physica status solidi (b)

ISSN 0370-1972 · SUBJECT AND AUTHOR INDEX (b) VOLUMES 127 TO 132 · 1985

ISSN 0031-8965 · SUBJECT AND AUTHOR INDEX (a) VOLUMES 87 TO 92 · 1985

physica status solidi (a)

THEORIA CUM PRAXI

AKADEMIE-VERLAG BERLIN

Classification Scheme

1. Structure of Crystalline Solids
   1.1 Perfectly Periodic Structure
   1.2 Solid-State Phase Transformations
   1.3 Alloys. Metallurgy
   1.4 Microstructure (Magnetic Domains See 18; Ferroelectric Domains See 14.4.1)
   1.5 Films
   1.6 Surfaces
2. Non-Crystalline State
3. Crystal Growth
4. Bonding Properties
5. Mössbauer Spectroscopy
6. Lattice Dynamics. Phonons
7. Acoustic Properties
8. Thermal Properties
9. Diffusion
10. Defect Properties (Irradiation Defects See 11)
    10.1 Metals
    10.2 Non-Metals
11. Irradiation Effects (X-Ray Diffraction Investigations See 1 and 10)
12. Mechanical Properties (Plastic Deformations See 10)
    12.1 Metals
    12.2 Non-Metals
13. Electron States
    13.1 Band Structure
    13.2 Fermi Surfaces
    13.3 Surface and Interface States
    13.4 Impurity and Defect States
    13.5 Elementary Excitations (Phonons See 6)
        13.5.1 Excitons
        13.5.2 Plasmons
        13.5.3 Polarons
        13.5.4 Magnons
14. Electrical Properties. Transport Phenomena
    14.1 Metals. Semi-Metals
    14.2 Superconductivity. Superconducting Materials and Devices
    14.3 Semiconductors
        14.3.1 Films
        14.3.2 Surfaces and Interfaces
        14.3.3 Devices. Junctions (Contact Problems See 14.3.4)
        14.3.4 High-Field Phenomena, Space-Charge Effects, Inhomogeneities, Injected Carriers (Electroluminescence See 20.3; Junctions See 14.3.3)
    14.4 Dielectrics
        14.4.1 Ferroelectrics
15. Thermoelectric and Thermomagnetic Properties
16. Photoconductivity. Photovoltaic Effects
17. Emission of Electrons and Ions
    17.1 Field Emission Microscope Investigations
18. Magnetic Properties
    18.1 Paramagnetic Properties
    18.2 Ferromagnetic Properties
        18.2.1 Ferromagnetic Films
    18.3 Ferrimagnetic Properties
    18.4 Antiferromagnetic Properties

(Continued on cover three)

physica status solidi (b)

basic research

physica status solidi (a)

applied research

SUBJECT AND AUTHOR INDEX

Volumes 127 to 132, 1985 — Volumes 87 to 92, 1985

ISSN 0370-1972/0031-8965

Edited by
E. BORCHARDT

AKADEMIE-VERLAG BERLIN

Subscriptions and orders for this Subject and Author Index
or single copies of physica status solidi (b) and physica status solidi (a) should be directed
in the GDR:

to the Postzeitungsvertrieb or
to the Akademie-Verlag Berlin, DDR-1086 Berlin, Leipziger Str. 3—4;

in the other socialist countries:

to a book-shop for foreign language literature or
to the competent news-distributing agency;

in the FRG and BERLIN (WEST):

to a book-shop or
to the wholesale distributing agency Kunst und Wissen, Erich Bieber OHG, Wilhelmstr. 4—6, D-7000 Stuttgart 1;

in the other Western European countries:

to Kunst und Wissen Erich Bieber GmbH, General Wille-Str. 4, CH-8002 Zürich;

in USA and CANADA:

to VCH Publishers, Inc., 303 NW 12 th Avenue, Deerfield Beach, FL 33442—1705, USA;

in other countries:

to the international book- and journal-selling trade,
DDR-7010 Leipzig, Postfach 160, or
to the Akademie-Verlag Berlin, DDR-1086 Berlin, Leipziger Straße 3—4.
to Buchexport, Volkseigener Außenhandelsbetrieb der Deutschen Demokratischen Republik,

Schriftleiter und verantwortlich für den Inhalt der Zeitschriften:

Professor Dr. Dr. h. c. P. Görlich †; amtierend: Professor Dr. E. Gutsche, DDR-1086 Berlin, Leipziger Str. 3—4, Postfach 1233.

Herausgeber und verantwortlich für den Inhalt des Registers:

Dipl.-Phys. E. Borchardt, DDR-1086 Berlin, Leipziger Str. 3—4, Postfach 1233.

Verlag:

Akademie-Verlag Berlin, DDR-1086 Berlin, Leipziger Str. 3—4; Fernruf: 2236221 und 2236229; Telex-Nr.: 114420; Bank: Staatsbank der DDR, Berlin, Kto.-Nr.: 6836-26-20712.

Veröffentlicht unter der Lizenznummer 1310 und 1620 des Presseamtes beim Vorsitzenden des Ministerrates der Deutschen Demokratischen Republik.

Gesamtherstellung:

VEB Druckerei „Thomas Müntzer", DDR-5820 Bad Langensalza.

Bestellnummer des Registers: 1068/85 R18.

Printed in the German Democratic Republic.

AN (EDV) 20831/20735

Preface

The eighteenth index volume contains volumes 127 to 132 of "physica status solidi (b) – basic research" as well as volumes 87 to 92 of "physica status solidi (a) – applied research", i.e. all volumes having appeared in 1985 containing 1461 papers.

Beginning from January 1986 the International Classification System for Physics (Condensed Matter, items 60 and 70) of the International Council for Scientific and Technical Information, Abstracting Board completed by a Substance Classification will be used for quoting papers in physica status solidi. Both systems are reproduces on pages 4 and 5 and page 599 to 604 of this index volume. However, in the present index volume the general classification scheme of the last index volume was maintained as it was the basis for quoting of papers in the respective issues having appeared in 1985. Nevertheless, before taking over these subject classifications into the index volume they were once more carefully revised and in some cases improved and completed in order to get a most uniform and clear classification for all papers. The average number of quotations per paper in this index volume is nearly the same as in the last one, namely about 3.0.

Again I have to express my thanks to the readers for their interest, and I will always readily take up any critical remarks and proposals for improvements. I am especially obliged to Prof. Dr. P. Görlich, Prof. Dr. E. Gutsche, Dr. H.-J. Hänsch, Dr. H. Lange, Dr. S. Oberländer, and Dipl. Phys. K. Müller for giving me advice and unselfish help. Furthermore I have to thank Mrs. E. Schwan for her exact and reliable secretary's work at the manuscript.

Berlin, October 1986

Eva Borchardt

Classification Scheme

(In use up to December 1, 1985;
the new Classification Scheme used from January 1, 1986 is reproduced on page 599 to 604)

1. Structure of Crystalline Solids
   1.1 Perfectly Periodic Structure
   1.2 Solid-State Phase Transformations
   1.3 Alloys. Metallurgy
   1.4 Microstructure (Magnetic Domains See 18; Ferroelectric Domains See 14.4.1)
   1.5 Films
   1.6 Surfaces
2. Non-Crystalline State
3. Crystal Growth
4. Bonding Properties
5. Mössbauer Spectroscopy
6. Lattice Dynamics. Phonons
7. Acoustic Properties
8. Thermal Properties
9. Diffusion
10. Defect Properties (Irradiation Defects See 11)
    10.1 Metals
    10.2 Non-Metals
11. Irradiation Effects (X-Ray Diffraction Investigations See 1 and 10)
12. Mechanical Properties (Plastic Deformations See 10)
    12.1 Metals
    12.2 Non-Metals
13. Electron States
    13.1 Band Structure
    13.2 Fermi Surfaces
    13.3 Surface and Interface States
    13.4 Impurity and Defect States
    13.5 Elementary Excitations (Phonons See 6)
        13.5.1 Excitons
        13.5.2 Plasmons
        13.5.3 Polarons
        13.5.4 Magnons
14. Electrical Properties. Transport Phenomena
    14.1 Metals. Semi-Metals
    14.2 Superconductivity. Superconducting Materials and Devices
    14.3 Semiconductors
        14.3.1 Films
        14.3.2 Surfaces and Interfaces
        14.3.3 Devices Junctions (Contact Problems See 14.3.4)
        14.3.4 High-Field Phenomena, Space-Charge Effects, Inhomogeneities, Injected Carriers (Electroluminescence See 20.3; Junctions See 14.3.3)
    14.4 Dielectrics
        14.4.1 Ferroelectrics
15. Thermoelectric and Thermomagnetic Properties
16. Photoconductivity. Photovoltaic Effects
17. Emission of Electrons and Ions
    17.1 Field Emission Microscope Investigations

18. Magnetic Properties
   18.1 Paramagnetic Properties
   18.2 Ferromagnetic Properties
      18.2.1 Ferromagnetic Films
   18.3 Ferrimagnetic Properties
   18.4 Antiferromagnetic Properties

19. Magnetic Resonance

20. Optical Properties
   20.1 Spectra. Optical Constants (X-Ray Spectra See 20)
   20.2 Lasers
   20.3 Luminescence (X-Ray Spectra See 20)

21. Metals. Alloys
   21.1 Transition Metals (Group Sc to Zn) and Their Alloys
      21.1.1 Fe and Fe Alloys
   21.2 Alkali-Metals
   21.3 Alkali-Earth Metals
   21.4 Rare-Earth Metals
   21.5 Actinides
   21.6 Noble Metals
   21.7 Semi-Metals (Semiconducting Àlloys See 22.7)

22. Semiconductors and Ionic Crystals
   22.1 Elements
      22.1.1 Germanium
      22.1.2 Silicon
      22.1.3 Group VI Elements
   22.2 III–V Compounds
      22.2.1 Arsenides
      22.2.2 Phosphides
      22.2.3 Antimonides
      22.2.4 Mixed Crystals
   22.3 IV–IV, II–V, and III–VI Compounds
      22.3.1 Mixed Crystals
   22.4 II–VI Compounds
      22.4.1 Sulphides
      22.4.2 Selenides
      22.4.3 Tellurides
      22.4.4 Mixed Crystals
   22.5 Halides
      22.5.1 Silver Halides
      22.5.2 Alkali Halides
      22.5.3 Alkali-Earth Halides
      22.5.4 Mixed Crystals
   22.6 Simple Oxides (CdO, ZnO See 22.4)
      22.6.1 Mixed Crystals
   22.7 Semiconducting Intermetallic Compounds
      22.7.1 Mixed Crystals
   22.8 Tertiary and Higher Compounds
      22.8.1 Salts of Oxy-Acids (Carbonates, Phosphates, Silicates Including Mica etc).
      22.8.2 Spinels, Garnets, Ferrites
   22.9 Organic Semiconductors

23. Solidified Gases

For comments on the use of the classification scheme see p. 6.

Some Comments on the Use of the Subject and Author Index

The classification system[1]) comprises 20 broad sections each devoted to a major field of solid state physics and 3 broad sections devoted to the main groups of substances. A section may contain a number of subsections dealing with aspects of this field of special substance, respectively, which tend to receive particular attention. A paper dealing with this field will either be classified under one or more of the subsections or, if none of these are relevant, it will just appear under the general section heading (i.e. sections and subsections are equivalent in this sense). For example, a paper dealing with problems of paramagnetism would appear under 18.1 but not 18, whereas a paper on diamagnetism would be quoted under 18; a paper dealing with GaP would appear under 22.2.2, not under 22 or 22.2, whereas a paper on AlN would be quoted under 22.2, and a paper on a II–IV or a IV–VI compound would be quoted under 22.

All papers will be quoted under each relevant section. There is made a distinction between one or two principal entries (refering to the main subject matter of the paper) and secondary entries (refering to subsidiary subject matter and substance, respectively).In the principal entries the title of the paper is given in its original language and in English. In addition, every principal entry is marked by * in front of the title of the paper. In the secondary entries the title is only given in English.

On the line following the title the subject classification is given (in parantheses) in the same manner as in the volumes. Each entry gives the volume, the number of the issue, the initial and final page numbers, and the year in which the paper appeared. A paper of physica status solidi (b) or physica status solidi (a) is marked by (b) or (a) in front of the line, respectively. The letter "K" before the page numbers indicates a short note published in the offset part of the journal.

Review articles appear, at the head of the list of entries under a given subject and are given the subtitle "Review Articles". These articles are listed in the chronological order of their publication, irrespective of being a paper of physica status solidi (b) or physica status solidi (a). The subject classification for a review article consisting of several parts is based on the article as a whole rather than on the individual parts.

The entries for original papers, short notes, and errata are made together in a single list for each subject and this follows the entries for review articles. These are again ordered according to data of publication, irrespective of being a paper of physica status solidi (b) or physica status solidi (a). Original papers are sometimes published in several parts under the same principal title (in general with special subtitles). The various parts of papers of this type are treated separately and each part is classified according to its particular contents.

In quoting errata the following rule is obeyed. If the erratum refers to a paper quoted in the preceding index, it is listed in the chronological order of its publication otherwise the erratum is quoted together with the paper to which it refers.

---

[1]) See cover two and three, and pages 4 and 5. For the case of losing the cover by bookbinding we repeat the classification scheme on pages 4 and 5.

In using the subject index attention should be given to the following details:

subject	quoted in	remarks
representations of space groups	1 and 13.1	
Alfen waves, helicon waves	13 and 20	
hot electrons	13 and 14.3.4	
transport phenomena	14 or subsection of 14	if not clearly to be assigned to 8 or 15
galvanomagnetic effects	14 or subsection of 14	not in 18
electro-optical effects	14 and 20 (or subsection of 14 and 20)	
piezoelectricity, electrostriction	14 or subsection of 14	not in 12 or subsection of 12; different electromechanical effects are quoted in 12 or subsection of 12
ionic conduction	14.4	not in 14.3
magneto-optical effects	18 and 20 (or subsection of 18 and 20)	possibly in additional sections, e.g. Faraday effect in 13
ferromagnetic films	18.2.1	not additionally in 1.5
X-ray spectra	20	but not X-ray studies of structure
MIS and MOS structures	the section assigned to the semiconductor	not in the section assigned to the oxide or the metal
MIM structures	the section assigned to the insulator	generally not in the section assigned to the metal
epitaxial layers	the section assigned to the layer	not in the section assigned to the substrate

The author index gives the names of the authors in alphabetical order. Each name is followed by the bibliographic details of the paper. A paper of physica status solidi (b) or physica status solidi (a) is again marked by (b) or (a) in front of the number of the volume, respectively. In the last column the subject classification of the paper is presented in the same manner as in the volumes. Behind the principal subject classification(s) the page number is given (in parantheses) on which the paper appears in the subject index. When several papers have been published by the same author they are arranged according to the dates of publication, irrespective of being a paper of physica status solidi (b) or physica status solidi (a).

Subject Index

1. Structure of Crystalline Solids

Original Papers and Short Notes

T. OHKAWA and H. HASHIMOTO
*An X-Ray Wave Theory for Heavily Distorted Crystals. I. On the Modified Monochromatic Spherical Waves
(1)
(a) Vol. **87**, No. 1, p. 59–71 (1985).

R. S. VARDANIAN and A. KH. KHACHATRIAN
*Application of the Invariance Principle in the Dynamic Theory of X-Ray Scattering
(1; 20)
(a) Vol. **87**, No. 1, p. 73–78 (1985).

T. OHKAWA and H. HASHIMOTO
*An X-Ray Wave Theory for Heavily Distorted Crystals. II. On the Resonance Shift Function and the Dispersion Relation in Distorted Crystals
(1)
(a) Vol. **87**, No. 2, p. 417–423 (1985).

V. B. MOLODKIN, S. I. OLIKHOVSKII, M. E. OSINOVSKII, A. N. GUREEV, L. I. DATSENKO, A. I. NIZKOVA, and B. F. ZHURAVLEV
*The Integrated Intensities of the Laue-Diffracted X-Rays for Monocrystals Containing Macroscopically Homogeneously Distributed Defects
(1 and 10)
(a) Vol. **87**, No. 2, p. 597–608 (1985).

C. MALGRANGE and J. GRONKOWSKI
*Erratum to: Fraunhofer Diffraction of X-Ray Beams at Bragg Incidence in Distorted Crystals
(1)
(a) Vol. **87**, No. 2, p. 741 (1985).

Z. JANÁČEK
*Coherence Description of a Double Crystal Diffractometer
(1; 10)
(a) Vol. **88**, No. 1, p. 49–59 (1985).

F. N. CHUKHOVSKII and YU. P. KHAPACHEV
*Exact Solution of the Takagi-Taupin Equation for Dynamical X-Ray Bragg Diffraction by a Crystal with a Transition Layer
(1 and 10)
(a) Vol. **88**, No. 1, p. 69–76 (1985).

P. A. BEZIRGANYAN, K. T. GABRIELYAN, and G. O. DEMIRCHYAN
*An X-Ray Interferometer with an Elastically Deformed Crystal
(1 and 10)
(a) Vol. **88**, No. 1, p. 143–149 (1985).

F. EGGERT and W. SCHOLZ
*A Rapid Deconvolution Method Based on the Bayesian Theorem Applied to Energy Dispersive X-Ray Emission Analysis
(1 and 20)
(a) Vol. **88**, No. 2, p. K123–K125 (1985).

A. A. KOROLEV and V. P. SMIRNOV
*Coordinate Transformations Induced by the Elements of the Permutation-Inversion Symmetry Group of a Nonrigid Molecular Crystal
(1; 22.8.1)
(b) Vol. **129**, No. 1, p. 41–47 (1985).

S. ADAM, GH. ADAM, and A. CORCIOVEI
*One-Particle Crystal Field Hamiltonians
(1 and 13)
(b) Vol. **129**, No. 2, p. 763–774 (1985).

T. TAKAI, T. HALICIOĞLU, and W. A. TILLER
*A Parametric Study of Crystal Stability and Surface Energy for Diamond Cubic Structures
(1 and 4; 22.1)
(b) Vol. **130**, No. 1, p. 131–138 (1985).

H. S. KIM and S. S. SHEININ
*Asymmetry in Bright Field Images of Stacking Faults
(1 and 10)
(a) Vol. **90**, No. 1, p. 231–240 (1985).

H.-K. HYVÄRINEN, V.-P. TANNINEN, A. GREKULA, and P.-L. KALLIOMÄKI
*Profile Fitting Analysis of the X-Ray Powder Diffraction Pattern of a Complicated Metal Aerosol
(1; 22.8)
(a) Vol. **90**, No. 1, p. K15–K18 (1985).

T. TAKAI, T. HALICIOĞLU, and W. A. TILLER
*Calculation of Potential Energy Parameters of the Silicon–Carbon System
(1 and 4; 22.1; 22.1.2; 22.3)
(b) Vol. **130**, No. 2, p. 475–481 (1985).

R. Somashekar, Ch. Satyanarayana, and M. S. Madhava
*Mesomorphic Behaviour of Mixtures of Cetyl Alcohol and Dimethyl Sulfoxide
(1 and 2; 22.9)
(a) Vol. **90**, No. 2, p. 449–456 (1985).

Yu. A. Babanov and V. R. Shvetsov
*EXAFS: Bond Length Determination for Multicomponent Systems
(1 and 20; 21.1)
(b) Vol. **131**, No. 1, p. K1–K4 (1985).

L. T. Wille and J. Vennik
*X-Ray Diffraction by Vibrating Crystals
(1 and 7)
(b) Vol. **131**, No. 2, p. K97–K100 (1985).

I. R. Entin
On the Suppression of X-Ray Anomalous Transmission by Acoustic Oscillations
(7; 1)
(b) Vol. **132**, No. 2, p. 355–364 (1985).

J. Ganster and D. Geiss
*Polycrystalline Simple Average of Mechanical Properties in the General (Triclinic) Case
(1 and 12.2; 22.9)
(b) Vol. **132**, No. 2, p. 395–407 (1985).

Yu. G. Chukalkin, V. V. Petrov, V. R. Shtirts, and B. N. Goshchitskii
*Effects of Radiation Disorder in Chromium Spinels
(1 and 11; 22.8.2)
(a) Vol. **92**, No. 2, p. 347–354 (1985).

R. G. Gabrielyan, A. R. Mkrtchyan, H. A. Aslanyan, and Kh. V. Kotandyan
*On the Theory of X-Ray Diffraction in Oscillating Piezocrystals
(1 and 10)
(a) Vol. **92**, No. 2, p. 361–368 (1985).

1.1 Perfectly Periodic Structure

Original Papers and Short Notes

H.-G. Brühl and H. Rhan
*On the Extension of the Theory of Umweganregung with Respect to the Use of Divergent White X-Ray Radiation
(1.1; 22.2.1)
(a) Vol. **87**, No. 1, p. 121–126 (1985).

U. Pietsch
*X-Ray Structure Amplitudes for GaAs and InP
(1.1; 22.2.1; 22.2.2)
(a) Vol. **87**, No. 1, p. 151–156 (1985).

W. Coene, H. Bender, F. C. Lovey, D. van Dyck, and S. Amelinckx
*On the Influence of Crystal Orientation on the High Resolution Image Contrast of Polytypes
(1.1; 21.1; 22.3; 22.4.1)
(a) Vol. **87**, No. 2, p. 483–497 (1985).

D. Stephan and V. Geist
*Investigation of Lattice Strain in Proton-Irradiated GaP by a Modified Auleytner Technique
(1.1 and 11; 22.2.2)
(a) Vol. **87**, No.2, p. 589–596 (1985).

V. G. Baryshevskii and S. V. Cherepitsa
*Neutron Spin Precession and Spin Dichroism of Nonmagnetic Unpolarized Single Crystals
(1.1 and 11; 22)
(b) Vol. **128**, No. 2, p. 379–387 (1985).

G. Mavrodiev, A. Koneska, and A. Szytuła
*Dilatometric and DDC Measurements of $NiMnGe_{1-x}Si_x$ and $NiMn_{1-t}Ti_tGe$ Systems
(1.1 and 1.2; 21.1)
(a) Vol. **88**, No. 2, p. K105–K109 (1985).

B. Jenichen, R. Köhler, and W. Möhling
*Double Crystal Topography Compensating for the Strain in Processed Samples
(1.1 and 1.4; 10.2; 22.2.4)
(a) Vol. **89**, No. 1, p. 79–87 (1985).

R. Oberschmid
Absorption Centers of $Bi_{12}GeO_{20}$ and $Bi_{12}SiO_{20}$ Crystals
(13.4 and 20.1; 1.1; 16; 22.6)
(a) Vol. **89**, No. 1, p. 263–270 (1985).

J. Przedmojski, S. Gierlotka, and B. Pałosz
*X-Ray Thermal Investigations of Cadmium Iodide Single Crystals
(1.1 and 8; 22.5)
(a) Vol. **89**, No. 1, p. K9–K11 (1985).

B. Kozarzewski
*Calculation of Lattice Sums
(1.1)
(b) Vol. **129**, No. 2, p. 459—463 (1985)

V. V. Aristov, W. Graeff, V. I. Polovinkina, and A. A. Snigirev
*Dynamical Diffraction of X-Ray Spherical Waves in Perfect Crystals Using a Narrow Slit as Radiation Source
(1.1; 10.2; 22.1.2)
(a) Vol. **90**, No. 1, p. 69—75 (1985).

H. Kido, T. Hoshikawa, M. Shimada, and M. Koizumi
Crystals Structure Magnetic and Electrical Properties of Pseudo-Ternary $DyCu_{1-x}Zn_xSi$ Compounds
(14.1 and 18.1; 1.1; 18.2; 21)
(a) Vol. **90**, No. 1, p. 307—311 (1985)

H. R. Wenk, D. O'Brien, and Z. You
*Spherical Representation of Orientation Distribution Functions
(1.1; 22.8)
(a) Vol. **90**, No. 1, p. K19—K24 (1985).

V. Holý, J. Hlávka, and J. Kuběna
On the Correlation of X-Ray Generated Photoconductivity with the X-Ray Reflection Curve in Silicon
(16; 1.1; 22.1.2)
(a) Vol. **90**, No. 1, p. K87—K89 (1985).

N. A. Razik
*Precise Lattice Constant Determination of Hexagonal, Rhombohedral, and Tetragonal Crystals from X-Ray Powder Diffractometric Data
(1.1; 22.6)
(a) Vol. **90**, No. 2, p. K125—K128 (1985).

D. K. Chakrabarty, M. K. Jha, and S, Unnikrishnan
Preparation and Properties of Barium Substituted Gadolinium Cobaltite
(18.2; 1.1; 14.3; 22.8.1)
(a) Vol. **90**, No, 2, p. K175—K177 (1985).

M. Schenk, H. Berger, C. Kleint, N. S. Golovanova, O. I. Tananaeva, and V. P. Zlomanov
The Influence of Bi on Lattice Parameter and Carrier Concentration of PbTe
(10.2; 1.1; 14.3; 22.7)
(a) Vol. **91**, No. 1, p. K35—K39 (1985).

A. Ben Fredj et G. Nihoul
*Observation en microscopie électronique à haute résolution d'oxydes à grande maille (c-Dy_2O_3) et interprétation des images obtenues
(High-Resolution Electron Microscopic Observation of Large Cell Oxides (c-Dy_2O_3) and Interpretation of the Obtained Images)
(1.1 and 1.4, 22.6)
(a) Vol. **91**, No. 2, p. 351—369 (1985).

P. Kh. Bezirganyan and Kh. R. Drmeyan
*Fine Structure of X-Ray Interferograms. III. Complex Interference Systems

1.2 Solid-State Phase Transformations

Review Articles

S. Sugai
Lattice Vibrations in the Charge-Density-Wave States of Layered Transition Metal Dichalcogenides
(6 and 13; 1.2; 20.1; 22)
(b) Vol. **129**, No. 1, p. 13–39 (1985).

S. A. Shabalovskaya
*Phase Transitions in the Intermetallic Compound TiNi with Charge-Density Wave Formation
(1.2 and 13.1; 6; 14.1; 21.1)
(b) Vol. **132**, No. 2, p. 327–344 (1985).

Original Papers and Short Notes

M. M. Szostak, A. V. Belushkin, and I. Natkaniec
Neutron Spectroscopy and Model Calculations of Phonon Dynamics in Crystalline p-Chloroaniline
(6; 1.2; 22.9)
(b) Vol. **127**, No. 1, p. K1–K4 (1985).

S. Åsbrink, M. Wołcyrz, and S.-H. Hong
*X-Ray Bond-Type Diffractometer Investigations on V_3O_5 in the Temperature Interval 298 to 480 K Including the Phase Transition Temperature $T_t = 428$ K
(1.2; 8; 22.6)
(a) Vol. **87**, No. 1, p. 135–140 (1985).
Erratum
(a) Vol. **89**, No. 1, p. 415 (1985).

J. A. Cusidó, A. Isalgué, and J. Tejada
*On the Amorphous to Crystalline Transformation of $Fe_{80}B_{20}$ by Means of Electrical and Thermal Conductivity, X-Ray Diffraction, and Mössbauer Measurements
(1.2 and 2; 5; 8; 14.1; 21.1.1)
(a) Vol. **87**, No. 1, p. 169–174 (1985).

S. N. Vaidya
*Theory of Melting. II. Relaxion Model
(1.2 and 6; 8; 23)
(a) Vol. **87**, No. 1, p. 181–186 (1985).

P. Saint-Grégoire, A. Pérez, R. Almairac, and M. Lopez
*Some Anomalies Connected with the Structural Transition of $KFeF_4$ near 400 K
(1.2 and 8; 22.8)
(a) Vol. **87**, No. 1, p. K1–K6 (1985).

G. A. Bordovskii and A. B. Zarkoi
*Polymorphic Phase Transition in $Pb_5Bi_8O_{17}$
(1.2; 22.6.1)
(a) Vol. **87**, No. 1, p. K7–K12 (1985).

Z. M. Alekseeva, V. S. Vikhnin, A. P. Zdebskii, and I. M. Tsivileva
Acoustical and Electrochemical Behaviour of Proustite Parameters in the Incommensurate Phase
(7; 1.2; 10.2; 22.8)
(a) Vol. **87**, No. 1, p. K35–K39 (1985).

P. Enders
Structure and Bonding in Cubic IV–VI Crystals. IV. Evidence for Covalent Bonding in the IV–VI Family from the Extended Mulliken Scheme
(4 and 13; 1.2; 22)
(b) Vol. **127**, No. 2, p. 449–457 (1985).

F. Bolduan, H. D. Hochheimer, and H. P. Beck
Optical and Energy-Dispersive X-Ray Investigations of the Mixed Halide SmFCl
(20.1; 1.2; 13.1; 22.5.4)
(b) Vol. **127**, No. 2, p. 673–678 (1985).

E. M. Smolyarenko, V. M. Trukhan, and A. P. Bologa
*Phase Diagram of the ZnP_2-ZnAs_2 System
(1.2; 22.3.1)
(a) Vol. **87**, No. 2, p. K123–K126 (1985).

H. Oshima, A. Tanaka, and T. Sukegawa
*Calculation of the Ga–Al–Sb Phase Diagram Using the Redlich-Kister Expression
(1.2; 3; 22.2.4)
(a) Vol. **87**, No. 2, p. K131–K134 (1985).
Erratum
(a) Vol. **92**, No. 1, p. K87 (1985).

G. Blasse and G. J. Dirksen
*The Influence of the Phase Transformation on the Luminescence of $NbPO_5$
(1.2 and 20.3; 22.8)
(a) Vol. **87**, No. 2, p. K181–K184 (1985).

G. Üffinger and A. Kasten
*Specific Heat and Magnetocaloric Effects in $TbPO_4$. I. Experiments
(1.2 and 18.4; 8; 22.8.1)
(b) Vol. **128**, No. 1, p. 201–207 (1985).

H. D. Hochheimer, E. Gmelin, Ch. Mensing, L. Viczian, K. D. Kreuer, and H. G. von Schnering
*AC Conductivity, Calorimetric, X-Ray, and Raman Study of the Hexagonal Phase $TlInS_2$-III
(1.2; 6; 8; 14.3; 20.1; 22.8)
(a) Vol. **88**, No. 1, p. 43–48 (1985).

M. R. Nagy, M. S. Sakr, M. M. Mostafa, and R. Kamel
Effect of Structure Transformation on Creep and Creep Recovery of Al–14 wt% Zn
(10.1; 1.2; 21; 21.1)
(a) Vol. **88**, No. 1, p. 61–67 (1985).

A. G. Bondarenko and A. B. Khmelinin
*Phase Transition in the System Solid–Gas Admixture Due to the Macroscopic Defect Formation and Growth
(1.2; 10)
(a) Vol. **88**, No. 1, p. 121–127 (1985).

P. Rochegude and J. Foct
*The Transformation $\varepsilon \rightarrow \alpha''$ in Iron–Nitrogen Solid Solutions Studied by Mössbauer Spectrometry
(1.2 and 5; 21.1.1)
(a) Vol. **88**, No. 1, p. 137–142 (1985).

R. Kuśka
*The Application of Landau's Theory for Studying First-Order Phase Transitions by the Internal Friction Method
(1.2; 10.1; 12.1; 21)
(a) Vol. **88**, No. 1, p. 151–156 (1985).

A. Kasten and G. Üffinger
Specific Heat and Magnetocaloric Effects in $TbPO_4$. II. Mean Field Theory
(1.2 and 18.4; 8; 22.8.1)
(b) Vol. **128**, No. 2, p. 525–535 (1985).

V. L. Aksenov and M. Bobeth
*Structural Glass in Solid Solutions with Competing Interactions
(1.2; 14.4; 22.8.1)
(b) Vol. **128**, No. 2, p. K105–K109 (1985).

A. V. Dobromyslov, A. E. Ermakov, N. I. Taluts, and M. A. Uimin
*Electron Microscopy Investigation of Phase Transformations in Mn–Al–C Alloy
(1.2; 1.4; 18.2; 21.1)
(a) Vol. **88**, No. 2, p. 443–454 (1985).

A. H. Phillips, N. K. Gobran, and R. Kamel
*Structure Analysis of the Metallic Glass $Fe_{40}Ni_{40}B_{20}$ Using the Tangent Lattice Model
(1.2; 14.1; 21.1; 21.1.1)
(a) Vol. **88**, No. 2, p. 463–467 (1985).

M. A. Uimin, A. E. Ermakov, O. B. Andreeva, E. I. Teitel, and V. M. Gundyrev
*Phase Transformations and Magnetic Properties of MnAl–C Single Crystals
(1.2 and 18.2; 21.1)
(a) Vol. **88**, No. 2, p. 587–594 (1985).

C. Julien, M. Eddrief, M. Balkanski, E. Hatzikraniotis, and K. Kambas
Electrical Transport Properties of In_2Se_3
(14.3; 1.2; 22.3)
(a) Vol. **88**, No. 2, p. 687–695 (1985).

G. Mavrodiev, S. Koneska, and A. Szytuła
*Dilatometric and DDC Measurements of $NiMnGe_{1-x}Si_x$ and $NiMn_{1-t}Ti_tGe$ Systems
(1.1 and 1.2; 21.1)
(a) Vol. **88**, No. 2, p. K105–K109 (1985).

A. F. Goncharov, I. N. Makarenko, and N. M. Gasanly
*Effect of Pressure on the Optical Absorption Edge in InS Monocrystals
(1.2 and 20.1; 13.1; 22.3)
(b) Vol. **129**, No. 1, p. K79–K81 (1985).

R. KAWASHIMA, K. KATSUKI, and K. SUZUKI
*Temperature Dependence of Optical Absorption near the Structural Phase Transition of Potassium Nitrate
(1.2 and 13.1; 14.4.1; 20.1; 22.8.1)
(b) Vol. **129**, No. 2, p. 697–705 (1985).

S. GARCÍA and E. ALTSHULER
*Mössbauer Study of the Reaction Kinetics of Hexagonal M-Phase Ferrites
(1.2 and 5; 22.8.2)
(a) Vol. **89**, No. 2, p. 427–436 (1985).

M. PETROV, L. KOMITOV, P. SIMOVA, and G. HAUCK
*Intermediate Smectic A-State at Nematic–Smectic C Phase Transition at Tilted Orientation
(1.2; 2; 22.9)
(a) Vol. **89**, No. 2, p. 451–456 (1985).

M. PIOTROWSKI and A. MURASIK
Exchange and Dipolar Interactions in TbF_3
(18; 1.2; 22.5)
(a) Vol. **89**, No. 2, p. 571–580 (1985).

I. H. ISMAILZADE and O. A. SAMEDOV
*The Nature of the Intermediate Phase of $PbZrO_3$
(1.2 and 14.4.1; 22.8.1)
(a) Vol. **89**, No. 2, p. K133–K136 (1985).

M. S. SAKR, A. E. E. ABDEL-RAAHEIM. and A. A. EL-DALLY
*Effect of Structure Transformation on the Transient Creep Characteristics of Zn–40 wt% Al and Zn–0.5 wt% Al Alloys
(1.2 and 10.1; 21.1)
(a) Vol. **89**, No. 2, p. K157–K162 (1985).

A. N. DIDENKO, O. KH. ASAINOV, V. P. KRIVOBOKOV, E. I. LOGACHEV, A. D. POGREBNYAK, L. N. PUCHKAREVA, G. E. REMNEV, and SH. M. RUZIMOV
Positron Annihilation at Defects in Surface Layers of Steel Irradiated by Supercurrent Ion Beams
(10.1 and 11; 1.2; 21.1.1)
(a) Vol. **89**, No. 2, p. K167–K172 (1985).

I. O. BASHKIN, YU. M. DERGACHEV, and E. G. PONYATOVSKII
*Pressure-Induced Phase Transition in RbH
(1.2; 22.5)
(b) Vol. **130**, No. 1, p. 73–77 (1985).

D. M. HATCH and H. T. STOKES
*Active Representations in Crystals of Cubic Symmetry
(1.2)
(b) Vol. **130**, No. 1, p. 79–86 (1985).

P. J. BECKER, H. G. KAHLE, and E. KELLER
*The Phase Diagram of $TbPO_4$ Measured by Linear Optical Birefringence
(1.2 and 18.4; 20; 22.8.1)
(b) Vol. **130**, No. 1, p. 191–196 (1985).

D. I. UZUNOV, E. R. KORUTCHEVA, and Y. T. MILLEV
*Random-Field Correlations and Quantum Critical Phenomena
(1.2 and 13)
(b) Vol. **130**, No. 1, p. 243–253 (1985).

R. STEARNS, R. BRAUNSTEIN, and L. MULDAWER
Wavelength-Modulated Spectra of the Optical Properties of β'-Cu_xZn_{1-x} from 1.5 to 5.1 eV at the β' ⇄ α + β' Phase Transition
(13.1 and 20.1; 1.2; 21.1)
(b) Vol. **130**, No. 2, p. 595 – 606 (1985).

I. H. ISMAILZADE and O. A. SAMEDOV
*Influence of Permanent Electric Fields on the Successive Phase Transitions in the Solid Solutions of the System $(Pb_{1-x}Ba_x)ZrO_3$
(1.2 and 14.4.1; 22.8.1)
(a) Vol. **90**, No. 2, p. 445 – 448 (1985).

O. BOSTANJOGLO, E. ENDRUSCHAT, and W. TORNOW
*Kinetics of Laser-Induced Liquid Metal Etching of a-Si Films
(1.2, 1.4; 22.1.2)
(a) Vol. **90**, No. 2, p. 457 – 462 (1982).

P. URWANK, E. WIESER, A. HÄSSNER, CH. KAUFMANN, H. LIPPMANN, and I. MELZER
*Formation of $MoSi_2$ by Light Pulse Irradiation
(1.2 and 11, 14.1; 21)
(a) Vol. **90**, No. 2, p. 463 – 468 (1985).

J. MAYER and K. URBAN
*Observation of Ni_8Mo Ordered Phase in Ni–Mo Alloys
(1.2, 1.3, 11, 21.1)
(a) Vol. **90**, No. 2, p. 469 – 475 (1985).

F. REYNAUD
*Relation between the Pair Potentials V_{AA} and V_{BB} in Classical Statistical Physics of the Order–Disorder Transitions in Binary Substitutional Alloys
(1.2 and 1.3, 21)
(a) Vol. **90**, No. 2, p. 477 – 481 (1985).

A. I. GUSEV and A. A. REMPEL
*Order Parameter Functional Method in the Theory of Atomic Ordering
(1.2; 22)
(b) Vol. **131**, No. 1, p. 43 – 51 (1985).

S. IHARA and K. SUZUKI
Molecular Dynamical Study of β-$AgI_{1-x}Cl_x$
(10.2 and 14.4, 1.2; 22.5.4)
(b) Vol. **131**, No. 1, p. 97 – 103 (1985).

B. MARTIN, J. M. PASTOR, and J. A. DE SAJA
Ferroelectric Behaviour of TGS and TGSe from Microhardness Measurements
(12.2; 1.2; 22.9)
(b) Vol. **131**, No. 1, p. K5 – K9 (1985).

T. SOMA, M. TAKAHASHI, and M.-MATSUO KAGAYA
*Compression Effect on the α–β Phase Transition of Cu-Based Binary Alloys
(1.2 and 12.1, 1.3, 21.1)
(b) Vol. **131**, No. 1, p. K11 – K14 (1985).

K. URBAN, N. MOSER, and H. KRONMÜLLER
*Phase Transitions between the Quasicrystalline, Crystalline and Amorphous Phases Al–14 at% Mn
(1.2; 2; 21; 21.1)
(a) Vol. **91**, No. 2, p. 411–422 (1985).

S. ÅSBRINK, L. GERWARD, and J. STAUN OLSEN
*Investigations of the Phase Transition in V_3O_5 Using Energy Dispersive X-Ray Diffraction and Synchrotron Radiation White Beam X-Ray Topography
(1.2; 22.6)
(a) Vol. **91**, No. 2, p. 423–432 (1985).

G. VAN TENDELOO, S. AMELINCKX, C. MANOLIKAS, and WEN SHULIN
*The Direct Observation of "Discommensurations" in Barium Sodium Niobate (BSN) and Its Homologues
(1.2 and 1.4; 22.8.1)
(a) Vol. **91**, No. 2, p. 483–501 (1985).

J. CORBERO, D. J. PUSIOL, A. E. WOLFENSON, and A. H. BRUNETTI
*Possible Incommensurate Structure in 4,4'-Dichlorobiphenyl Sulphone as Studied by the ^{35}Cl Nuclei NQR
(1.2 and 19; 22.9)
(a) Vol. **91**, No. 2, p. K97–K99 (1985).

E. M. SMOLYARENKO, V. M. TRUKHAN, and A. P. BOLOGA
*Phase Diagram of the ZnP_2–CdP_2 System
(1.2; 22.3.1)
(a) Vol. **91**, No. 2, p. K101–K102 (1985).

F. BOLDUAN, W. HÖNLE, H. D. HOCHHEIMER, and W. HENKEL
*Temperature and High Pressure Raman Study of M_3P_7 (M = Li, Na, K, Rb, Cs) Compounds
(1.2; 20.1; 21.2)
(b) Vol. **132**, No. 1, p. 41–50 (1985).

D. DUVAL and R. A. CONDRATE, SR.
*Temperature Dependence of the Raman Spectra of Calcium Oxalate Monohydrate
(1.2; 6; 20.1; 22.9)
(b) Vol. **132**, No. 1, p. 83–92 (1985).

PAN XIAO-QING, HU MEI-SHEN, YAO MING-HUI, and FENG DUAN
*A TEM Study of the Incommensurate Phase and Related Phase Transitions in Barium Sodium Niobate
(1.2 and 1.4; 22.8.1)
(a) Vol. **92**, No. 1, p. 57–68 (1985).

O. BOSTANJOGLO, E. ENDRUSCHAT, E. I. GIVARGIZOV, and W. TORNOW
*Metal-Assisted Grain Growth in Beam Crystallized Amorphous Ge/Au Films
(1.2 and 3; 21.6; 22.1.1)
(a) Vol. **92**, No. 1, p. K1–K4 (1985).

R. A. DUNLAP
*The Formation of Icosahedral Al–Transition Metal Alloys
(1.2 and 1.3; 21; 21.1)
(a) Vol. **92**, No. 1, p. K11–K13 (1985).

S. L. Chaplot
*Phase Transition in Tetracyanoethylene at High Pressure by X-Ray Diffraction
(1. 2; 12.2; 22.9)
(a) Vol. **92**, No. 1, p. K23–K27 (1985).

J. Narain, N. K. Dwivedi, G. G. Agrawal, and J. Shanker
*Structural Phase Transformation and the Equation of State of CsCl, CsBr, and CsI Crystals
(1.2 and 4; 22.5.2)
(b) Vol. **132**, No. 2, p. 389–393 (1985).

V. P. Meiklyar, A. E. Usachev, Yu. V. Yablokov, and V. A. Shustov
*EPR and ENDOR in $CsCaCl_3:Gd^{3+}$ Crystals
(1.2 and 19; 22.8)
(b) Vol. **132**, No. 2, p. K73–K76 (1985).

N. Lawerence and T. M. Haridasan
Mean Square Amplitude of Transition Metal Dichalcogenide 1T-$TiSe_2$ in the Normal and Commensurate Charge Density Wave Phases
(6; 1.2; 22)
(b) Vol. **132**, No. 2, p. K77–K79 (1985).

M. Wagner, H.-D. Geiler, and G. Götz
Time-Resolved Investigation of Large-Area Explosive Crystallization of Amorphous Silicon Layers
(2 and 3; 1.2; 11; 22.1.2)
(a) Vol. **92**, No. 2, p. 413–420 (1985).

K. H. Heinig and H.-D. Geiler
Penomenological Theory of Explosive Solid Phase Crystallization of Amorphous Silicon. I. Stationary Solutions
(2 and 3; 1.2; 22.1.2)
(a) Vol. **92**, No. 2, p. 421–430 (1985).

B. Martin, H. Jex, and F. Wittmann
*Elastic and Inelastic Scattering from NH_4Br Measured by Mössbauer γ-Ray Diffraction
(1.2; 5; 6; 22.5)
(a) Vol. **92**, No. 2, p. 457–465 (1985).

R. Mukhopadhyay, S. L. Chaplot, and K. R. Rao
*Phase Transition in Tetracyanoethylene
(1.2; 22.9)
(a) Vol. **92**, No. 2, p. 467–474 (1985).

G. Hübner and H. Löffler
On the Quench-Sensitivity of the Decomposition Rate of Al–Zn Alloys
(1.3 and 10.1; 1.2; 21; 21.1)
(a) Vol. **92**, No. 2, p. 517–524 (1985).

Z. A. Matysina, A. M. Ovrutskii, and L. M. Chuprina
*Influence of Correlation and Anisotropy on the Molecular Roughness of the Crystal Face
(1.2 and 1.6; 21.7)
(a) Vol. **92**, No. 2, p. K93–K97 (1985).

A. U. Sheleg and V. V. Zaretskii
*The Relationship of the "Invar Effect" with Polytypism and Modulation in α-ZnP_2
(1.2; 22.3)
(a) Vol. **92**, No. 2, p. K99–K104 (1985).

C. Dauwe and K. Motoko
*Three-Quantum Annihilation in the Plastic Crystal Norbornane
(1.2 and 10.2; 22.9)
(a) Vol. **92**, No. 2, p. K117–K121 (1985).

1.3 Alloys. Metallurgy

Original Papers and Short Notes

B. A. Men and M. L. Katsnelson
*The Ground State Superstructures in Ordered Binary S.C., B.C.C., and F.C.C. Alloys with Pair Interactions up to Third Neighbours
(1.3; 21)
(a) Vol. **87**, No. 1, p. 93–108 (1985).

J. E. Frąckowiak
*Determination of the Long-Range Order Parameter in the Transition Metal Alloys Using Mössbauer Spectroscopy
(1.3 and 5; 21.1; 21.1.1)
(a) Vol. **87**, No. 1, p. 109–119 (1985).

A. E. Curzon and O. S. Rajora
*An Electron Diffraction Study of the Ordering of Silver in Ag_xNbSe_2 for $0 < x < 0.33$
(1.3; 22.8)
(a) Vol. **87**, No. 1, p. 157–163 (1985).

M. Singh, M. Barkei, G. Inden, and S. Bhan
*High-Temperature X-Ray Diffraction Study on $Co_{75}Sn_{25}$ Alloy
(1.3; 21.1)
(a) Vol. **87**, No. 1, p. 165–168 (1985).

J. F. Dinhut, P. Bouillaud, and J. P. Rivière
Ion Irradiation of Thin Films of Fe(65 at%)–Co (35 at%) Alloy. Production of a Metallic Fe(50 at%)-Co(30 at%)–P (20 at%) Glass
(11 and 14.1; 1.3; 21.1; 21.1.1)
(a) Vol. **87**, No. 1, p. 199–205 (1985).

A. Prasad, R. M. Singru, and A. K. Biswas
Study of the Roasting of Pyrite Minerals by Mössbauer Spectroscopy
(5; 1.3; 21.1.1)
(a) Vol. **87**, No. 1, p. 267–271 (1985).

H. Kido, T. Hoshikawa, M. Shimada, and M. Koizumi
Crystal Structure and Magnetic and Electrical Properties of Pseudo-Ternary $HoCu_{1-x}Zn_xSi$ and $NdCu_{0.6}Zn_{0.4}Si$ Compounds
(14.1 and 18.1; 1.3; 18.2; 22.8)
(a) Vol. **87**, No. 1, p. 273–277 (1985).

E. A. Vasilev and V. A. Virchenko
*Effect of Interstitial Atoms on the Quadrupole Interaction at ^{119}Sn Nuclei in Some B8 Type Solid Solutions
(1.3 and 5; 4; 21.1.1)
(a) Vol. **87**, No. 1, p. K19–K24 (1985).

H. KIDO, T. HOSHIKAWA, M. SHIMADA, and M. KOIZUMI
Preparation of Magnetic Properties of $YMn_2(Si_{1-x}Ge_x)_2$ Compounds
(18.1 and 18.2; 1.3; 21.1; 21.4)
(a) Vol. **87**, No. 1, p. K61—K64 (1985).

D. SCHRYVERS, G. VAN TENDELOO, and S. AMELINCKX
*On the Ordering Mechanism of Ni_3Mo. A High Resolution Electron Microscopy Study
(1.3 and 1.4; 21; 21.1)
(a) Vol. **87**, No. 2, p. 401—415 (1985).

V. G. TSIRELSON, M. YU. ANTIPIN, R. G. GERR, R. P. OZEROV, and YU. T. STRUCHKOV
*Ruby Structure Peculiarities Derived from X-Ray Diffraction Data. Localization of Chromium Atoms and Electron Deformation Density
(1.3 and 10.2; 13.4; 22.6)
(a) Vol. **87**, No. 2, p. 425—433 (1985).

K. URBAN
*Determination of Long-Range Order Parameter in Alloys by Means of Electron Diffraction in the Electron Microscope
(1.3; 21.1; 21.1.1)
(a) Vol. **87**, No. 2, p. 459—471 (1985).

L. DĄBROWSKI
*Short- and Long-Range Ordering in Binary Alloys
(1.3; 21.1)
(b) Vol. **128**, No. 2, p. 371—378 (1985).

J. P. ZHANG, H. Q. YE, K. H. KUO, and S. AMELINCKX
*A HREM Study of the Crystal Structure of Cu_4Ti
(1.3 and 1.4; 21.1)
(a) Vol. **88**, No. 2, p. 475—482 (1985).

H. Q. YE and S. AMELINCKX
*Long-Period and Hexagonal Shear Domain Structures Co-Existing with the Cu_3Ti_2 Phase
(1.3 and 1.4; 21.1)
(a) Vol. **88**, No. 2, p. 483—495 (1985).

F. H. SÁNCHEZ, R. C. MERCADER, A. F. PASQUEVICH, A. G. BIBILONI, and A. R. LÓPEZ-GARCÍA
*The Influence of Particle Sizes on the Oxidation Kinetics of AgSn Alloys Studied by Mössbauer Spectroscopy
(1.3 and 5; 21.6)
(a) Vol. **89**, No. 1, p. K17—K21 (1985).

M. TAKAHASHI, K. NUSHIRO, and S. ISHIO
*Formation of the F.C.C. Phase in Fe–C Alloys by Rapid Quenching
(1.3; 21.1.1)
(a) Vol. **89**, No. 1, p. K27—K29 (1985).

D. J. GONZÁLEZ, C. M. ZAPATA, and J. A. ALONSO
*Density Functional Pseudopotential Calculation of the Cohesive Properties of Disordered Solid Alloys of Alkaline-Earth Metals. Influence of the Ionic Pseudopotential
(1.3 and 13; 4; 21.3)
(b) Vol. **129**, No. 2, p. 483—488 (1985).

F. REYNAUD
*Relation between the Pair Potentials V_{AA} and V_{BB} in Classical Statistical Physics of the Order–Disorder Transitions in Binary Substitutional Alloys
(1.2 and 1.3; 21)
(a) Vol. **90**, No. 2, p. 477–481 (1985).

G. WENDROCK and H. LÖFFLER
*Two-Step Ageing Treatments as a Means to Estimate the Zinc Content of Guinier-Preston Zones at Elevated Temperatures in an Al–15 at% Zn Alloy
(1.3; 21; 21.1)
(a) Vol. **90**, No. 2, p. 483–487 (1985).

H. NEUMANN, W. MATZ, W. HOYER, and M. WOBST
Investigation of the Structure of Molten $Ge_{15}Te_{85}$ in Dependence on Temperature
(2; 1.3; 22.7)
(a) Vol. **90**, No. 2, p. 489–498 (1985).

M. J. MARCINKOWSKI
*Strengthening of Alloys by Atomic Order
(1.3 and 10.1; 21.1.1)
(a) Vol. **90**, No. 2, p. 621–630 (1985).

T. SOMA, M. TAKAHASHI, and M.-MATSUO KAGAYA
Compression Effect on the α–β Phase Transition of Cu-Based Binary Alloys
(1.2 and 12.1; 1.3; 21.1)
(b) Vol. **131**, No. 1, p. K11–K14 (1985).

D. S. HAROUTUNIAN, A. G. KARAGYOZYAN, and G. N. KARAJIAN
*Anomalous Paramagnetic State in AgMn Spin Glass and the Kondo Effect
(1.3 and 14.1; 18.1; 21)
(b) Vol. **131**, No. 1, p. K87–K90 (1985).

F. C. LOVEY, G. VAN TENDELOO, J. VAN LANDUYT, L. DELAEY, and S. AMELINCKX
*Erratum to: On the Nature of Various Stacking Defects in 18R Martensite in Cu–Al Alloys. A Study by High Resolution Electron Microscopy
(1.3; 10.1; 21; 21.1)
(a) Vol. **91**, No. 1, p. 301 (1985).

T. SATO, N. MORITA, and M. SHIMADA
*Synthesis and Electrical Properties of $Li_xMg_{1-x}M_2O_4$ (M = Ti, V) ($0 \leqq x \leqq 1.0$)
(1.3 and 14.3; 22.8)
(a) Vol. **91**, No. 1, p. K81–K85 (1985).

L. DĄBROWSKI
*Atomic Ordering in Multicomponent Alloys
(1.3; 21.1)
(b) Vol. **131**, No. 2, p. 437–442 (1985).

V. I. BOGDANOV and D. L. FUKS
Many-Body Contribution to the Energy of Metals with B.C.C. Lattice
(4; 1.3; 21.1; 21.2)
(b) Vol. **131**, No. 2, p. 495–503 (1985).

K. Bortel
Electronic Contribution to the Ordering Energy of a Binary Alloy
(13; 1.3; 21.1)
(b) Vol. **131**, No. 2, p. 595–604 (1985).

D. Colaitis, D. van Dyck, and S. Amelinckx
*Electron Microscopic Study of the Domains and Interfaces in MV_4S_8 (M = Fe, Co, Ni)
(1.3 and 1.4; 21.1; 21.1.1)
(a) Vol. **91**, No. 2, p. 385–406 (1985).

R. Schülbe, and U. Schmidt
*On the Influence of the Pre-Ageing Conditions on the Reversion Behaviour of an Al–1.87 at% Cu Alloy
(1.3; 12.1; 21; 21.1)
(a) Vol. **91**, No. 2, p. 433–438 (1985).

G. Wendrock, H. Löffler, O. Kabisch, and Chu Thien Truong
*On the Position of the Critical Temperature for the Dissolution of Guinier-Preston Zones in an Al–15 at% Zn Alloy
(1.3 and 10.1; 21; 21.1)
(a) Vol. **91**, No. 2, p. 453–460 (1985).

L. Uray
Investigation of Dilute Tungsten Alloys by Electrical Transport Measurements
(14.1 and 15; 1.3; 21.1; 21.1.1)
(a) Vol. **91**, No. 2, p. 729–735 (1985).

W. M. Stobbs
*The Short Range Order Exhibited by Au–Fe Alloys
(1.3; 1.4; 21.1.1; 21.6)
(a) Vol. **92**, No. 1, p. 69–81 (1985).

Minh Hoang, J.-C. Niepce et C. Mutin
*Mise en évidence du rôle chimique joué par l'oxyde de niobium Nb_2O_5 lors du frittage d'un mélange des perovskites $BaTiO_3$ et $CaZrO_3$
(Evidence of the Chemical Role Played by Niobium Oxide Nb_2O_5 during Sintering of Mixtures of $BaTiO_3$ and $CaZrO_3$ Perovskites)
(1.3 and 14.4; 22.8.1)
(a) Vol. **92**, No. 1, p. 83–93 (1985).

W. J. Jiang, Z. Zhang, Dz. K. Hei, and K. H. Kuo
*High Resolution Electron Microscopy of Translation Twins in NiZr
(1.3 and 1.4; 21.1)
(a) Vol. **92**, No. 1, p. K5–K6 (1985).

R. A. Dunlap
*The Formation of Icosahedral Al–Transition Metal Alloys
(1.2 and 1.3; 21; 21.1)
(a) Vol. **92**, No. 1, p. K11–K13 (1985).

H. NEUMANN, W. HOYER, W. MATZ, and M. WOBST
*Neutron Scattering Investigations on Molten Eutectic Ag–Ge Alloys
(1.3 and 2; 21.6; 22.1.1)
(a) Vol. **92**, No. 1, p. K19–K22 (1985).

C. OTERO AREÁN and M. C. TROBAJO-FERNANDEZ
Cation Distribution in $Mg_xNi_{1-x}Ga_2O_4$ Oxide Spinels
(1.1; 1.3; 22.8.2)
(a) Vol. **92**, No. 2, p. 443–447 (1985).

G. HÜBNER and H. LÖFFLER
*On the Quench-Sensitivity of the Decomposition Rate of Al–Zn Alloys
(1.3 and 10.1; 1.2; 21; 21.1)
(a) Vol. **92**, No. 2, p. 517–524 (1985).

1.4 Microstructure
(Magnetic Domains See 18; Ferroelectric Domains See 14.4.1)

Original Papers and Short Notes

M. TAKEDA, H. YOSHIDA, and H. HASHIMOTO
*Local Tetragonality and Atomic Structure in Nb_3Sn Superconductor Studied by High Resolution Electron Microscopy
(1.4 and 14.2; 21)
(a) Vol. **87**, No. 2, p. 473–482 (1985).

W. GRAEFF and K. WIETESKA
*Observation of Domain Structure in a (111) Oriented Grain of Fe(Si) by X-Ray White Beam Topography
(1.4; 18.2; 21.1.1)
(a) Vol. **87**, No. 2, p. 517–525 (1985).

M. BEN SALEM, B. YANGUI, G. SCHIFFMACHER, and C. BOULESTEIX
*Twinning of the Hexagonal (A) Structure of Rare Earth Sesquioxides
(1.4; 22.6)
(a) Vol. **87**, No. 2, p. 527–536 (1985).

R. ŁYŻWA
*Connection between the Thermodynamic Potential and the Interaction Function for Systems with Bilinear Hamiltonians in Molecular Field Approximation
(1.4)
(b) Vol. **128**, No. 1, p. K1–K3 (1985).

L. SZCZEŚNIAK, K.-P. MEYER, H. BLUMTRITT, B. HILCZER, R. LE BIHAN, and E. H. BOUDJEMA
*Shape and Arrangement of the "Hexagonal Domain" and Distribution of Ferroelectric Domains in GASH Single Crystals
(1.4 and 14.4.1; 22.9)
(a) Vol. **88**, No. 1, p. 93–101 (1985).

G. MARTENS, P. RABE, and P. WENCK
*Destructive Interference and Multiple Scattering Effects Observed in Ca K-Edge EXAFS Spectra
(1.4 and 20; 22.6; 22.8.1)
(a) Vol. **88**, No. 1, p. 103–111 (1985).

M. J. MARCINKOWSKI
*The Structure of Twist Boundaries in Terms of Surface Dislocations
(1.4; 10)
(a) Vol. **88**, No. 1, p. 129–136 (1985).

M. ARITA, H.-U. NISSEN, and W. SCHAUER
*Structure Images of the Hexagonal Nb_5Ge_3 ($D8_8$) Structure
(1.4; 14.2; 21)
(a) Vol. **88**, No. 1, p. K1–K4 (1985).

P. A. BEZIRGANYAN, A. G. SEDRAKYAN, A. A. PAPOYAN, and K. G. TRUNI
*Investigation of the Influence of Deformation on Displacement Lines
(1.4)
(a) Vol. **88**, No. 2, p. 407–414 (1985).

L. BENEŠ, J. VOTINSKÝ, P. LOŠŤÁK, J. KALOUSOVÁ, and J. KLIKORKA
*Cobaltocene Intercalate of the Layered $SnSe_2$
(1.4; 22)
(a) Vol. **89**, No. 1, p. K1–K4 (1985).

B. GŁOWACKI and M. HOROBIOWSKI
*Influence of Temperature of Diffusion Growth and Morphology of Nb_3Sn Superconducting Layer on the Value of the Pinning Force
(1.4 and 14.2; 21)
(a) Vol. **89**, No. 1, p. K13–K16 (1985).

J. KUBĚNA and J. HLÁVKA
*Space Correlation of Microdefects with Recombination of Excess Carriers in CZ-Si
(1.4 and 16; 22.1.2)
(a) Vol. **89**, No. 1, p. K23–K25 (1985).

J. SITEK and M. MIGLIERINI
*Mössbauer Spectroscopy on Amorphous $Fe_xNi_{80-x}B_{20}$ after Neutron Irradiation
(1.4; 5; 11; 21.1; 21.1.1)
(a) Vol. **89**, No. 1, p. K31–K33 (1985).

C. YU. SHUNYAEV, M. T. VARSHAVSKII et A. N. MEN
*Influence de l'ordre à courte distance sur le moment magnétique des solutions solides quasibinaires à la structure du grenat avec la substitution dans un sous-réseau
(Influence of Short-Range Order on the Magnetic Moment of Quasibinary Granat-Type Solid Solutions with Sublattice Substitution)
(1.4 and 18; 22.8.2)
(b) Vol. **129**, No. 2, p. K131–K134 (1985).

L. DELAEY, G. GUÉNIN, Y. MURAKAMI, P. F. GOBIN, G. VAN TENDELOO, and J. VAN LANDUYT
Strain Field Observation and Lattice Stability at the Tip of Thermoelastic Cu–Zn–Al Martensite Plates
(1.3; 1.4; 12.1; 21.1)
(a) Vol. **89**, No. 2, p. 457–465 (1985).

W. GRUSCHEL and M. WILKENS
*Focus Dependence of the Black–White Contrast of Small Point Defect Clusters on Transmission Electron Microscope Images of Crystalline Specimens
(1.4; 10)
(a) Vol. **89**, No. 2, p. 467–482 (1985).

L. T. MIYADA-NABORIKAWA, R. DE BATIST, and P. DELAVIGNETTE
*Transmission Electron Microscopy Observations of Dislocations and Twins in Polycrystalline Zirconium
(1.4; 10.1; 21)
(a) Vol. **89**, No. 2, p. 521–531 (1985).

M. G. BLANCHIN, L. A. BURSILL, and DAVID J. SMITH
Extended Defects in Deformed Rutile
(10.2; 1.4; 22.6)
(a) Vol. **89**, No. 2, p. 559–570 (1985).

A. Ben Fredj et G. Nihoul
*Observation en microscopie électronique γ haute résolution d'oxydes γ grande maille (c-Dy_2O_3) et interprétation des images obtenues
(High-Resolution Electron Microscopic Observation of Large Cell Oxides (c-Dy_2O_3) and Interpretation of the Obtained Images)
(1.1 and 1.4; 22.6)
(a) Vol. **91**, No. 2, p. 351—369 (1985).

D. Colaitis, D. van Dyck, and S. Amelinckx
*Electron Microscopic Study of the Domains and Interfaces in MV_4S_8 (M = Fe, Co, Ni)
(1.3 and 1.4; 21.1; 21.1.1)
(a) Vol. **91**, No. 2, p. 385—406 (1985).

K. Scheerschmidt and R. Hillebrand
Calculated Defocus Diffraction Contrast and Lattice Fringe Imaging of Small Crystal Defects
(1.1; 1.4; 10)
(a) Vol. **91**, No. 2, p. 465—482 (1985).

G. van Tendeloo, S. Amelinckx, C. Manolikas, and Wen Shulin
*The Direct Observation of "Discommensurations" in Barium Sodium Niobate (BSN) and Its Homologues
(1.2 and 1.4; 22.8.1)
(a) Vol. **91**, No. 2, p. 483—501 (1985).

I. Gaal and J. M. Neugebauer
Effect of Grain Boundaries on the Out-Diffusion Rate of Volatile Solutes from Thin Wires
(9; 1.4; 21)
(a) Vol. **91**, No. 2, p. 573—581 (1985).

V. Yu. Gertsman, R. Z. Valiev, and O. A. Kaibyshev
*On the Energy Change during the Grain Boundary Structure Recovery from Non-Equilibrium State to Equilibrium State
(1.4 and 10)
(a) Vol. **91**, No. 2, p. K119—K124 (1985).

Pan Xiao-qing, Hu Mei-shen, Yao Ming-hui, and Feng Duan
*A TEM Study of the Incommensurate Phase and Related Phase Transitions in Barium Sodium Niobate
(1.2 and 1.4; 22.8.1)
(a) Vol. **92**, No. 1, p. 57—68 (1985).

W. M. Stobbs
The Short Range Order Exhibited by Au–Fe Alloys
(1.3; 1.4; 21.1.1; 21.6)
(a) Vol. **92**, No. 1, p. 69—81 (1985).

S. A. Demin, A. I. Ustinov, and K. V. Chuistov
*Structural Aspects of Reversible Plastic Deformation of Martensitic Mn–Cu Alloys
(1.4 and 10.1; 21.1)
(a) Vol. **92**, No. 1, p. 101—107 (1985).

1.5 Films

Original Papers and Short Notes

R. P. Goyal, Raviendra D., and B. R. K. Gupta
*Electroless Deposition of In_2O_3 and In_2O_3:Sn (ITO)
(1.5 and 14.3.1; 20.1; 22.6)
(a) Vol. 87, No. 1, p. 79–83 (1985).

V. Kraševec, A. Prodan, M. Hudomalj, and S. Sulčič
*A New Form of SnO_2 Found during Oxidation of α-SnO Thin Films
(1.5; 22.6)
(a) Vol. 87, No. 1, p. 127–133 (1985).

H.-J. Fitting, K. König, and D. Hecht
Exoelectron Emission Due to Alkali Ions on Insulating Layers
(17; 1.5; 21.2; 22.1.2; 22.6)
(a) Vol. 87, No. 1, p. K77–K79 (1985).

G. D. Bagratishvili, R. B. Dzhanelidze, T. V. Eterashvili, D. A. Jishiashvili, and E. R. Kutelia
*Properties and Structure of Silicon Oxynitride Films Obtained in a Hydrazine Plasma
(1.5 and 20.1; 22.8)
(a) Vol. 87, No. 2, p. 435–440 (1985).

A. Strigazzi
*A New Method for Measuring the Mixed Splay–Bend Elasticity and the Anchoring Strength in a Nematic Film
(1.5 and 12.2; 22)
(a) Vol. 87, No. 2, p. 537–541 (1985).

D. Stock, H.-D. Geiler, and K. Hehl
*Theoretical Evidence for Opposite Moving Phase Fronts during Ultrafast Solidification Processes
(1.5 and 3; 11; 22.1.2)
(a) Vol. 87, No. 2, p. K115–K118 (1985).

B. Rother and C. Weissmantel
*Structure and Chemical Composition of RF-Sputtered Boron Nitride Films
(1.5; 3; 22)
(a) Vol. 87, No. 2, p. K119–K121 (1985).

W. Wierzchowski
*Double-Crystal Topographic Investigation of Misfit Dislocations in Silicon Epitaxial Layers
(1.5; 22.1.2)
(a) Vol. 88, No. 1, p. 77–87 (1985).

Z. Ivanov, B. Todorov, and V. Gancheva
*Thin ε-Phase Films Prepared by Alloy Evaporation
(1.5; 14.2)
(a) Vol. 88, No. 1, p. 89–92 (1985).

C. H. Ling, C. Y. Kwok, and K. Prasad
*Relative Hydrogen Content in Plasma-Enhanced CVD Silicon Nitride Films: Substrate Temperature Dependence and Effect of Thermal Annealing
(1.5 and 10.2; 22)
(a) Vol. **89**, No. 1, p. K39–K43 (1985).

L. N. Aleksandrov
Simulation of the Influence of Mechanical Stresses on the Kinetics of Crystallization of Ion-Implanted Silicon Layers under Pulse Heating
(2 and 3; 1.5; 11; 22.1.2)
(a) Vol. **89**, No. 2, p. 443–449 (1985).

V. A. Pilipovich, B. A. Budkevich, I. M. Romanov, G. D. Ivlev, I. A. Ges, and S. P. Zhvavyi
Laser Annealing Effects on Electrochromic Properties of Amorphous Evaporated WO_3 Films
(14.3.1 and 20.1; 1.5; 22.6)
(a) Vol. **89**, No. 2, p. 709–717 (1985).

Th. Moritz and P. Schäfer
*Preparation of Pb(Te, Se) Films by Simultaneous Evaporation of PbTe and Se
(1.5 and 3; 22.8)
(a) Vol. **89**, No. 2, p. K119–K121 (1985).

U. Rensch and G. Eichhorn
*Investigation of the State of Strain and of the Electrical Properties of AlN Layers on Monocrystalline Si
(1.5 and 14.3.1; 22.1.2; 22.2)
(a) Vol. **90**, No. 1, p. 135–141 (1985).

P. D. Cikmach, J. J. Kleperis, A. R. Lusis, and G. M. Ramāns
*Influence of Ordering of Structure on the Optical Properties of WO_3 Thin Film
(1.5 and 20.1; 22.6)
(a) Vol. **90**, No. 1, p. K1–K4 (1985).

R. M. Candea, D. Dărdâlat, R. Turcu, and E. Indrea
*Properties of PbSe Films Prepared by Chemical „Anorganic" Deposition
(1.5 and 14.3.1; 16; 20.1; 22)
(a) Vol. **90**, No. 1, p. K91–K95 (1985).

N. N. Koren, V. F. Gremenok, and V. V. Kindyak
*On the Crystal Structure of Thin CdS Films Grown by Pulse Laser Evaporation
(1.5; 22.4.1)
(a) Vol. **90**, No. 2, p. K121–K123 (1985).

B. V. Vasilev, A. P. Dolgarev, A. A. Potseluiko, and V. G. Pynko
*Transitional Epitaxial Layers in Nickel Films Deposited in Vacuum on Complex LiF/NaCl Substrates
(1.5 and 3; 21.1)
(a) Vol. **90**, No. 2, p. K129–K130 (1985)

O. Bostanjoglo and E. Endruschat
*Kinetics of Laser-Induced Crystallization of Amorphous Germanium Films
(1.5 and 3; 2; 22.1.1)
(a) Vol. **91**, No. 1, p. 17–28 (1985).

1.6 Surfaces

Original Papers and Short Notes

V. M. GENKIN and A. A. FRAERMAN
*The Rate of Advance of a Monoatomic Step on the Surface of a Growing Kossel Crystal
(1.6 and 3)
(a) Vol. **87**, No. 1, p. 85–91 (1985).

V. O. ESIN and V. I. DANILYUK
Cluster Structure of a Two-Dimensional Lattice Gas Model
(1.4; 1.6)
(b) Vol. **127**, No. 2, p. 443–448 (1985).

H. DÜSTERHÖFT, R. DAUGS, and R. PIPPIG
A New Method for the Interpretation of Energy Distributions Measured on Secondary Ions
(17; 1.6; 11; 21.1; 21.1.1; 22.1.2)
(a) Vol. **87**, No. 2, p. 499–506 (1985).

I. MOJZES, YU. VAULIN, N. GERASIMENKO, V. OBODNIKOV, B. KOVÁCZ, and R. VERESEGYHÁZY
*Spatial Distribution of the Components in Au/GaAs System after Scanning Laser Annealing
(1.6 and 11; 22.2.1)
(a) Vol. **87**, No. 2, p. K127–K130 (1985).

I. URSU, V. CRĂCIUN, L. NANU, D. CRĂCIUN, I. N. MIHĂILESKU, AL. POPA, A. PANTELICĂ, and M. ŠĂLĂGEAN
*On the Behaviour of GaAs Single Crystals Irradiated with a Free-Running Ruby Laser
(1.6; 11; 22.2.1)
(a) Vol. **88**, No. 2, p. 415–419 (1985).

M. A. ANDREEVA, K. ROCETE, and YU. P. KHAPACHEV
*Matrix Analog of the Takagi Equations for Grazing-Incidence Diffraction
(1.6)
(a) Vol. **88**, No. 2, p. 455–462 (1985).

A. L. GOLOVIN, R. M. IMAMOV, and E. A. KONDRASHKINA
*Potentialities of New X-Ray Diffraction Methods in Structural Studies of Ion-Implanted Silicon Layers
(1.6; 11; 22.1.2)
(a) Vol. **88**, No. 2, p. 505–514 (1985).

O. G. MÜLLER, W. KÄTZEL, and F. STORBECK
*Determination of Sulphur Coverages of Fe(III) by Means of Quantitative Auger-Electron Spectrometry
(1.6; 21.1.1)
(a) Vol. **89**, No. 1, p. 89–94 (1985).

A. L. GOLOVIN, R. M. IMAMOV, and E. A. KONDRASHKINA
*Absolute Measurements of Lattice Spacings in Surface Layers of Crystals
(1.6; 22.2.3)
(a) Vol. **89**, No. 1, p. K5–K7 (1985).

L. T. Wille and J. Vennik
*Kinetics of Surface Segregation in Binary Alloys
(1.6; 21.1)
(b) Vol. **131**, No. 2, p. 443–449 (1985).

Th. Nattermann
The Random Field Ising Model in 2 + ε Dimensions. Critical Behaviour from Interface Fluctuations
(18; 1.6)
(b) Vol. **131**, No. 2, p. 563–571 (1985).

W. Frentrup, M. Griepentrog, and U. Müller-Jahreis
Negative Secondary Ion Emission Influenced by Alkali Atoms
(17; 1.6; 22.1.1; 22.1.2; 22.2.1; 22.2.2)
(a) Vol. **91**, No. 2, p. 447–452 (1985).

F. Appel, H. Bethge, and M. Krohn
*Direct Evidence of Post-Nucleation Decoration
(1.6; 9; 22.5.2)
(a) Vol. **91**, No. 2, p. 461–464 (1985).

S. Trimper
*On a Random Surface Tension Model
(1.6)
(b) Vol. **132**, No. 1, p. K1–K4 (1985).

A. M. Afanasev, R. M. Imamov, E. Kh. Mukhamedzhanov, and Le Cong Qui
*Photoemission Curves Obtained in Conditions of X-Ray Grazing-Incidence Laue Diffraction from Crystals with a Disturbed Surface Layer
(1.6; 22.5.2)
(a) Vol. **92**, No. 2, p. 355–360 (1985).

G. P. Yablonskii
*Light Figures on the Surfaces of Semiconductors
(1.6; 22.4.1)
(a) Vol. **92**, No. 2, p. 431–441 (1985).

Z. A. Matysina, A. M. Ovrutskii, and L. M. Chuprina
*Influence of Correlation and Anisotropy on the Molecular Roughness of the Crystal Face
(1.2 and 1.6; 21.7)
(a) Vol. **92**, No. 2, p. K93–K97 (1985).

R. Burkova, S. Takova, and G. Dimitrov
*S-Pits on (111) Silicon Surfaces
(1.6 and 10.2; 22.1.2)
(a) Vol. **92**, No. 2, p. K107–K108 (1985).

2. Non-Crystalline State

Review Articles

Original Papers and Short Notes

S. G. SAXENA and K. B. GARG
*EXAFS Study of Some Iron Based Metallic Glasses
(2 and 4; 21.1.1)
(a) Vol. **87**, No. 1, p. K25–K28 (1985).

S. K. J. AN-ANI and C. A. HOGARTH
*Optical Absorption Edge in Amorphous $In_{30}Se_{70}$ Films
(2 and 20.1; 22.3)
(a) Vol. **87**, No. 1, p. K65–K67 (1985).

H. KRONMÜLLER
*The Rôle of Two-Level Systems in Amorphous Metallic Alloys. II. Micromagnetic Theory of Magnetic After-Effects in Amorphous Ferromagnets
(2 and 18.2; 21.1.1)
(b) Vol. **127**, No. 2, p. 531–541 (1985).

D. A. FUSHMAN
*Kinetic Properties of the Two-Level Systems in Glasses with Account of Tunneling
(2; 13)
(b) Vol. **127**, No. 2, p. 679–688 (1985).

W. SCHEIBNER and M. JÄCKEL
Thermal Conductivity and Specific Heat of an Epoxy Resin/Epoxy Resin Composite Material at Low Temperatures
(8; 2; 22.9)
(a) Vol. **87**, No. 2, p. 543–547 (1985).

N. I. NOSKOVA, N. F. VILDANOVA, YU. I. FILIPPOV, and A. P. POTAPOV
Strength, Plasticity, and Fracture of Ribbons of $Fe_5Co_{70}Si_{15}B_{10}$ Amorphous Alloy
(10.1 and 12.1; 2; 21.1; 21.1.1)
(a) Vol. **87**, No. 2, p. 549–557 (1985).

M. VÁZQUEZ, W. FERNENGEL, and H. KRONMÜLLER
Changes of the Magnetic Properties of the Nearly Non-Magnetostrictive Amorphous Alloy $Co_{58}Fe_5Ni_{10}Si_{11}B_{16}$ by Annealing under Tensile Stress
(10.1 and 18.2; 2; 21.1; 21.1.1)
(a) Vol. **87**, No. 2, p. 609–615 (1985).

P. KOŠTIAL and J. SLABEYCIUS
*The Determination of the Distribution Function of Activation Energies of Structural Changes in Amorphous Se
(2 and 7; 22.1.3)
(a) Vol. **87**, No. 2, p. K109–K113 (1985).

T. LUCIŃSKI and J. BASZYŃSKI
Thermopower in Amorphous $Fe_{1-x}Si_x$ Films
(15; 2; 21.1.1)
(a) Vol. **87**, No. 2, p. K191–K193 (1985).

P. I. Stoianova and R. E. Andreichin
Coordinate Sensitive Photoelement on the Basis of the Photoelectret State in Vitreous Semiconductors
(14.3.3 and 16; 2; 22)
(a) Vol. **88**, No. 1, p. K91–K93 (1985).

T. Bhattacharya and A. K. Ghosh
*Pair-Correlation Functions for Liquid Noble Metals through Percus-Yevick Theory
(2; 4; 21.1; 21.6)
(b) Vol. **128**, No. 2, p. 419–423 (1985).

W. Ebeling and W. Richert
*Thermodynamic Properties of Liquid Hydrogen Metal
(2; 8; 21; 23)
(b) Vol. **128**, No. 2, p. 467–474 (1985).

A. Bath, J. Hugel, and R. Kleim
*Thermoelectric Power of Liquid Germanium and Tin
(2 and 15; 14.1; 22.1; 22.1.1)
(b) Vol. **128**, No. 2, p. 761–767 (1985).

H. Bergner, V. Brückner, F. Kerstan, and W. Nowick
*Picosecond Processes of Laser-Excited Carriers in Silicon on Sapphire
(2 and 14.3.1; 16; 20.1; 22.1.2; 22.6)
(b) Vol. **128**, No. 2, p. 769–777 (1985).

J. G. Gasser, J. L. Bretonnet, and A. Bruson
*Temperature Dependence of the Liquid Tin Resistivity
(2 and 14.1; 21.7)
(b) Vol. **128**, No. 2, p. 789–796 (1985).

H. Böttger, P. Szyler, and D. Wegener
*Numerical Investigation of Non-Ohmic Hopping Conduction at Low Temperatures
(2 and 14.3)
(b) Vol. **128**, No. 2, p. K179–K182 (1985).

G. Suryanarayana Reddy, H. C. Gupta, and B. B. Tripathi
Non-Linear Dependence of the Temperature Coefficient of Electrical Resistivity on Electron–Phonon Interaction Constant in the Case of Transition Metal Amorphous Alloys
(14.1; 2; 21.1)
(b) Vol. **128**, No. 2, p. K183–K186 (1985).

P. Allia, E. Bonetti, E. Evangelista, L. Turroni, and F. Vinai
*Reversible and Irreversible Processes of Structural Relaxation and Dynamic Young Modulus Behaviour in the $Fe_{40}Ni_{38}Mo_4B_{18}$ Amorphous Alloy
(2; 12.1; 21.1; 21.1.1)
(a) Vol. **88**, No. 2, p. 521–527 (1985).

X. B. Liao, J. Y. Lian, and G. L. Kong
Analysis for Steady State Field Effect Measurements on a-Si:H Films
(14.3.1; 2; 13.3; 22.1.2)
(a) Vol. **88**, No. 2, p. 697–704 (1985).

J. Krištofik, J. J. Mareš, and V. Šmíd
The Effect of Pressure on Conductivity and Permittivity of As_2Te_3-Based Glasses
(14.3; 2; 22)
(a) Vol. **89**, No. 1, p. 333–345 (1985).

S. A. Y. Al-Ismail, K. Arshak, and C. A. Hogarth
Electron Spin Resonance Measurements and Electrical Characteristics, before and after Electroforming, of Thin Films of SiO/Nb_2O_5 and Nb_2O_5
(14.3.1 and 14.3.4; 2; 19; 22.6; 22.6.1)
(a) Vol. **89**, No. 1, p. 363–374 (1985).

A. V. Ivanov and A. I. Melker
*Interatomic Bond Rupture in the System of Two Coupled Anharmonic Chains with a Weak Interaction
(2 and 10.2; 4; 22.9)
(a) Vol. **89**, No. 1, p. K51–K54 (1985).

A. Fontana, G. Mariotto, and F. Rocca
*Low-Frequency Light Scattering in Superionic Glasses $(AgI)_x(Ag_2O\ 1B_2O_3)_{1-x}$
(2 and 14.4; 20.1; 22.8)
(b) Vol. **129**, No. 2, p. 489–496 (1985).

P. Blaudeck
*Shell Model for Clusters of Dence-Packed s-Scatterers and Calculation of the Density of States
(2 and 13)
(b) Vol. **129**, No. 2, p. 729–734 (1985).

S. Yu. Potapenko and A. M. Satanin
The Resistance of a One-Dimensional System with a Stationary Random Potential which is a Markov Process
(13 and 14; 2)
(b) Vol. **129**, No. 1, p. 805–811 (1985).

M. I. Katsnelson and A. S. Shcherbakov
*Temperature Dependence of Conductivity at Small Electron Mean Free Path
(2 and 14.1; 21.1)
(b) Vol. **129**, No. 2, p. 813–822 (1985).

P. Singh, Ch. Pal, and K. N. Khanna
Flory's Formula for the Entropy of Mixing of LiNa Alloy
(1.3 and 8; 2; 21.2)
(b) Vol. **129**, No. 2, p. K111–K115 (1985).

V. Hietschold and G. Seifert
Soft X-Ray Spectra of Amorphous Hydrogenated Silicon
(13 and 20; 2; 22.1.2)
(b) Vol. **129**, No. 2, p. K163–K166 (1985).

L. N. Aleksandrov
*Simulation of the Influence of Mechanical Stresses on the Kinetics of Crystallization of Ion-Implanted Silicon Layers under Pulse Heating
(2 and 3; 1.5; 11; 22.1.2)
(a) Vol. **89**, No. 2, p. 443–449 (1985).

Nguyen Chi Thanh and R. Krishnan
Anistropy of Magnetostriction in Coevaporated $Co_{1-x}Zr_x$ Amorphous Thin Films
(18.2.1; 2; 21.1)
(a) Vol. **90**, No. 1, p. 313–323 (1985).

K. Eidner, G. Mayer, R. Schuster, L. Komitov, and A. G. Petrov
*Influence of Finite Sample Thickness and Distortion on the Critical Angle of Total Reflection
(2 and 20)
(a) Vol. **90**, No. 1, p. 349–353 (1985).

K. Yasuda, A. Yoshida, M. Takeda, H. Masuda, and I. Akasaki
*Boron Nitride Films Prepared by RF Sputtering
(2 and 3; 6; 20.1; 22)
(a) Vol. **90**, No. 1, p. K7–K9 (1985).

A. A. Bahgat, E. E. Shaisha, A. I. Sabry, and N. A. Eissa
*Mössbauer Analysis of Valence and Coordination of Fe Cations in TeO_2–Fe_2O_3 Glasses
(2 and 4; 5; 22.6.1)
(a) Vol. **90**, No. 1, p. K25–K29 (1985).

Y. Hoshino and T. Nishida
*Transient Photocurrent Wave Form Deformation in Amorphous Silicon by Successive Light Pulse Irradiations
(2 and 16; 14.3.4; 22.1.2)
(a) Vol. **90**, No. 1, p. K101–K102 (1985).

Y. Takano, Y. Tandoh, H. Ozaki, and N. Mori
*XPS Study on a-GeO_x ($1.3 < x < 2$) Films Prepared from GeO_2
(2; 17; 22.6)
(b) Vol. **130**, No. 2, p. 431–436 (1985).

B. R. Sood
Magnetoconductance in Metallic Granular Gallium Films
(14.1 and 18; 2; 21)
(b) Vol. **130**, No. 2, p. 801–806 (1985).

R. Somashekar, Ch. Satyanarayana, and M. S. Madhava
*Mesomorphic Behaviour of Mixtures of Cetyl Alcohol and Dimethyl Sulfoxide
(1 and 2; 22.9)
(a) Vol. **90**, No. 2, p. 449–456 (1985).

H. Neumann, W. Matz, W. Hoyer, and M. Wobst
*Investigation of the Structure of Molten $Ge_{15}Te_{85}$ in Dependence on Temperature
(2; 1.3; 22.7)
(a) Vol. **90**, No. 2, p. 489–498 (1985).

W. Fernengel, M. Rapp, and H. Kronmüller
*The Coercive Field of the Partially Crystallized Amorphous Alloy $Fe_{40}Ni_{40}P_{14}B_6$
(2 and 18.2; 21.1; 21.1.1)
(a) Vol. **90**, No. 2, p. 637–642 (1985).

M. M. Aleksandrova, V. D. Blank, V. I. Larchev, S. V. Popova, and G. G. Skrotskaya
*Transition Temperature and Heat of Crystallization of Amorphous Bulk Gallium Antimonide Obtained by Rapid Quenching from the Melt at High Pressure
(1.2 and 2; 22.2.3)
(a) Vol. **91**, No. 1, p. K5–K8 (1985).

I. N. Shabanova
*Chemical Bond Peculiarities of Elements in Amorphous Alloy Surface Layers on the Basis of Fe
(1.6 and 2; 17; 21.1; 21.1.1)
(a) Vol. **91**, No. 1, p. K13–K17 (1985).

R. Fromageau, A. Magnouche, and N. Gerard
*The Resistivity of Dissolved Hydrogen in Palladium and Amorphous $Pd_{80}Si_{20}$
(2 and 14.1; 21.6)
(a) Vol. **91**, No. 1, p. K75–K80 (1985).

L. M. Sandler, V. N. Pushkar, I. V. Filonchuk, and T. S. Korsunskaya
*Crystallization of Amorphous Fe–Si Films
(1.2 and 2; 1.5; 21.1.1)
(a) Vol. **91**, No. 2, p. 371–378 (1985).

J. Čermák, M. Černanský, P. Duhaj, and J. Horváth
*Crystallization of the Metallic Glass $Fe_{40}Ni_{40}B_{20}$ during Isochronal Annealing in Argon and Hydrogen
(2 and 3; 21.1; 21.1.1)
(a) Vol. **91**, No. 2, p. 407–410 (1985).

K. Urban, N. Moser, and H. Kronmüller
Phase Transitions between the Quasicrystalline, Crystalline, and Amorphous Phases in Al–14 at% Mn
(1.2; 2; 21; 21.1)
(a) Vol. **91**, No. 2, p. 411–422 (1985).

M. M. El-Zaidia, A. A. Ammar, and R. A. El-Mallwany
*Infra-Red Spectra, Electron Spin Resonance Spectra, and Density of $(TeO_2)_{100-x}(WO_3)_x$ and $(TeO_2)_{100-x}(ZnCl_2)_x$ Glasses
(2 and 20.1; 19; 22.6)
(a) Vol. **91**, No. 2, p. 637–642 (1985).

S. E. Abdulgafarov, V. L. Averyanov, B. T. Kolomiets, and V. M. Lyubin
*Photoconductivity of Vitreous As_2Se_3 Films Modified with Nickel
(2 and 16; 22)
(a) Vol. **92**, No. 1, p. 257–262 (1985).

A. K. George
*Anomalous Ultrasonic Absorption in a Nematic and a Polymesomorphic Liquid Crystal
(2 and 7; 22.9)
(a) Vol. **92**, No. 1, p. K15–K17 (1985).

H. Neumann, W. Hoyer, W. Matz, and M. Wobst
*Neutron Scattering Investigations on Molten Eutectic Ag–Ge Alloys
(1.3 and 2; 21.6; 22.1.1)
(a) Vol. **92**, No. 1, p. K19–K22 (1985).

3. Crystal Growth

Original Papers and Short Notes

V. M. GENKIN and A. A. FRAERMAN
*The Rate of Advance of a Monoatomic Step on the Surface of a Growing Kossel Crystal
(1.6 and 3)
(a) Vol. **87**, No. 1, p. 85–91 (1985).

N. RYŠAVÁ, L. TICHÝ, Č. BARTA, A. TŘÍSKA, and H. TICHÁ
*Kinetics of Recrystallization of Sb_2S_3 in the Glassy $(GeS_2)_{0.3}(Sb_2S_3)_{0.7}$
(2 and 3; 22)
(a) Vol. **87**, No. 1, p. K13–K17 (1985).

D. STOCK, H.-D. GEILER, and K. HEHL
*Theoretical Evidence for Opposite Moving Phase Fronts during Ultrafast Solidification Processes
(1.5 and 3; 11; 22.1.2)
(a) Vol. **87**, No. 2, p. K115–K118 (1985).

B. BOTHER and C. WEISSMANTEL
Structure and Chemical Composition of RF-Sputtered Boron Nitride Films
(1.5; 3; 22)
(a) Vol. **87**, No. 2, p. K119–K121 (1985).

H. OSHIMA, A. TANAKA, and T. SUKEGAWA
Calculation of the Ga–Al–Sb Phase Diagram Using the Redlich-Kister Expression
(1.2; 3; 22.2.4)
(a) Vol. **87**, No. 2, p. K131–K134 (1985).
Erratum
(a) Vol. **92**, No. 1, p. K87 (1985).

H. ATMANI
*Determination of Thermodynamic Parameters of the Amorphous Selenium Film Crystallization
(1.5 and 3; 22.1.3)
(a) Vol. **88**, No. 1, p. 113–120 (1985).

YU. B. BOLKHOVITYANOV and R. I. BOLKHOVITYANOVA
*Stress-Induced Isothermal Formation of Films during LPE Growth of III–V Compounds with $\Delta a/a \neq 0$
(3; 1.5; 22.2.1; 22.2.2; 22.2.4)
(a) Vol. **88**, No. 1, p. K5–K8 (1985).

D. RAVIENDRA, SUNDEEP, and J. K. SHARMA
In_2O_3 and In_2O_3:Sn (ITO) Films by Post-Oxidation of Metal Films
(14.3.1; 3; 20.1; 22.6)
(a) Vol. **88**, No. 1, p. K83–K86 (1985).

S. CHANDRA, O. N. SRIVASTAVA, and S. N. SAHU
Electrodeposited Tungsten Selenide Films. I. Preparatory Technique and Structural Characterization
(1.5; 3; 22)
(a) Vol. **88**, No. 2, p. 497–503 (1985).

S. E. Zabolotskii, V. P. Kalinushkin, T. M. Murina, M. G. Ploppa, and K. Tempelhoff
*The Influence of Growth Rate on the Impurity Cloud Parameters in Pure Silicon
(3 and 10.2; 22.1.2)
(a) Vol. **88**, No. 2, p. 539–542 (1985).

Z. Cimpl, F. Kosek, J. J. Janoš, andM. Mikhailov
*Heterogeneity of Evaporated As_2S_3 Films
(3; 1.5; 22)
(a) Vol. **88**, No. 2, p. K111–K113 (1985).

D. Stock, H.-D. Geiler, and K. Hehl
*A Model of Crystallization Processes Controlled by Temperature Pulses in Amorphous Semiconductors. I. Step-Like Temperature Pulse
(3; 1.2; 2; 22.1.2)
(a) Vol. **89**, No. 1, p. 57–63 (1985).

V. A. Zhorin, D. P. Shashkin, and N. S. Enikolopian
*Dynamics of Solid Solution Formation in Cu–Ni Mixture under Plastic Flow at High Pressure
(1.3 and 3; 10.1; 21.1)
(a) Vol. **89**, No. 2, p. 437–442 (1985).

L. N. Aleksandrov
*Simulation of the Influence of Mechanical Stresses on the Kinetics of Crystallization of Ion-Implanted Silicon Layers under Pulse Heating
(2 and 3; 1.5; 11; 22.1.2)
(a) Vol. **89**, No. 2, p. 443–449 (1985).

Th. Moritz and P. Schäfer
*Preparation of Pb(Te, Se) Films by Simultaneous Evaporation of PbTe and Se
(1.5 and 3; 22.8)
(a) Vol. **89**, No. 2, p. K119–K121 (1985).

K. Yasuda, A. Yoshida, M. Takeda, H. Masuda, and I. Akasaki
*Boron Nitride Films Prepared by RF Sputtering
(2 and 3; 6; 20.1; 22)
(a) Vol. **90**, No. 1, p. K7–K9 (1985).

V. O. Esin and L. P. Tarabaev
*A Generalized Criterion for the Transition from the Layer to the Continuous Mechanism of Crystal Growth
(3; 1.6)
(a) Vol. **90**, No. 2, p. 425–437 (1985).

B. V. Vasilev, A. P. Dolgarev, A. A. Potseluiko, and V. G. Pynko
*Transitional Epitaxial Layers in Nickel Films Deposited in Vacuum on Complex LiF/NaCl Substrates
(1.5 and 3; 21.1)
(a) Vol. **90**, No. 2, p. K129–K130 (1985).

O. Bostanjoglo and E. Endruschat
*Kinetics of Laser-Induced Crystallization of Amorphous Germanium Films
(1.5 and 3; 2; 22.1.1)
(a) Vol. **91**, No. 1, p. 17—28 (1985).

J. Čermák, M. Čerňanský, P. Duhaj, and J. Horváth
*Crystallization of the Metallic Glass $Fe_{40}Ni_{40}B_{20}$ during Isochronal Annealing in Argon and Hydrogen
(2 and 3; 21.1; 21.1.1)
(a) Vol. **91**, No. 2, p. 407—410 (1985).

L. N. Aleksandrov, A. N. Kogan, and N. P. Tikhonova
*Study of the Peculiarities of Film Growth from Molecular Beams by the Method of Simulation
(1.5 and 3; 22.1.2)
(a) Vol. **92**, No. 1, p. 109—113 (1985).

O. Bostanjoglo, E. Endruschat, E. I. Givargizov, and W. Tornow
*Metal-Assisted Grain Growth in Beam Crystallized Amorphous Ge/Au Films
(1.2 and 3; 21.6; 22.1.1)
(a) Vol. **92**, No. 1, p. K1—K4 (1985).

M. Wagner, H.-D. Geiler, and G. Götz
*Time-Resolved Investigation of Large-Area Explosive Crystallization of Amorphous Silicon Layers
(2 and 3; 1.2; 11; 22.1.2)
(a) Vol. **92**, No. 2, p. 413—420 (1985).

K. H. Heinig and H.-D. Geiler
*Phenomenological Theory of Explosive Solid Phase Crystallization of Amorphous Silicon. I. Stationary Solutions
(2 and 3; 1.2; 22.1.2)
(a) Vol. **92**, No. 2, p. 421—430 (1985).

4. Bonding Properties

Original Papers and Short Notes

H. RAUH and R. GEICK
Crystal-Independent Interionic Potentials and the Lattice Dynamics of the Perovskite-Type Layer Material K_2ZnF_4
(6; 4; 22.8)
(b) Vol. **127**, No. 1, p. 55–65 (1985).

N. DUTT, O. P. SHARMA, and J. SHANKER
*Analysis of Photoelastic Behaviour of Fluorite-Structure Crystals
(4 and 12.2; 14.4; 22.5.3)
(b) Vol. **127**, No. 1, p. 67–71 (1985).

W. N. BHENDE, P. S. BAKHSHI, and J. SHANKER
*Thermal Energy and the Equation of State for Alkali Halide Crystals
(4; 22.5.2)
(b) Vol. **127**, No. 1, p. 73–77 (1985).

E. A. VASILEV and V. A. VIRCHENKO
Effect of Interstitial Atoms on the Quadrupole Interaction at ^{119}Sn Nuclei in Some B8 Type Solid Solutions
(1.3 and 5; 4; 21.1.1)
(a) Vol. **87**, No. 1, p. K19–K24 (1985).

S. G. SAXENA and K. B. GARG
*EXAFS Study of Some Iron Based Metallic Glasses
(2 and 4; 21.1.1)
(a) Vol. **87**, No. 1, p. K25–K28 (1985).

P. ENDERS
*Structure and Bonding in Cubic IV–VI Crystals. IV. Evidence for Covalent Bonding in the IV–VI Family from the Extended Mulliken Scheme
(4 and 13; 1.2; 22)
(b) Vol. **127**, No. 2, p. 449–457 (1985).

R. K. GUPTA, A. SHIROMANY, P. S. BAKHSHI, and J. SHANKER
*Crystalline Properties of Alkali Halides Using the Generalized Huggins-Mayer Potentials
(4 and 8; 12.2; 22.5.2)
(b) Vol. **127**, No. 2, p. 473–479 (1985).

A. K. M. A. ISLAM and E. HUQ
*On Cohesive Energy and Grüneisen Parameter of Silver Halide Crystals
(4 and 8; 6; 22.5.1)
(b) Vol. **127**, No. 2, p. K109–K111 (1985).

S. C. AGRAWAL
*Analysis of the Equation of State of Ammonium Halides
(4; 22.5)
(b) Vol. **128**, No. 1, p. 69–73 (1985).

P. Enders
*Structure and Bonding in Cubic IV–VI Crystals. V. Lattice Repulsion. Comparison of Lattice and Vibronic Anharmonicity
(4 and 13; 22)
(b) Vol. **128**, No. 1, p. 75–82 (1985).

B. R. K. Gupta
*Static and Thermophysical Properties of Mixed Crystal Systems
(4 and 8; 22.5.4)
(b) Vol. **128**, No. 1, p. 99–104 (1985).

W. N. Bhende, P. S. Bakhshi, and J. Shanker
*Thermoelastic Behaviour and Higher-Order Elastic Constants of Ionic Crystals
(4 and 12.2; 6; 8; 22.5.2; 22.6)
(b) Vol. **128**, No. 1, p. 105–112 (1985).

V. K. Bashenov, V. E. Gorbachev, and D. I. Marvakov
Electron Density Distribution and Bond Charges in Covalent Crystals
(13; 4; 22.1)
(b) Vol. **128**, No. 1, p. K59–K63 (1985).

V. Valvoda, V. Šíma, Z. Smetana, and P. Čapková
*Neutron Powder Diffraction Study of the Debye-Waller Factor in TiC and TiN Compounds and Comparison with X-Ray Results
(4; 6; 21.1)
(a) Vol. **88**, No. 1, p. 157–163 (1985).

T. Bhattacharya and A. K. Ghosh
Pair-Correlation Functions for Liquid Noble Metals through Percus-Yevick Theory
(2; 4; 21.1; 21.6)
(b) Vol. **128**, No. 2, p. 419–423 (1985).

U. Pietsch
*Static and Dynamical Valence Charge and Bond Charge Properties of Zincblende Structure Compounds
(4; 13; 22.2)
(b) Vol. **128**, No. 2, p. 439–451 (1985).

P. J. Pomonis
*Electron Density and Work Function Dependence of the Enthalpy of Formation of Metal Oxides
(4; 8; 13.1; 13.3; 22.6)
(a) Vol. **88**, No. 2, p. 515–519 (1985).

N. Dutt, G. G. Agrawal, and J. Shanker
*Investigation of the Equation of State and the Grüneisen-Anderson Parameters for Fluorite-Structure Crystals
(4; 6; 8; 12.2; 22.5.3; 22.8.1)
(b) Vol. **129**, No. 1, p. 73–79 (1985).

T. TAKAI, T. HALICIOĞLU, and W. A. TILLER
*A Parametric Study of Crystal Stability and Surface Energy for Diamond Cubic Structures
(1 and 4; 22.1)
(b) Vol. **130**, No. 1, p. 131–138 (1985).

XIE YING, L. SNYDER, S. N. SAHU, and J. W. CORBETT
*MNDO Calculation on Molecular Clusters Chosen to Model the Bonding of Oxygen in Silicon
(4 and 13; 22.1.2)
(b) Vol. **130**, No. 1, p. 333–338 (1985).

A. G. PELMENCHIKOV, V. I. PAVLOV, and G. M. ZHIDOMIROV
*Quantum-Chemical Comparison of the Coordination Abilities of B and Al in Oxides within the Framework of the Molecular Model
(1.2 and 4; 22.6)
(b) Vol. **130**, No. 1, p. K1–K4 (1985).

A. A. BAHGAT, E. E. SHAISHA, A. I. SABRY, and N. A. EISSA
*Mössbauer Analysis of Valence and Coordination of Fe Cations in TeO_2-Fe_2O_3 Glasses
(2 and 4; 5; 22.6.1)
(a) Vol. **90**, No. 1, p. K25–K29 (1985).

T. TAKAI, T. HALICIOĞLU, and W. A. TILLER
*Calculation of Potential Energy Parameters for the Silicon–Carbon System
(1 and 4; 22.1; 22.1.2; 22.3)
(b) Vol. **130**, No. 2, p. 475–481 (1985).

T. SOMA and H.-MATSUO KAGAYA
Mode Grüneisen Parameters for Long-Wave Phonons of III–V Covalent Compounds under Pressure
(6; 4; 12.2; 22.2.1; 22.2.3)
(b) Vol. **130**, No. 2, p. 497–501 (1985).

P. ENDERS
*Structure and Bonding on Cubic IV–VI Crystals. VII. On the Influence of Defects on the Stability of the Cubic Structure
(4 and 13; 1.2; 10.2; 22)
(b) Vol. **130**, No. 2, p. 511–516 (1985).

L. SKALA
Electron Transfer in Chemisorption
(13.3; 4)
(b) Vol. **130**, No. 2, p. K157–K162 (1985).

U. PIETSCH, J. MAHLBERG, and K. UNGER
*Investigation of Dynamical Bond Charge Transfer in GaAs by Changing X-Ray Reflection Power under High Electric Field
(4; 13; 22.2.1)
(b) Vol. **131**, No. 1, p. 67–73 (1985).

A. A. VASILYEV, V. V. SIROTINKIN, and A. I. MELKER
*Small Vacancy Clusters in Nickel
(4 and 10.1; 21.1)
(b) Vol. **131**, No. 2, p. 537–544 (1985).

F. BECHSTEDT, D. REICHARDT, and R. ENDERLEIN
Self-Consistent Tight-Binding Method for Total Energy Calculations of Tetrahedral Semiconductors Including Surfaces and Defects
(13.1; 4; 22.1; 22.1.1; 22.1.2)
(b) Vol. **131**, No. 2, p. 643–657 (1985).

R. J. WATTS-TOBIN
*Critical Lengths for Formation of Kink Structures in Finite Trans-Polyacetylene Chains
(4; 10.2; 14.2; 22.9)
(b) Vol. **131**, No. 2, p. 683–698 (1985).

W. KUCHARCZYK
*On the Nature of the Quadratic Electrooptic Effect in LiF
(4 and 20.1; 22.5.2)
(b) Vol. **131**, No. 2, p. K107–K109 (1985).

A. N. GUSATINSKII, A. A. LAVRENTEV, M. A. BLOKHIN, and V. YU. SLIVKA
X-Ray Spectroscopic Investigation of the Electronic Structure and Chemical Bond of the Semiconductors $Sn_2P_2S_6$ and Tl_3PS_4
(13.1 and 20; 4; 22.8)
(b) Vol. **131**, No. 2, p. K139–K143 (1985).

B. N. MUKASHEV, M. F. TAMENDAROV, S. ZH. TOKMOLDIN, and V. V. FROLOV
Hydrogen Implantation into Silicon. Infra-Red Absorption Spectra and Electrical Properties
(10.2 and 11; 4; 14.3; 20.1; 22.1.2)
(a) Vol. **91**, No. 2, p. 509–522 (1985).

N. DUTT, G. G. AGRAWAL, and J. SHANKER
*Evaluation of the First- and Higher-Order Volume and Pressure Derivatives of Dielectric Constants of Alkali Halide Crystals
(4; 6; 8; 22.5.2)
(b) Vol. **132**, No. 1, p. 99–107 (1985).

B. DUTTA, S. K. SARKAR, and D. ROY
*Three-Body Interaction in H.C.P. Metals
(4 and 6; 12.1; 21.3)
(b) Vol. **132**, No. 2, p. 345–354 (1985).

T. SOMA, H. HIGUCHI, and H.-MATSUO KAGAYA
Effect of Lattice Anharmonicity on Specific Heat of Alkali Alloys
(6; 4; 8; 21.2)
(b) Vol. **132**, No. 2, p. 383–388 (1985).

5. Mössbauer Spectroscopy

Review Article

K. H. FISCHER
Spin Glasses (II)
(18.2 and 18.4; 2; 5; 7; 8; 13; 14; 21; 21.1; 21.6; 22.6; 22.8)
(b) Vol. **130**, No. 1, p. 13–71 (1985).

Original Papers and Short Notes

V. D. GOROBCHENKO and F. J. LITTERST
*Mössbauer Relaxation Spectra Due to Electric Field Gradient Reorientation and Nuclear Hopping for an $I = 5/2$ to 7/2 Nuclear Transition
(5; 22.1.2)
(b) Vol. **127**, No. 1, p. 351–358 (1985).

J. E. FRĄCKOWIAK
*Determination of the Long-Range Order Parameter in the Transition Metal Alloys Using Mössbauer Spectroscopy
(1.3 and 5; 21.1; 21.1.1)
(a) Vol. **87**, No. 1, p. 109–119 (1985).

J. A. CUSIDÓ, A. ISALGUÉ, and J. TEJADA
On the Amorphous to Crystalline Transformation of $Fe_{80}B_{20}$ by Means of Electrical and Thermal Conductivity, X-Ray Diffraction, and Mössbauer Measurements
(1.2 and 2; 5; 8; 14.1; 21.1.1)
(a) Vol. **87**, No. 1, p. 169–174 (1985).

A. PRASAD, R. M. SINGRU, and A. K. BISWAS
*Study of the Roasting of Pyrite Minerals by Mössbauer Spectroscopy
(5; 1.3; 21.1.1)
(a) Vol. **87**, No. 1, p. 267–271 (1985).

E. A. VASILEV and V. A. VIRCHENKO
*Effect of Interstitial Atoms on the Quadrupole Interaction at ^{119}Sn Nuclei in Some B8 Type Solid Solutions
(1.3 and 5; 4; 21.1.1)
(a) Vol. **87**, No. 1, p. K19–K24 (1985).

V. KUMAR, R. M. SINGRU, O. PARKASH, and D. BAHADUR
*Mössbauer Studies of the System $La_{1-x}Sr_xCo_{1-y}Ti_yO_3$
(5; 22.8)
(b) Vol. **128**, No. 1, p. 223–230 (1985).

P. ROCHEGUDE and J. FOCT
*The Transformation $\varepsilon \rightarrow \alpha''$ in Iron-Nitrogen Solid Solutions Studied by Mössbauer Spectrometry
(1.2 and 5; 21.1.1)
(a) Vol. **88**, No. 1, p. 137–142 (1985).

A. M. VAN DER KRAAN, D. B. DE MOOIJ, and K. H. J. BUSCHOW
Magnetic Properties and ^{57}Fe Mössbauer Effect in $V_{1-x}Fe_x$ Alloys
(18.2; 5; 21.1; 21.1.1)
(a) Vol. **88**, No. 1, p. 231—237 (1985).

F. H. SÁNCHEZ, R. C. MERCADER, A. F. PASQUEVICH, A. G. BIBILONI, and A. R. LÓPEZ-GARCÍA
*The Influence of Particle Sizes on the Oxidation Kinetics of AgSn Alloys Studied by Mössbauer Spectroscopy
(1.3 and 5; 21.6)
(a) Vol. **89**, No. 1, p. K17—K21 (1985).

J. SITEK and M. MIGLIERINI
Mössbauer Spectroscopy on Amorphous $Fe_xNi_{80-x}B_{20}$ after Neutron Irradiation
(1.4; 5; 11; 21.1; 21.1.1)
(a) Vol. **89**, No. 1, p. K31—K33 (1985).

E. V. ZOLOTOYABKO and N. S. OVANESYAN
*Mössbauer Diffraction in the Superionic Conductor $RbAg_4I_5$
(1.2 and 5; 9; 22.5.4)
(b) Vol. **129**, No. 2. p. 587—594 (1985).

S. GARCÍA and E. ALTSHULER
*Mössbauer Study of the Reaction Kinetics of Hexagonal M-Phase Ferrites
(1.2 and 5; 22.8.2)
(a) Vol. **89**, No. 2, p. 427—436 (1985).

J. A. CUSIDÓ and A. ISALGUÉ
A Simple Generalized Model for the Kinetics of Crystallization in Metallic Glasses
(1.2 and 2; 5; 8; 21.1.1)
(a) Vol. **90**, No. 1, p. 127—133 (1985).

C. MICHALK and W. THIEL
*Garnet Formation of Al-Substituted Yttrium Iron Garnet Ceramics Studied by Mössbauer Effect and Magnetization Measurements
(1.2 and 5; 18.3; 22.8.2)
(a) Vol. **90**, No. 1, p. 325—330 (1985).

A. A. BAHGAT, E. E. SHAISHA, A. I. SABRY, and N. A. EISSA
Mössbauer Analysis of Valence and Coordination of Fe Cations in TeO_2–Fe_2O_3 Glasses
(2 and 4; 5; 22.6.1)
(a) Vol. **90**, No. 1, p. K25—K29 (1985).

P. C. M. GUBBENS, A. M. VAN DER KRAAN, and K. H. J. BUSCHOW
*^{161}Dy Mössbauer Effect in $Dy_2Fe_{14}B$
(5 and 18.3; 22.8)
(b) Vol. **130**, No. 2, p. 575—580 (1985).

F. S. NASREDINOV, L. V. PROKOFIEVA, P. P. SEREGIN, S. V. ZARUBO, A. V. ERMOLAEV, and A. N. KURMANTAEV
*Mössbauer Study of Two-Electron Exchange between Neutral and Ionized Tin Donor Centres in Solid Solutions on the Basis of PbS and PbSe
(5 and 13.4; 22)
(b) Vol. **130**, No. 2, p. 727—735 (1985).

M. Krause, C. Michalk, M. Benedix, and J. Suwalski
*Low-Temperature Mössbauer Measurement of a Polymeric Oxalate (2,2′-Bipyridine) Iron Compound
(5; 18; 22.9)
(a) Vol. **90**, No. 2, p. 657–660 (1985).

E. E. Shaisha and A. A. Bahgat
Magnetic and Thermal Behaviour of the Amorphous Ferromagnet $Fe_{79}B_{16}Si_5$
(13.5.4 and 18.2; 5; 8; 21.1.1)
(a) Vol. **90**, No. 2, p. 661–668 (1985).

N. Suárez, J. L. Sánchez, and S. García
*Cd_2–W: a New Material for Permanent Magnets?
(5 and 18; 22.8.2)
(a) Vol. **90**, No. 2, p. K171–K174 (1985).

A. M. Afanasev and V. E. Sedov
*Mössbauer Absorption Spectra under Continuous Localized Diffusion
(5)
(b) Vol. **131**, No. 1, p. 299–308 (1985).

K. Choudhury, B. Das, C. K. Majumdar, K. R. P. M. Rao, and M. Adhikari
Mössbauer Studies of Corrosion and Inhibition
(1.6; 5; 21.1.1)
(a) Vol. **91**, No. 1, p. 29–35 (1985).

Y. Kreislerová, T. Zemčík, and H. Hemschik
*Mössbauer, Electrical, and Magnetic Studies of the Sn-Doped Ni_3Fe Alloy
(5 and 18.2; 14.1; 21.1; 21.1.1)
(a) Vol. **91**, No. 1, p. 115–122 (1985).

N. N. Delyagin, V. I. Krylov, G. T. Mujiri, V. I. Nesterov, and S. I. Reiman
*The Magnetic Hyperfine Field for ^{119}Sn Atoms in GdMnSi Intermetallic Compound. Zero Contribution from 3d Moments
(5 and 18.3; 21.1; 21.4)
(b) Vol. **131**, No. 2, p. 555–561 (1985).

V. I. Nikolaev, V. S. Rusakov, and N. I. Chistyakova
*Covalency Effects and the Pressure Dependence of the Mössbauer Line Shifts for Spinel-Type Ferrites
(5; 22.8.2)
(a) Vol. **91**, No. 2, p. K139–K142 (1985).

G. Weyer, H. Andreasen, and H. de Waard
The Electronic Configuration of ^{121}Sb in Semiconductors
(13.4; 5; 22.2; 22.4)
(b) Vol. **132**, No. 1, p. 219–223 (1985).

B. K. Das, A. K. Tewari, and D. L. Bhattacharya
*Study of Pressure on Metallic Tin Foil by Mössbauer Spectroscopy
(5 and 12.1; 21.7)
(b) Vol. **132**, No. 1, p. K23–K25 (1985).

6. Lattice Dynamics. Phonons

Review Articles

H. W. STREITWOLF
Physical Properties of Polyacetylene
(13.5 and 13.5.3; 6; 18; 19; 20.1; 22.9)
(b) Vol. **127**, No. 1, p. 11–54 (1985).

S. SUGAI
*Lattice Vibrations in the Charge-Density-Wave States of Layered Transition Metal Dichalcogenides
(6 and 13; 1.2; 20.1; 22)
(b) Vol. **129**, No. 1, p. 13–39 (1985).

YU. E. PERLIN and A. A. KAMINSKII
Nonradiative Transitions of the Trivalent Lanthanides in Insulating Laser Crystals
(13.4 and 20.3; 6; 20.2; 22.8.1)
(b) Vol. **132**, No. 1, p. 11–40 (1985).

S. A. SHABALOVSKAYA
Phase Transitions in the Intermetallic Compound TiNi with Charge-Density Wave Formation
(1.2 and 13.1; 6; 14.1; 21.1)
(b) Vol. **132**, No. 2, p. 327–344 (1985).

Original Papers and Short Notes

H. RAUH and R. GEICK
*Crystal-Independent Interionic Potentials and the Lattice Dynamics of the Perovskite-Type Layer Material K_2ZnF_4
(6; 4; 22.8)
(b) Vol. **127**, No. 1, p. 55–65 (1985).

P. C. MAITHANI and M. D. TIWARI
*Lattice Specific Heat and Superconducting Transition Temperature of a V: W Alloy
(6 and 8; 14.2; 21.1)
(b) Vol. **127**, No. 1, p. 79–88 (1985).

H.-MATSUO KAGAYA and T. SOMA
*Calculation of Phonon Dispersion Curves for AlP, AlAs, and AlSb
(6; 12.2; 22.2)
(b) Vol. **127**, No. 1, p. 89–93 (1985).

P. KLEINERT
*Self-Consistent Field Approximation for the Vibrational Spectra of Quaternary Alloys
(6; 20.1; 22.2.4)
(b) Vol. **127**, No. 1, p. 109–119 (1985).

S. N. Vaidya
*Theory of Melting. II. Relaxion Model
(1.2 and 6; 8; 23)
(a) Vol. **87**, No. 1, p. 181–186 (1985).

T. Tarnóczi
Structural Relaxation and Curie Point of a Metallic Glass
(2 and 18.2; 6; 21.1.1)
(a) Vol. **87**, No. 1, p. 283–291 (1985).

K. R. Allakhverdiev, U. A. Aleshenko, N. A. Bakhyshov, L. K. Vodopyanov, F. M. Gashimzade, R. M. Sardarly, and V. Ya. Shteinshraiber
*Vibrational Spectrum of TlS_xSe_{1-x} Solid Solutions
(6 and 20.1; 22.3.1)
(b) Vol. **127**, No. 2, p. 459–464 (1985).

G. V. Kozlov, E. B. Kryukova, B. Wyncke, F. Brehat, and M. E. Sherif
*Low-Frequency Lattice Dynamics of $NaNO_2$ Crystals
(6 and 20.1; 14.4.1; 22.8.1)
(b) Vol. **127**, No. 2, p. 465–472 (1985).

W. Wasserbäch
*Electron–Phonon Scattering and Low-Temperature Thermal Conductivity of Niobium and Tantalum Single Crystals
(6 and 8; 21)
(b) Vol. **127**, No. 2, p. 481–492 (1985).

M. A. Ivanov and L. B. Kvashnina
Multiple Scattering of Axially Channeled Particles by Thermal Vibrations and Electrons
(11; 6; 13)
(b) Vol. **127**, No. 2, p. 587–597 (1985).

K. Dieterich and M. Wagner
On the Temperature Behaviour of DC Conductivity in One -Dimensional Conductors above T_c
(13 and 14; 6)
(b) Vol. **127**, No. 2, p. 715–726 (1985).

K. B. Petrosyan and K. M. Pokhsraryan
*Stimulated Raman Scattering in a Potassium Pentaborate Crystal
(6 and 20.1; 22.8.1)
(b) Vol. **127**, No. 2, p. K105–K108 (1985).

A. K. M. A. Islam and E. Huq
On Cohesive Energy and Grüneisen Parameter of Silver Halide Crystals
(4 and 8; 6; 22.5.1)
(b) Vol. **127**, No. 2, p. K109–K111 (1985).

J. B. Radunović
The Influence of Large-Scale Fluctuations on the Anomalous Conductivity in Piezoelectric Semiconductors
(14.3; 6; 22)
(b) Vol. **127**, No. 2, p. K181–K185 (1985).

L. P. Mokhracheva, Yu. G. Listopadov, and P. V. Geld
*Frequency Ratio Change of Localized Hydrogen Atom Vibrations in Nb and Ta on Isotope Substitution
(6; 21)
(b) Vol. **128**, No. 1, p. K9–K14 (1985).

S. K. Pal and A. Kumar
*Resonant Scattering of Phonons in GaInSb
(6 and 8; 22.2.4)
(b) Vol. **128**, No. 1, p. K15–K19 (1985).

D. V. Kapor, D. Lj. Mirjanić, S. D. Stojanović, J. P. Šetrajčić, and B. S. Tošić
On the Possibility of the Realization of Superfluid Excitation Transfer in Molecular Crystals
(13.5.1; 6)
(b) Vol. **128**, No. 1, p. K49–K52 (1985).

U. F. Kozmidis-Luburić, D. Lj. Mirjanić, J. P. Šetrajčić, and B. S. Tošić
The Influence of Phonons on Two-Particle Exciton States in Molecular Chains
(13.5.1; 6)
(b) Vol. **128**, No. 1, p. K53–K57 (1985).

H. D. Hochheimer, E. Gmelin, Ch. Mensing, L. Viczian, K. D. Kreuer, and H. G. von Schnering
AC Conductivity, Calorimetric, X-Ray, and Raman Study of the Hexagonal Phase $TlInS_2$-III
(1.2; 6; 8; 14.3; 20.1; 22.8)
(a) Vol. **88**, No. 1, p. 43–48 (1985).

V. Valvoda, V. Šíma, Z. Smetana, and P. Čapková
Neutron Powder Diffraction Study of the Debye-Waller Factor in TiC and TiN Compounds and Comparison with X-Ray Results
(4; 6; 21.1)
(a) Vol. **88**, No. 1, p. 157–163 (1985).

X. Gerbaux, A. Hadni, and A. Kitade
*Surface Layers and Far-Infrared Quasi-Harmonic Oscillators in Triglycine Sulphate at 4 K
(6 and 20.1; 14.4.1; 22.9)
(a) Vol. **88**, No. 1, p. 165–174 (1985).

V. M. Burlakov, A. P. Litvinchuk, V. N. Pyrkov, G. G. Tarasov, and N. I. Vitrikhovskii
*Optical Properties of the Quaternary II–VI Mixed Crystals in the Far Infrared Region
(6 and 20.1; 22.4.4)
(b) Vol. **128**, No. 2, p. 389–400 (1985).

A. G. Abdullaev, K. R. Allakhverdiev, T. D. Ibragimov, and R. M. Sardarly
*Two-Phonon Processes in TlSe and $TlInSe_2$
(6 and 20.1; 22.3; 22.3.1)
(b) Vol. **128**, No. 2, p. 401–409 (1985).

A. Rajput
*Phonon Dispersion in Platinum and Palladium
(6; 8; 12.1; 21)
(b) Vol. **128**, No. 2, p. 411–417 (1985).

H.-Matsuo Kagaya and T. Soma
*Temperature Dependence of the Linear Thermal Expansion Coefficient for Si and Ge
(6 and 8; 22.1.1; 22.1.2)
(b) Vol. **129**, No 1, p. K5–K8 (1985).

Anil Kumar and D. Pal
*Lattice Thermal Conductivity of Boron Nitride Crystals at Temperatures 1.5 to 300 K
(6 and 8; 22.2)
(b) Vol. **129**, No. 1, p. K9–K12 (1985).

N. A. Abdullaev, L. N. Alieva, and R. A. Suleimanov
*Thermal Expansion and Temperature Dependence of Raman Active Modes in Layer Semiconductors
(6 and 8; 22.3)
(b) Vol. **129**, No. 1, p. K13–K16 (1985).

V. Frei
*Born-von Kármán Force Constants of Eight Coordination Spheres of Atoms in Face-Centred Cubic Elements Studied by Means of Phonon Frequency Sums
(6; 21; 23)
(b) Vol. **129**, No. 1, p. K17–K20 (1985).

D. A. Fushman
Saturation of Tunneling State in Glasses in the Presence of the Phonon Bottleneck
(2; 6; 13)
(b) Vol. **129**, No. 1, p. K21–K25 (1985).

A. K. Abass, F. Y. M. Al-Eithan, and R. H. Misho
Indirect Electronic Transitions in Single Crystals of Triglycine Sulfate
(20.1; 6; 13.1; 22.9)
(a) Vol. **89**, No. 1, p. 225–229 (1985).

N. Gopi Krishna, D. B. Sirdeshmukh, R. Rama Rao, B. J. Beaudry, and K. A. Gschneidner, Jr.
*X-Ray Debye Temperature of Ytterbium
(6 and 8; 21.4)
(a) Vol. **89**, No. 1, p. K37–K38 (1985).

K. R. Rao and S. L. Chaplot
*Dynamics of Paraelectric and Ferroelectric SbSI
(6 and 14.4.1; 22.8)
(b) Vol. **129**, No 2, p. 471–482 (1985).

A. K. Sood, E. Anastassakis, and M. Cardona
*Raman Piezospectroscopy in GaAs Revisited
(6 and 12.2; 22.2.1)
(b) Vol. **129**, No. 2, p. 505–512 (1985).

L. Wendler
*Electron–Phonon Interaction in Dielectric Bilayer Systems. Effect of the Electronic Polarizability
(6 and 13; 13.5.3; 14.4; 22.2.1; 22.2.2)
(b) Vol. **129**, No. 2, p. 513—530 (1985).

V. M. POLYANOVSKII
*Harmonics of Cyclotron and Cyclotron–Phonon Resonances in Semiconductors with Superlattice
(6 and 13; 19; 20.1; 22)
(b) Vol. **130**, No. 1, p. 95—101 (1985).

W. WOJDOWSKI
*Vibrational Modes in $Bi_{12}GeO_{20}$ and $Bi_{12}SiO_{20}$ Crystals
(6 and 20.1; 4; 22.8)
(b) Vol. **130**, No. 1, p. 121—130 (1985).

M. WÖHLECKE, U. SCHERZ, and D. K. MARETIS
*Invariant Harmonic Oscillator Functions. I. The Cubic Groups T_d, O, and O_h
(6 and 13.4)
(b) Vol. **130**, No. 1, p. 139—150 (1985).

H. KOHL
*Unambiguous Determination nf Interatomic Force Constants of Crystals by Coherent Inelastic Neutron Scattering
(6)
(b) Vol. **130**, No. 1, p. 151—159 (1985).

E. ANASTASSAKIS, J. S. RAPTIS, and W. RICHTER
*Raman and Infrared Spectra of Tellurium Sub-Bromide (Te_2Br)
(6 and 20.1; 22.5)
(b) Vol. **130**, No. 1, p. 161—168 (1985).

M. GRINBERG, S. ŁĘGOWSKI, and H. MĘCZYŃSKA
The Energy of Shallow Donor States in CdTe, GaAs, AgCl, AgBr Semiconductors
(13.4; 6; 22.2.1; 22.4.3; 22.5.1)
(b) Vol. **130**, No. 1, p. 325—331 (1985).

T. SOMA, S. NEHASHI, and H.-MATSUO KAGAYA
Specific Heat and Thermal Expansion Coefficient of Al
(8; 6; 21)
(b) Vol. **130**, No. 1, p. K11—K14 (1985).

T. SOMA, H.-MATSUO KAGAYA, and H. OHSUGI
Thermal Properties for the Alloy System Composed of Alkali Metal Elements
(8; 6; 21.2)
(b) Vol. **130**, No. 1, p. K15—K18 (1985).

K. R. ZHDANOV, V. R. BELOSLUDOV, and F. S. RAKHMENKULOV
Entropy and Heat Capacity of an Anharmonic Crystal
(8; 6; 21)
(b) Vol. **130**, No. 1, p. K19—K22 (1985).

P. K. SRIVASTAVA and O. P. KULSHRESTHA
*Lattice Dynamics of Lanthanum by Using a Model Pseudopotential Approach
(6; 21.4)
(b) Vol. **130**, No. 1, p. K23—K25 (1985).

B. MARI, A. SEGURA, and A. CHEVY
*Free Carrier Absorption in n-Type Indium Selenide
(6 and 20.1; 14.3; 22.3)
(b) Vol. **130**, No. 2, p. 793–799 (1985).

V. V. KRASNOGOLOVETS and P. M. TOMCHUK
The Coulomb Correlations and the Electric Field Local Heterogeneities on the Proton Conductivity of Hydrogen Bonded Chains
(14.4; 6; 22)
(b) Vol. **130**, No. 2, p. 807–817 (1985).

J. J. SINAI and S. Y. WU
*Mixed Mode Behavior of $Ga_{1-x}In_xAs$
(6; 22.2.4)
(b) Vol. **130**, No. 2, p. K91–K93 (1985).

H.-MATSUO KAGAYA, Y. KITANI, and T. SOMA
Calculation of Specific Heat for the Si–Ge System
(8; 6; 22.1.1; 22.1.2)
(b) Vol. **130**, No. 2, p. K95–K96 (1985).

S. MAHMOUD, E. M. KASSEM, H. MÜLLER, and E. HEGENBARTH
Thermal Conductivity of $LiKSO_4$ at Low Temperatures
(8 and 14.4.1; 6; 22.8.1)
(a) Vol. **90**, No. 2, p. K135–K138 (1985).

F. BECHSTEDT and R. ENDERLEIN
*Dielectric Screening, Polar Phonons, and Longitudinal Electronic Excitations of Quantum Well Double Heterostructures. Application to Light Scattering from Charge Density Fluctuations
(6 and 13.3; 14.4; 20; 22.2.4)
(b) Vol. **131**, No. 1, p. 53–66 (1985).

A. V. SHERMAN
Exciton Absorption Spectrum
(13.5.1 and 20.1; 6)
(b) Vol. **131**, No. 1, p. 225–233 (1985).

S. GHOSH and S. DIXIT
Stimulated Raman Scattering and Raman Instability of an Intense Helicon Wave in Longitudinally Magnetized n-Type Piezoelectric Semiconducting Plasma
(13.5.2; 6; 22.2.3)
(b) Vol. **131**, No. 1, p. 255–265 (1985).

I. A. MARUSHKO and O. P. BUGAETS
*Hellmann-Feynman Theorem and the Theory of Force Constants
(4 and 6)
(b) Vol. **131**, No. 1, p. K15–K17 (1985).

A. SCHENK, R. ENDERLEIN, and D. SUISKY
Electroabsorption for Deep Level to Band Transition Accompanied by Multiphonon Processes
(13.4 and 20.1; 6; 14.3.4; 22)
(b) Vol. **131**, No. 2, p. 729–739 (1985).

E. ANACHKOVA and I. SAVATINOVA
*Isotopic Shift and Mode Interaction in $K(D_xH_{1-x})_2PO_4$ Crystals
(6 and 20.1; 14.4.1; 22.8.1)
(b) Vol. **131**, No. 2, p. K101–K105 (1985).

V. V. KRASNOGOLOVETS and P. M. TOMCHUK
The Anharmonicity Influence on the Proton Conductivity of an Hydrogen-Bonded Chain
(14.4; 6; 22.8.1)
(b) Vol. **131**, No. 2, p. K177–K179 (1985).

M. BALKANSKI, M. JOUANNE, and M. KANEHISA
*Bound Phonons
(6 and 13.4; 22.1.2; 22.4.3)
(b) Vol. **132**, No. 1, p. 57--67 (1985).

V. M. FOMIN and E. P. POKATILOV
*Phonons and the Electron–Ponon Interaction in Multi-Layer Systems
(6 and 13.3; 22)
(b) Vol. **132**, No. 1, p. 69–82 (1985).

D. DUVAL and R. A. CONDRATE, SR.
Temperature Dependence of the Raman Spectra of Calcium Oxalate Monohydrate
(1.2; 6; 20.1; 22.9)
(b) Vol. **132**, No. 1, p. 83–92 (1985).

G. IRMER, A. HEINRICH, and J. MONECKE
*First Order Raman Scattering in $CdSnP_2$
(6 and 20.1; 22.8)
(b) Vol. **132**, No. 1, p. 93–98 (1985).

N. DUTT, G. G. AGRAWAL, and J. SHANKER
Evaluation of the First- and Higher-Order Volume and Pressure Derivatives of Dielectric Constants of Alkali Halide Crystals
(4; 6; 8; 22.5.2)
(b) Vol. **132**, No. 1, p. 99–107 (1985).

F. F. CHARFI and M. ZOUAGHI
*Hot-Exciton Orientation in Indirect-Gap $GaAs_{1-x}P_x$ Alloys
(6 and 13.5.1; 20.3; 22.2.4)
(b) Vol. **132**, No. 1, p. 109–115 (1985).

P. ENDERS
Acoustic and Optical Deformation Potentials in Cubic IV–VI Compounds. Strong Dependence upon the Lattice Distortion
(13.1; 6; 22)
(b) Vol. **132**, No. 1, p. 165–172 (1985).

N. Lawerence and T. M. Haridasan

*Mean Square Amplitude Transition Metal Dichalcogenide 1T-$TiSe_2$ in the Normal and Commensurate Charge Density Wave Phases
(6; 1.2; 22)
(b) Vol. **132**, No. 2, p. K77—K79 (1985).

K. K. Danilevich-Tovtsyuk and B. A. Lukiyanets

*On the Electron–Phonon Interaction in Layered Crystals
(6; 12.2; 13)
(b) Vol. **132**, No. 2, p. K81—K85 (1985).

H.-Matsuo Kagaya, Y. Kitani, and T. Soma

*Linear Thermal Expansion Coefficient of $Si_{1-x}Ge_x$ Solid Solution
(6 and 8; 22.1.1; 22.1.2)
(b) Vol. **132**, No. 2, p. K87—K90 (1985).

R. H. Enns and S. S. Rangnekar

Solitary Wave Solutions of the Driven and Damped Three-Wave Problem
(13.5; 6; 20.2)
(b) Vol. **132**, No. 2, p. K121—K126 (1985).

L. G. Midodashvili and A. I. Tugushi

*Saturation of the Inhomogeneously Broadened EPR Line at Low Temperatures under Bottleneck Conditions
(6 and 19)
(b) Vol. **132**, No. 2, p. K127—K130 (1985).

B. Martin, H. Jex, and F. Wittmann

Elastic and Inelastic Scattering from NH_4Br Measured by Mössbauer γ-Ray Diffraction
(1.2; 5; 6; 22.5)
(a) Vol. **92**, No. 2, p. 457–465 (1985).

7. Acoustic Properties

Review Article

K. H. Fischer
Spin Glasses (II)
(18.2 and 18.4; 2; 5; 7; 8; 13; 14; 21; 21.1; 21.6; 22.6; 22.8)
(b) Vol. **130**, No. 1, p. 13–71 (1985).

Original Papers and Short Notes

S. Ghosh and S. Dixit
Effect of Relativistic Mass Variation of the Electron on the Parametric Instability of Acoustic Waves in Transversely Magnetised Piezoelectric Semiconducting Plasmas
(13.5.2; 7; 22.2.3)
(b) Vol. **127**, No. 1, p. 245–252 (1985).

Z. M. Alekseeva, V. S. Vikhnin, A. P. Zdebskii, and I. M. Tsivileva
*Acoustical and Electrochemical Behaviour of Proustite Parameters in the Incommensurate Phase
(7; 1.2; 10.2; 22.8)
(a) Vol. **87**, No. 1, p. K35–K39 (1985).

E. M. Smokotin, O. P. Kvashnina, and A. M. Kapitonov
*Acoustic Properties of Terbium Ferrite Garnet in the Region of the Compensation Temperature Point
(7 and 18.3; 22.8.2)
(a) Vol. **87**, No. 1, p. K53–K56 (1985).

S. Ghosh and R. B. Saxena
Stimulated Brillouin Scattering of High-Power Helicon Waves in Longitudinally Magnetized Cubic Piezoelectric Semiconducting Plasmas
(13.5.2; 7; 22.2.3)
(b) Vol. **127**, No. 2, p. 557–566 (1985).

H. Cerva and W. Graeff
*Contrast Investigations of Surface Acoustic Waves by Stroboscopic Topography. II. Wavefield Deviation Contrast
(7; 14.4.1; 22.8.1)
(a) Vol. **87**, No. 2, p. 507–516 (1985).

P. Koštial and J. Slabeycius
*The Determination of the Distribution Function of Activation Energies of Structural Changes in Amorphous Se
(2 and 7; 22.1.3)
(a) Vol. **87**, No. 2, p. K109–K113 (1985).

V. L. Bonch-Bruevich and T. Z. Kakhlishvili
*On the Nonlinear Light and Sound Absorption in Materials with Weakly Bound Atoms
(7 and 20.1; 22)
(b) Vol. **128**, No. 1, p. 39–48 (1985).

Z. ŠIMŠA, Y. KAWAI, and V. A. M. BRABERS
Anealisticity and Relaxation Effects in Single Crystal Mn Ferrites
(12.2 and 18.3; 7; 22.8.2)
(a) Vol. **88**, No. 2, p. 611–617 (1985).

P. KOŠTIAL
*The Influence of Non-Isothermic Thermal Treatment on the Attenuation of Surface Waves in Electrographic Se Layers
(1.5 and 7; 2; 22.1.3)
(a) Vol. **88**, No. 2, p. K115–K118 (1985).

S. GHOSH and M. S. PATHAK
Transverse Phonon–Helicon Interactions in Semiconducting Plasmas
(6 and 13.5.2; 7; 22.2.3)
(b) Vol. **129**, No. 1, p. 81–87 (1985).

E. L. ALBUQUERQUE, N. S. ALMEIDA, and M. C. OLIVEROS
Light Scattering by Surface Ripples in a Spatially Dispersive Medium
(13.5.1 and 20; 7)
(b) Vol. **129**, No. 1, p. 177–186 (1985).

S. GHOSH and D. K. SINHA
*Stimulated Brillouin Scattering and Brillouin Instability in Dissipative Piezosemiconducting Plasmas
(7; 22.2.3)
(b) Vol. **129**, No. 2, p. K117–K122 (1985).

A. G. BAGDOEV and A. V. SHEKOYAN
*Focusing of Nonlinear Ultrasonic Waves in Viscous Thermoelastic Materials with Spherical Inclusions
(7)
(a) Vol. **89**, No. 2, p. 499–507 (1985).

S. GHOSH and S. DIXIT
*Effect of Relativistic Mass Variation of the Electron on the Modulational Instability of Laser Beams in Transversely Magnetised Piezoelectric Semiconducting Plasmas
(7 and 13.5.2; 22.2.3)
(b) Vol. **130**, No. 1, p. 219–224 (1985).

J. BERDOWSKI and M. HAJDUK
*The Influence of Isotope Effects on the Interaction of Light and Surface Acoustic Waves in DADA Crystals
(7 and 20.1; 14.4.1; 22.8.1)
(a) Vol. **90**, No. 1, p. 157–165 (1985).

S. SHIGETOMI, T. IKARI, Y. KOGA, and S. SHIGETOMI
*Annealing Behavior of Photoacoustic Spectra of Undoped InSe
(7 and 13.4; 16; 22.3)
(a) Vol. **90**, No. 1, p. K61–K64 (1985).

M. OMRI and A. VINCENT
*Attenuation Stress Hysteresis Loops and Low-Frequency Internal Friction Associated with Thermomechanical Breakaway of Dislocations
(7 and 10.1; 12.1; 21)
(b) Vol. **130**, No. 2, p. 547–560 (1985).

A. Pawełek, W. Stryjewski, W. Bochniak, and H. Dybiec
Mobile Dislocation Density Variation during Strain Rate Change Evidenced by Acoustic Emission
(10.1; 7; 21.1)
(a) Vol. **90**, No. 2, p. 531–536 (1985).

L. T. Wille and J. Vennik
*X-Ray Diffraction by Vibration Crystals
(1 and 7)
(b) Vol. **131**, No. 2, p. K97–K100 (1985).

L. Voicu, L. Stamatescu, Al. Hening, V. Raeţchi, I. N. Mihăilescu, and L. Nanu
*On the Signals Generated by Lead Zirconium Titanate (PZT) Ceramics when Irradiated with Microsecond Pulsed TEA-CO_2 Laser Pulses
(7 and 11; 22.9)
(a) Vol. **91**, No. 2, p. K103–K106 (1985).

G. Urushadze
*Acoustic Phenomena in Tunnel Junctions
(7 and 14.1; 13.2; 21)
(b) Vol. **132**, No. 1, p. 51–56 (1985).

E. L. Albuquerque and C. A. Dos Santos
Deby-Waller Factor for Mössbauer Resonant Surface Atoms: A Linear Response Theory
(6; 7)
(b) Vol. **132**, No. 1, p. K5–K10 (1985).

A. K. George
*Anomalous Ultrasonic Absorption in a Nematic and a Polymesomorphic Liquid Crystal
(2 and 7; 22.9)
(a) Vol. **92**, No. 1, p. K15–K17 (1985).

I. R. Entin
*On the Suppression of X-Ray Anomalous Transmission by Acoustic Oscillations
(7; 1)
(b) Vol. **132**, No. 2, p. 355–364 (1985).

V. F. Kitaeva, E. V. Zharikov, and I. L. Chistyi
The Properties of Crystals with Garnet Structure
(12.2; 7; 22.8.2)
(a) Vol. **92**, No. 2, p. 475–488 (1985).

8. Thermal Properties

Review Article

K. H. Fischer
Spin Glasses (II)
(18.2 and 18.4; 2; 5; 7; 8; 13; 14; 21; 21.1; 21.6; 22.6; 22.8)
(b) Vol. **130**, No. 1, p. 13–71 (1985).

Original Papers and Short Notes

P. C. Maithani and M. D. Tiwari
*Lattice Specific Heat and Superconducting Transition Temperature of a V: W Alloy
(6 and 8; 14.2; 21.1)
(b) Vol. **127**, No. 1, p. 79–88 (1985).

N. Dass and M. Kumari
*Derivation of Some Equations of State for Solids. A New Approach
(8; 12.2; 22.5.2)
(b) Vol. **127**, No. 1, p. 103–108 (1985).

J. Schreiber and D. Schulz
Self Consistent Dynamical Theory for Heisenberg Spin Glasses
(2 and 18.1; 8; 18.2; 18.4)
(b) Vol. **127**, No. 1, p. 139–147 (1985).

E. A. Goremychkin, E. Mühle, B. Lippold, O. D. Chistyakov, and E. M. Savitskii
Crystal Field Study of $HoNi_5$ by Neutron Scattering
(13.4; 8; 18; 21.4)
(b) Vol. **127**, No. 1, p. 371–375 (1985).
Erratum
(b) Vol. **128**, No. 2, p. 807 (1985).

D. Sen and S. K. Sen
*Transport in Simple Metals in the Framework of a Unified Pseudopotential Study
(8 and 14.1; 15; 21; 21.2)
(b) Vol. **127**, No. 1, p. 377–383 (1985).

G. Ch. Christakudis, S. K. Plachkova, E. S. Avilov, L. E. Shelimova, and N. Kh. Abrikosov
Transport Phenomena of Ferroelectric Solid Solutions
$(GeTe)_{1-x}((Ag_2Te)_{1-y}(Sb_2Te_3)_y)_x$ with $y = 0.60$, $0 \leqq x \leqq 1$ at 300 K
(14.3; 8; 14.4.1; 15; 22.7.1)
(b) Vol. **127**, No. 1, p. 419–426 (1985).

H.-Matsuo Kagaya and T. Soma
*Specific Heat of Si and Ge
(6 and 8; 22.1.1; 22.1.2)
(b) Vol. **127**, No. 1, p. K5–K7 (1985).

A. K. M. A. ISLAM and E. HUQ
*On Cohesive Energy and Grüneisen Parameter of Silver Halide Crystals
(4 and 8; 6; 22.5.1)
(b) Vol. **127**, No. 2, p. K109–K111 (1985).

W. SCHEIBNER and M. JÄCKEL
*Thermal Conductivity and Specific Heat of an Epoxy Resin/Epoxy Resin Composite Material at Low Temperatures
(8; 2; 22.9)
(a) Vol. **87**, No. 2, p. 543–547 (1985).

P. ČAPKOVÁ, L. DOBIÁŠOVÁ, and J. MAIXNER
*The Effect of Non-Stoichiometry on Thermal Vibrations in TiC_x
(6 and 8; 21.1)
(a) Vol. **87**, No. 2, p. K139—K142 (1985).

B. A. VAID and K. C. SHARMA
Sructural Disorder Model of Amorphous Semiconductors. Thermodynamics of a-As
(2; 8; 22.1)
(b) Vol. **128**, No. 1, p. 93–98 (1985).

B. R. K. GUPTA
*Static and Thermophysical Properties of Mixed Crystal Systems
(4 and 8; 22.5.4)
(b) Vol. **128**, No. 1, p. 99–104 (1985).

W. N. BHENDE, P. S. BAKHSHI, and J. SHANKER
Thermoelastic Behaviour and Higher-Order Elastic Constants of Ionic Crystals
(4 and 12.2; 6; 8; 22.5.2; 22.6)
(b) Vol. **128**, No. 1, p. 105–112 (1985).

G. ÜFFINGER and A. KASTEN
Specific Heat and Magnetocaloric Effects in $TbPO_4$. I. Experiments
(1.2 and 18.4; 8; 22.8.1)
(b) Vol. **128**, No. 1, p. 201–207 (1985).

S. K. PAL and A. KUMAR
*Resonant Scattering of Phonons in GaInSb
(6 and 8; 22.2.4)
(b) Vol. **128**, No. 1, p. K15–K19 (1985).

M. MONDAL and K. P. GHATAK
*Theoretical Analysis of Oscillatory Heat Capacity in *n*-Channel Inversion Layers on Ternary Semiconductors
(8; 13.3; 22.4.4)
(b) Vol. **128**, No. 1, p. K21–K26 (1985).

H. D. HOCHHEIMER, E. GMELIN, CH. MENSING, L. VICZIAN, K. D. KREUER, and H. G. von SCHNERING
AC Conductivity, Calorimetric, X-Ray, and Raman Study of the Hexagonal Phase $TlInS_2$-III
(1.2; 6; 8; 14.3; 20.1; 22.8)
(a) Vol. **88**, No. 1, p. 43–48 (1985).

M. LAZARIDOU
*On the Connection between the Melting Temperature and the Parameters Obtained from Thermally Stimulated Depolarization Experiments
(8; 10.2; 22.5.2)
(a) Vol. **88**, No. 1, p. K19–K21 (1985).

P. SILFSTEN, P. KETOLAINEN, and M. LEHKONEN
*On the Thermal Resistance and Its Possible Role in Thermoluminescence Measurements
(8 and 20.3; 22.5.2; 22.5.4)
(a) Vol. **88**, No. 1, p. K67–K70 (1985).

A. RAJPUT
Phonon Dispersion in Platinum and Palladium
(6; 8; 12.1; 21)
(b) Vol. **128**, No. 2, p. 411–417 (1985).

W WASSERBÄCH
*Phonon–Dislocation Scattering and Low-Temperature Lattice Thermal Conductivity of Niobium and Tantalum Single Crystals Deformed at Low Temperatures
(8; 6; 10.1; 21)
(b) Vol. **128**, No. 2, p. 453–466 (1985).

W. EBELING and W. RICHERT
Thermodynamic Properties of Liquid Hydrogen Metal
(2; 8; 21; 23)
(b) Vol. **128**, No. 2, p. 467–474 (1985).

A. KASTEN and G. ÜFFINGER
Specific Heat and Magnetocaloric Effects in $TbPO_4$. II. Mean Field Theory
(1.2 and 18.4; 8; 22.8.1)
(b) Vol. **128**, No. 2, p. 525–535 (1985).

P. J. POMONIS
Electron Density and Work Function Dependence of the Enthalpy of Formation of Metal Oxides
(4; 8; 13.1; 13.3; 22.6)
(a) Vol. **88**, No. 2, p. 515–519 (1985).

P. VAROTSOS, K. ALEXOPOULOS, C. VAROTSOS, and M. LAZARIDOU
Interconnection of Point Defect Parameters in BaF_2
(10.2; 8; 9; 22.5.3)
(a) Vol. **88**, No. 2, p. K137–K140 (1985).

N. DUTT, G. G. AGRAWAL, and J. SHANKER
Investigation of the Equation of State and the Grüneisen-Anderson Parameters for Fluorite-Structure Crystals
(4; 6; 8; 12.2, 22.5.3; 22.8.1)
(b) Vol. **129**, No. 1, p. 73–79 (1985).

H.-Matsuo Kagaya and T. Soma
*Temperature Dependence of the Linear Thermal Expansion Coefficient for Si and Ge
(6 and 8; 22.1.1; 22.1.2)
(b) Vol. **129**, No. 1, p. K5–K8 (1985).

Anil Kumar and D. Pal
*Lattice Thermal Conductivity of Boron Nitride Crystals at Temperatures 1.5 to 300 K
(6 and 8; 22.2)
(b) Vol. **129**, No. 1, p. K9–K12 (1985).

N. A. Abdullaev, L. N. Alieva, and R. A. Suleimanov
*Thermal Expansion and Temperature Dependence of Raman Active Modes in Layer Semiconductors
(6 and 8; 22.3)
(b) Vol. **129**, No. 1, p. K13–K16 (1985).

L. J. Gallego, J. A. Somoza, and J. A. Alonso
Liquidus Curves of Eutectic Na–K and Na–Cs Systems from Semiempirical Theories of Mixtures
(2; 8; 21.2)
(a) Vol. **89**, No. 1, p. 73–78 (1985).

A. Akroune, J. Claverie, A. Tazairt, G. Villeneuve and A. Casalot
Structural, Magnetic, and Electric Properties of $V_{1-x}M_xO_{2-2x}F_{2x}$ (M = Mg, Ni) Oxyfluorides
(1.2 and 14; 8; 18.1; 22.8)
(a) Vol. **89**, No. 1, p. 271–282 (1985).

J. Przedmojski, S. Gierlotka, and B. Pałosz
*X-Ray Thermal Investigations of Cadmium Iodide Single Crystals
(1.1 and 8; 22.5)
(a) Vol. **89**, No. 1, p. K9–K11 (1985).

N. Gopi Krishna, D. B. Sirdeshmukh, R. Rama Rao, B. J. Beaudry, and K. A. Gschneidner, Jr.
*X-Ray Debye Temperature of Ytterbium
(6 and 8; 21.4)
(a) Vol. **89**, No. 1, p. K37–K38 (1985).

T. Ito
*A Pseudopotential Approach to the Structural and Thermodynamical Properties of III–V Ternary Semiconductor Alloys
(1.2 and 8; 12.2; 22.2.4)
(b) Vol. **129**, No. 2, p. 559–568 (1985).

U. Pietsch
The Ionicity Dependence of the Interatomic Potential Parameters of Covalently Bonded Materials
(4; 8; 22)
(b) Vol. **129**, No. 2, p. K99–K104 (1985).

M. Kumari and N. Dass
*On a Simple Relation for the Adiabatic Anderson-Grüneisen Parameter in Solids
(8; 6; 22.5.2)
(b) Vol. **129**, No. 2, p. K105–K109 (1985).

A. Szász
*One Possible Analytical Approximation for the Critical Point of the Three-Dimensional Ising Model
(8)
(b) Vol. **130**, No. 2, p. K97 – K100 (1985).

E. A. Pamyatnykh and A. V. Poltavets
*Effect of Thermal Charge Density Fluctuations on the Specific Heat of Normal Metals
(8, 21)
(b) Vol. **130**, No. 2, p. K101 – K104 (1985).

C. Varotsos, M. Lazaridou, K. Alexopoulos, and P. Varotsos
Point Defect Entropies and Enthalpies in KCl
(10.2; 8; 22.5.2)
(b) Vol. **130**, No. 2, p. K105 – K107 (1985).

E. E. Shaisha and A. A. Bahgat
Magnetic and Thermal Behaviour of the Amorphous Ferromagnet $Fe_{79}B_{16}Si_5$
(13.5.4 and 18.2, 5; 8; 21.1.1)
(a) Vol. **90**, No. 2, p. 661 – 668 (1985).

S. Mahmoud, E. M. Kassem, H. Müller, and E. Hegenbarth
*Thermal Conductivity of $LiKSO_4$ at Low Temperatures
(8 and 14.4.1, 6; 22.8.1)
(a) Vol. **90**, No. 2, p. K135 – K138 (1985).

W. Scheibner, R. Häszler, M. Stephan, and M. Jäckel
On the Filler Induced Modification of the Glass Transition Temperature and Hardness of an Epoxy Resin Composite Material
(12.2; 2; 8; 22.9)
(a) Vol. **90**, No. 2, p. K139 – K141 (1985).

M. Lazaridou and K. Eftaxias
*On a Connection between Vacancy Formation Parameters and Melting Process in Rare Gas Solids
(8 and 10.2; 23)
(a) Vol. **90**, No. 2, p. K147 – K149 (1985).

D. Neogy and T. Purohit
The Behaviour of Active Centres in a Laser Host. A Crystal Field Investigation on Nd^{3+} and Er^{3+} in LaF_3 Single Crystals
(13.4; 8; 18; 19; 22.5)
(b) Vol. **131**, No. 1, p. 329 – 338 (1985).

Yu. L. Khait
*Calculations of the Narrow Temperature Interval for Premelting Phenomena
(1.2 and 8; 10)
(b) Vol. **131**, No. 1, p. K19 – K22 (1985).

M. A. Aldzhanov, K. K. Mamedov, and A. A. Abdurragimov
*Heat Capacity and Moments of the Phonon Spectrum in $TlGaTe_2$
(6 and 8; 22.8)
(b) Vol. **131**, No. 1, p. K35 – K37 (1985).

Sanjeev Kumar, P. K. Ahluwalia, K. C. Sharma, and R. S. Tripathi
*Metallic Systems Containing Crystal Field Split Impurities. Free Energy in the Precence of Magnetic Field – a Contribution from the Loop Diagrams
(8 and 18; 21)
(b) Vol. **131**, No. 1, p. K49–K53 (1985).

S. N. Vaidya
Theory of Melting. III. Nature of the Molten Glasses
(1.2 and 6; 2; 8; 22.6)
(a) Vol. **91**, No. 1, p. 37–43 (1985).

Ch. Kuhrt, Th. Schittny, and K. Bärner
Magnetic *B–T* Phase Diagram of Anion Substituted MnAs. Magnetocaloric Experiments
(18; 8; 21.1)
(a) Vol. **91**, No. 1, p. 105–113 (1985).

R. K. Veliev and M. A. Aldzhanov
*Low-Temperature Heat Capacity of $ZnGa_2S_4$, $ZnGa_2Se_4$, and $ZnGa_2Te_4$ Thiogallates
(8; 22.8)
(a) Vol. **91**, No. 1, p. K23–K24 (1985).

I. V. Bodnar and N. S. Orlova
*X-Ray Study of the Thermal Expansion Anisotropy in $AgGaS_2$ and $AgInS_2$ Compounds over the Temperature Range from 80 to 650 K
(8; 22.8)
(a) Vol. **91**, No. 2, p. 503–507 (1985).

N. Dutt, G. G. Agrawal, and J. Shanker
Evaluation of the First- and Higher-Order Volume and Pressure Derivatives of Dielectric Constants of Alkali Halide Crystals
(4; 6; 8; 22.5.2)
(b) Vol. **132**, No. 1, p. 99–107 (10985).

H.-Matsuo Kagaya and T. Soma
*Specific Heat at Constant Volume of III–V Covalent Crystals
(8; 6; 22.2.1; 22.2.3)
(b) Vol. **132**, No. 1, p. K11–K14 (1985).

P. N. Ram, R. D. Bijalwan, P. D. Semalty, and M. Israel
*Resonant Scattering of Phonons in ZnS: Fe^{2+} and CdTe: Fe^{2+}
(6 and 8; 22.4.1; 22.4.3)
(b) Vol. **132**, No. 2, p. 365–373 (1985).

V. B. Kokshenev
Spin–Phonon Excitations in Paramagnetic Crystals. Heat Transfer at High Temperatures
(6 and 13.5.4; 8)
(b) Vol. **132**, No. 2, p. 375–382 (1985).

T. Soma, H. Higuchi, and H.-Matsuo Kagaya
Effect of Lattice Anharmonicity on Specific Heat of Alkali Alloys
(6; 4; 8; 21.2)
(b) Vol. **132**, No. 2, p. 383–388 (1985).

K. Bärner and E. Gmelin
Coupling of a Structural and a High-Spin–Low-Spin Transition in $MnAs_{1-x}P_x$ Compounds
(18; 8; 12.1; 21.1)
(b) Vol. **132**, No. 2, p. 431–436 (1985).

H.-Matsuo Kagaya, Y. Kitani, and T. Soma
*Linear Thermal Expansion Coefficient of $Si_{1-x}Ge_x$ Solid Solution
(6 and 8; 22.1.1; 22.1.2)
(b) Vol. **132**, No. 2, p. K87–K90 (1985).

9. Diffusion

Original Papers and Short Notes

J. Tritt-Goc and N. Piślewski
*A Study of H_2O Lattice Diffusion in Sodium Nitroprusside Single Crystal
(9 and 19; 22.8)
(a) Vol. **87**, No. 1, p. K41 – K44 (1985).

S. Jetschke and K. Kehl
Diffusion and Recrystallization Processes during Annealing of N^+- and P^+ -Implanted $LiNbO_3$
(11; 9; 20.1; 22.8.1)
(a) Vol. **88**, No 1, p. 193 – 205 (1985).

L. Latie, G. Villeneuve, Ch. Cros, and P. Hagenmuller
Influence of Cationic Vacancies on the Mobility of Li^+ Ions in Some Cation-Deficient Materials. An NMR Study
(10.2; 9; 14.4; 19; 22.5.4)
(b) Vol. **128**, No. 2, p. 475 – 482 (1985).

P. Varotsos, K. Alexopoulos, and M. Lazaridou
Erratum to: On the Proportionality of the Migration Volume and Its Migration Enthalpy in Fluorides
(10.2; 9; 22.5.3)
(b) Vol. **128**, No. 2, p. K191 (1985).

E. F. Krimmel, R. Langfeld, and A. G. Lutsch
*Modelling of Dopant Profiles Modified by Diffusing Defects in Polycrystalline Silicon
(9 and 14.3.1; 11; 22.1.2)
(a) Vol. **88**, No. 2, p. 533 – 538 (1985).

P. Varotsos, K. Alexopoulos, C. Varotsos, and M. Lazaridou
Interconnection of Point Defect Parameters in BaF_2
(10.2; 8; 9; 22.5.3)
(a) Vol. **88**, No. 2, p. K137 – K140 (1985).

F. H. M. Spit, H. Albers, A. Lubbes, Q. J. A. Rijke, L. J. v. Ruijven, J. P. A. Westerveld, H. Bakker, and S. Radelaar
*Diffusion of Antimony (^{125}Sb) in Polycrystalline Silicon
(9; 10.2; 22.1.2)
(a) Vol. **89**, No. 1, p. 105 – 115 (1985).
Erratum
(a) Vol. **91**, No. 1, p. 302 (1985).

A. Bauberger and H. Wipf
*H and D Diffusion in V and Nb in the Precence of a Temperature Gradient
(9; 10.1; 21.1)
(a) Vol. **89**, No. 1, p. 123 – 132 (1985).

D. Shaw
*The Chemical Diffusion of In in $Hg_{0.8}Cd_{0.2}Te$
(9; 22.4.4)
(a) Vol. **89**, No. 1, p. 173—183 (1985).

V. F. Stelmakh, Yu. R. Suprun-Belevich, V. D. Tkachev, and A. R. Chelyadinskii
*Diffusion of Boron Implanted into Silicon
(9; 10.2; 11; 22.1.2)
(a) Vol. **89**, No. 1, p. K45—K49 (1985).

E. V. Zolotoyabko and N. S. Ovanesyan
Mössbauer Diffraction in the Superionic Conductor $RbAg_4I_5$
(1.2 and 5; 9; 22.5.4)
(b) Vol. **129**, No. 2, p. 587—594 (1985).

S. Solmi, A. Armigliato, D. Govoni, R. Lotti, P. Negrini, M. Servidori, and A. Zani
*Effect of Chlorine Implantation on Phosphorus Predeposition in Silicon
(9 and 10.2; 11; 22.1.2)
(a) Vol. **89**, No. 2, p. 533—539 (1985).

Dao Khac An
*Numerical Solution of the Nonlinear Diffusion Equation for the Anomalous Boron Diffusion in Silicon
(9; 22.1.2)
(a) Vol. **90**, No. 1, p. 173—180 (1985).

G. G. Mahr von Staszewski and N. E. Walsöe de Reca
*Use of Autoradiography in the Study of Aluminium Electromigration
(9; 21)
(a) Vol. **90**, No 1, p. 191—195 (1985).

A. V. Eletskii and E. V. Stepanov
*Mechanisms of Diffusion of a Small Impurity in Noble Gas Crystals
(9 and 10.2; 23)
(b) Vol. **130**, No. 2, p. 517—529 (1985).

R. Braun and M. Feller-Kniepmeier
*Diffusion of Chromium in α-Iron
(9; 18.2; 21.1.1)
(a) Vol. **90**, No. 2, p. 553—561 (1985).

C. M. Demanet and H. J. Strydom
An Ion Microprobe Mass Analysis of Indium Contacts on Mercury Cadmium Telluride
(1.6; 9; 22.4.4)
(a) Vol. **90**, No. 2, p. K131—K133 (1985).

H. Nitta
*Semiclassical Theory of Dechanneling and the Diffusion Coefficients
(9 and 11; 22.1.2)
(b) Vol. **131**, No. 1, p. 75—86 (1985).

10. Defect Properties (Irradiation Defects See 11)

Original Papers and Short Notes

G. Schoeck
*The Superposition of Thermal Activation in Dislocation Movement
(10)
(a) Vol. **87**, No. 2, p. 571 – 581 (1985).

V. B. Molodkin, S. I. Olikhovskii, M. E. Osinovskii, A. N. Gureev, L. I. Datsenko, A. I. Nizkova, and B. F. Zhuravlev
*The Integrated Intensities of the Laue-Diffracted X-Rays for Monocrystals Containing Macroscopically Homogeneously Distributed Defects
(1 and 10)
(a) Vol. **87**, No. 2, p. 597 – 608 (1985).

Z. Janáček
Coherence Description of a Double Crystal Diffractometer
(1; 10)
(a) Vol. **88**, No. 1, 49 – 59 (1985).

F. N. Chukhovskii and Yu. P. Khapachev
*Exact Solution of the Takagi-Taupin Equation for Dynamical X-Ray Bragg Diffraction by a Crystal with a Transition Layer
(1 and 10)
(a) Vol. **88**, No. 1, p. 69 – 76 (1985).

A. G. Bondarenko and A. B. Khmelinin
Phase Transition in the System Solid–Gas Admixture Due to the Macroscopic Defect Formation and Growth
(1.2; 10)
(a) Vol. **88**, No. 1, p. 121 – 127 (1985).

M. J. Marcinkowski
The Structure of Twist Boundaries in Terms of Surface Dislocations
(1.4; 10)
(a) Vol. **88**, No. 1, p. 129 – 136 (1985).

P. A. Bezirganyan, K. T. Gabrielyan, and G. O. Demirchyan
*An X-Ray Interferometer with an Elastically Deformed Crystal
(1 and 10)
(a) Vol. **88**, No. 1, p. 143 – 149 (1985).

W. Gruschel and M. Wilkens
Focus Dependence of the Black–White Contrast of Small Point Defect Clusters on Transmission Electron Microscope Images of Crystalline Specimens
(1.4; 10)
(a) Vol. **89**, No. 2, p. 467 – 482 (1985).

V. Yu. Gerstman, R. Z. Valiev, and O. A. Kaibyshev
*On the Energy Change during the Grain Boundary Structure Recovery from Non-Equilibrium State to Equilibrium State
(1.4 and 10)
(a) Vol. **91**, No. 2, p. K119–K124 (1985).

M. Wautelet, C. Antoniadis, and L. D. Laude
*A Criterion for Amorphization or Transition to Metastable States of Highly Defective Solids
(2 and 10)
(a) Vol. **92**, No. 1, p. K43–K46 (1985).

G. Ernst and H.-J. Burmeister
*Calculations Concerning the Propagation of a Persistent Slip Band Nucleus
(10)
(a) Vol. **92**, No. 1, p. K57–K60 (1985).

R. G. Gabrielyan, A. R. Mkrtchyan, H. A. Aslanyan, and Kh. V. Kotandyan
*On the Theory of X-Diffraction in Oscillating Piezocrystals
(1 and 10)
(a) Vol. **92**, No. 2, p. 361–368 (1985).

10.1 Metals

Review Article

Original Papers and Short Notes

R. KUŚKA
The Application of Landau's Theory for Studying First-Order Phase Transsitions by the Internal Friction Method
(1.2; 10.1; 12.1; 21)
(a) Vol. **88**, No. 1, p. 151–156 (1985).

D. O. NORTHWOOD and I. O. SMITH
*Steady-State Creep and Strain Transients for Stress Change Tests in an Al–0.4 wt% Li Solid Solution Alloy
(10.1 and 12.1; 21)
(a) Vol. **88**, No. 1, p. 181–191 (1985).

K. MIURA
*Dissociation of Superdislocations and Damping in the $L1_2$ Ordered Structure
(10.1; 21.1; 21.6)
(a) Vol. **88**, No. 1, p. 207–212 (1985).

A. NOHARA
*Dislocation-Structure Changes during the Dip Tests and the Dynamical Internal Stress in Al–Mg Alloys at High Temperatures
(10.1; 12.1; 21; 21.3)
(a) Vol. **88**, No. 1, p. 213–222 (1985).

R. TOGNATO
*On the Free Energies of Hexagonal Dislocation Loops at High Temperatures
(10.1; 21)
(a) Vol. **88**, No. 1, p. K23–K25 (1985).

W. WASSERBÄCH
Phonon–Dislocation Scattering and Low-Temperature Lattice Thermal Conductivity of Niobium and Tantalum Single Crystals Deformed at Low Temperatures
(8; 6; 10.1; 21)
(b) Vol. **128**, No. 2, p. 453–466 (1985).

K. MIURA and A. MARUYAMA
*Equilibrium Superdislocation Damping in Cu_3Au near the Transformation Temperature
(10.1; 21.1; 21.6)
(a) Vol. **88**, No. 2, p. 529–532 (1985).

A. VILASSAKDANONT and K. N. SUBRAMANIAN
*Effect of Temperature on Some of the Mechanical Properties of Cu–10Ni–6Sn Spinodal Alloy
(10.1 and 12.1; 21.1)
(a) Vol. **88**, No. 2, p. 553–560 (1985).

P. KWAPULIŃSKI, J. RASEK, and J. W. MOROŃ
*The Kinetics of Structural Defect Recovery in Pure Iron in the Temperature Interval 300 to 800 K
(10.1; 14.1; 21.1.1)
(a) Vol. **88**, No. 2, p. 561–573 (1985).

L. T. Miyada-Naborikawa, R. de Batist, and P. Delavignette
Transmission Electron Microscopy Observations of Dislocations and Twins in Polycrystalline Zirconium
(1.4; 10.1; 21)
(a) Vol. **89**, No. 2, p. 521–531 (1985).

F. Walz, H. J. Blythe, and M. Földeaki
Magnetic After-Effects in Nitrogen-Charged α-Fe Following Low-Temperature Electron- and Neutron Irradiation
(11 and 18.2; 10.1; 21.1.1)
(a) Vol. **89**, No. 2, p. 581–594 (1985).

M. S. Sakr, A. E. E. Abdel-Raaheim. and A. A. El-Dally
*Effect of Structure Transformation on the Transient Creep Characteristics of Zn–40 wt% Al and Zn–0.5 wt% Al Alloys
(1.2 and 10.1; 21.1)
(a) Vol. **89**, No. 2, p. K157–K162 (1985).

A. N. Didenko, O. Kh. Asainov, V. P. Krivobokov, E. I. Logachev, A. D. Pogrebnyak, L. N. Puchkareva, G. E. Remnev, and Sh. M. Ruzimov
*Positron Annihilation at Defects in Surface Layers of Steel Irradiated by Supercurrent Ion Beams
(10.1 and 11; 1.2; 21.1.1)
(a) Vol. **89**, No. 2, p. K167–K172 (1985).

I. G. Ratishvili
*Annihilation of Frenkel Pairs in B.C.C. Metals (α-Iron). III. The Case of a Hydrostatically Compressed Lattice
(10.1; 11; 21.1.1)
(b) Vol. **130**, No. 1, p. 169–174 (1985).

W. Herremans, G. van Tendeloo M. David, and R. Gevers
*TEM Images of a Translation Interface on Edge
(1.4 and 10.1; 21.1.1)
(a) Vol. **90**, No. 1, p. 91–105 (1985).

W. Püschl
*Reactions between Glide Dislocations and Forest Dislocations in Anistropic B.C.C. Metals
(10.1; 21; 21.1; 21.1.1; 21.2)
(a) Vol. **90**, No. 1, p. 181–189 (1985).

O. A. Kaibyshev, R. Z. Valiev, and A. K. Emaletdinov
*Deformation Mechanisms and the Theory of Structural Superplasticity of Metals
(10.1; 1.4; 21)
(a) Vol. **90**, No. 1, p. 197–206 (1985).

B. V. Petukhov
*Theory of Solid Solution Softening in Comparison with Experiments on α-Fe
(10.1; 21.1.1)
(a) Vol. **90**, No. 1, p. 225–229 (1985).

K. Z. BOTROS and S. S. SHEININ
*On the Applicability of Conventional Contrast Theory to Weak Beam Images of Stacking Faults
(1.4 and 10.1; 21.1)
(a) Vol. **90**, No. 2, p. 499–514 (1985).

Z. GIERAK, J. W. MOROŃ, and J. RASEK
*Phenomena of Stuctural Defect Recovery in Molybdenum
(1.4 and 10.1; 14.1; 21)
(a) Vol. **90**, No. 2, p. 515–521 (1985).

A. PAWEŁEK, W. STRYJEWSKI, W. BOCHNIAK, and H. DYBIEC
*Mobile Dislocation Density Variation during Strain Rate Change Evidenced by Acoustic Emission
(10.1; 7; 21.1)
(a) Vol. **90**, No. 2, p. 531–536 (1985).

D. O. NORTHWOOD and I. O. SMITH
*Work-Hardening Rates during the Steady-State Creep of Aluminium and Alloys Determined from Stress-Change Tests
(10.1 and 12.1; 21)
(a) Vol. **90**, No. 2, p. 563–572 (1985).

M. J. MARCINKOWSKI
*Strengthening of Alloys by Atomic Order
(1.3 and 10.1; 21.1.1)
(a) Vol. **90**, No. 2, p. 621–630 (1985).

H. J. BLYTHE, F. WALZ, and E. KLUGMANN
The Influence of Neutron Irradiation on the Magnetic Relaxation Spectrum of Commercial Non-Oriented Iron–Silicon Steel
(18.2; 10.1; 11; 21.1.1)
(a) Vol. **90**, No. 2, p. 643–650 (1985).

A. R. ALI
*On the Behaviour of Interaction of Copper Atoms with Lattice Defects in Quenched and Cold-Worked Ni–Cu Alloys
(10.1; 14.1; 21.1)
(a) Vol. **91**, No. 1, p. 51–56 (1985).

S. SHEKHAR, T. C. LEE, and K. N. SUBRAMANIAN
*Deformation Studies on Cu–Ni–Sn Spinodal Alloys
(10.1; 21.1)
(a) Vol. **91**, No. 1, p. 63–72 (1985).

F. C. LOVEY, G. VAN TENDELOO, J. VAN LANDUYT, L. DELAEY, and S. AMELINCKX
Erratum to: On the Nature of Various Stacking Defects in 18R Martensite in Cu–Al Alloys. A Study by High Resolution Electron Microscopy
(1.3; 10.1; 21; 21.1)
(a) Vol. **91**, No. 1, p. 301 (1985).

G. GRAISS and G. SAAD
*Recovery of Internal Friction in Cold-Worked Aluminium Containing 0.35 wt% Fe
(10.1 and 12.1; 21)
(a) Vol. **92**, No. 2, p. 489–493 (1985).

I. RETAT, TH. STEFFENS, and CH. SCHWINK
*Dislocation Arrangement in FeNi–Invar Single Crystals
(10.1; 18.2; 21.1; 21.1.1)
(a) Vol. **92**, No. 2, p. 507–510 (1985).

G. HÜBNER and H. LÖFFLER
*On the Quench-Sensitivity of the Decomposition Rate of Al–Zn Alloys
(1.3 and 10.1; 1.2; 21; 21.1)
(a) Vol. **92**, No. 2, p. 517–524 (1985).

S. KISS, R. SCHALLER, and W. BENOIT
*Heating-Rate Dependent Internal Friction Background in Aluminium
(10.1; 21)
(a) Vol. **92**, No. 2, p. K109–K112 (1985).

10.2 Non-Metals

Original Papers and Short Notes

G. I. Pilipenko, D. V. Oparin, O. I. Tyutyunnik, and F. F. Gavrilov
*Photostimulated Transformations of Complex Colour Centres in Lithium Hydride Single Crystals
(6 and 10.2; 20.1; 20.3; 22)
(b) Vol. **127**, No. 1, p. 299–306 (1985).

L. E. Nagli
*F-Centres in KCl–Tl Crystals
(10.2 and 20.3; 22.5.2)
(b) Vol. **127**, No. 1, p. 319—326 (1985).

Y. Madhavan, K. Ramachandran, and T. M. Haridasan
*Resonant Modes Associated with a Vacancy in CdTe
(6 and 10.2; 22.4.3)
(b) Vol. **127**, No. 1, p. K15–K18 (1985).

M. Herget, A. Hofstaetter, and A. Scharmann
EPR of the F_2^--Centre in $LiYF_4$
(19; 10.2; 22.5)
(b) Vol. **127**, No. 1, p. K83–K87 (1985).

Yu. S. Boyarskaya, D. Z. Grabko, and D. S. Pishkova
*Correlation between Yield Stress and Microhardness in Crystals Such as NaCl
(10.2 and 12.2; 22.5.2)
(a) Vol. **87**, No. 1, p. 175–179 (1985).

A. Armigliato, S. Solmi, C. Donolato, P. Negrini, E. Gabilli, A. Garulli, and M. Kittler
*Influence of Chlorine Implantation on Phosphorus Diffusivity and Oxidation-Induced Defects in Silicon
(10.2 and 11; 14.3.3; 22.1.2)
(a) Vol. **87**, No. 1, p. 207–216 (1985).

G. A. Andreev and M. Hartmanová
*Effect of Impurity State on Deformation-Induced Density Change in NaCl:Ca Crystals
(10.2; 22.5.2)
(a) Vol. **87**, No. 1, p. 217–223 (1985).

V. L. Indenbom and V. M. Kaganer
The Formation of Plane-Wave X-Ray Images of Microdefects
(1.4; 10.2; 22.1)
(a) Vol. **87**, No. 1, p. 253–265 (1985).

Z. M. Alekseeva, V. S. Vikhnin, A. P. Zdebskii, and I. M. Tsivileva
Acoustical and Electrochemical Behaviour of Proustite Parameters in the Incommensurate Phase
(7; 1.2; 10.2; 22.8)
(a) Vol. **87**, No. 1, p. K35–K39 (1985).

N. A. DROZDOV, N. I. ZYATKOVA, A. A. PATRIN, and V. D. TKACHEV
*Photoluminescence of Silicon Annealed at Low Temperatures (420 to 550 °C)
(10.2 and 20.3; 13.4; 13.5.1; 22.1.2)
(a) Vol. **87**, No. 1, p. K45–K48 (1985).

A. AKILBEKOV, A. DAULETBEKOVA, and A. ELANGO
*Photo- and Thermochemical Reactions with Participation of Br_3^- Centres in X-Rayed KBr
(10.2; 11; 20.1; 22.5.2)
(b) Vol. **127**, No. 2, p. 493–501 (1985).

S. V. NISTOR, D. SCHOEMAKER, I. URSU, and B. R. YANG
*The Study of $Tl_2^+\langle 110 \rangle$ Centres in NaCl and RbCl
(10.2 and 19; 11; 22.5.2)
(b) Vol. **127**, No. 2, p. 503–508 (1985).

B. R. YANG, A. BOUWEN, and D. SCHOEMAKER
Electron Spin Resonance Study of Fe^+ Centers in $NaCl:Fe^{++}$ and Motion of the Associated Vacancies
(13.4 and 19; 10.2; 22.5.2)
(b) Vol. **127**, No. 2, p. 657–662 (1985).

T. KURIHARA, I. KANAZAWA, H. KOIZUMI, H. MURAKAMI, and T. SUZUKI
*Mobility and Trapping States of Positrons in NaCl Single Crystal Containing Ca^{2+} Ions
(10.2; 22.5.2)
(b) Vol. **127**, No. 2, p. K113–K116 (1985).

A. REQUARDT and G. LEHMANN
*Impurity-Related Electron Centers in α-Berlinite ($AlPO_4$)
(10.2 and 19; 11; 22.8.1)
(b) Vol. **127**, No. 2, p. K117–K120 (1985).

V. G. TSIRELSON, M. YU. ANTIPIN, R. G. GERR, R. P. OZEROV, and YU. T. STRUCHKOV
*Ruby Structure Peculiarities Derived from X-Ray Diffraction Data. Localization of Chromium Atoms and Electron Deformation Density
(1.3 and 10.2; 13.4; 22.6)
(a) Vol. **87**, No. 2, p. 425–433 (1985).

W. MAYER, D. GRASSE, and J. PEISL
*Diffuse X-Ray Scattering from Neutron-Irradiated Silicon Doped with Boron
(10.2 and 11; 22.1.2)
(a) Vol. **87**, No. 2, p. 583–588 (1985).

V. S. KORTOV, T. S. BESSONOVA, M. S. AKSELROD, and I. I. MILMAN
Hole-Induced Exoelectron Emission and Luminescence of Corundum Doped with Mg
(17 and 20.3; 10.2; 13.4; 22.6)
(a) Vol. **87**, No. 2, p. 629–639 (1985).

V. V. KVEDER, YU. A. OSIPYAN, I. R. SAGDEEV, A. I. SHALYNIN, and M. N. ZOLOTUKHIN
*The Effect of Annealing and Hydrogenation on the Dislocation Conduction in Silicon
(10.2 and 14.3; 13.4; 22.1.2)
(a) Vol. **87**, No. 2, p. 657–665 (1985).

M. Lazaridou
*On the Connection between the Melting Temperature and the Parameters Obtained from Thermally Stimulated Depolarization Experiments
(8 and 10.2; 22.5.2)
(a) Vol. **88**, No. 1, p. K19–K21 (1985).

R. Assenov, V. A. Moshnikov, and D. A. Yaskov
*On the Behaviour of Iodine in PbTe and SnTe
(10.2; 14.3; 22)
(a) Vol. **88**, No. 1, p. K27–K30 (1985).

N. M. Nancheva, D. N. Popov, K. I. Terziisky, B. N. Zlatov, and G. Kh. Tumbev
*Doppler Broadening of the Annihilation Gamma Line in KCl and LiF
(10.2; 22.5.2)
(a) Vol. **88**, No. 1, p. K31–K34 (1985).

H. Bahadur and R. Parshad
*On Obtaining *f–T* Characteristics without Hysteresis in Quartz
(10.2 and 12.2; 22.6)
(a) Vol. **88**, No. 1, p. K35–K38 (1985).

V. A. Kalukhov and S. I. Chikichev
The Influence of Isoelectronic Impurities on Intrinsic Deep Levels in Liquid Phase Epitaxial Gallium Arsenide
(13.4; 10.2; 22.2.1)
(a) Vol. **88**, No. 1, p. K59–K61 (1985).
Erratum
(a) Vol. **89**, No. 1, p. K115 (1985).

E. A. Secco
*Electrical Conductivity Measurements to Test for Rotating Sulfate Ions in Fast Ion Conductors
(10.2 and 14.4; 22.8.1)
(a) Vol. **88**, No. 1, p. K75–K77 (1985).

L. Latie, G. Villeneuve, Ch. Cros, and P. Hagenmuller
*Influence of Cationic Vacancies on the Mobility of Li^+ Ions in Some Cation-Deficient Materials. An NMR Study
(10.2; 9; 14.4; 19; 22.5.4)
(b) Vol. **128**, No. 2, p. 475–482 (1985).

A. M. Basu, A. Choudhury, S. Chaudhuri, P. Mukhopadhyay, and B. B. Baliga
*The Influence of Impurities on the Range of 28 MeV α-Particles in NaCl Single Crystals
(10.2 and 11; 20.1; 22.5.2)
(b) Vol. **128**, No. 2, p. 483–488 (1985).

R. Mićić and B. Drašković
*Some Photoluminescence Properties of Mn^{2+} in Magnesium Bromide
(10.2 and 20.3; 22.5.3)
(b) Vol. **128**, No: 2, p. 489–494 (1985).

F. H. M. SPIT, H. ALBERS, A. LUBBES, Q. J. A. RIJKE, L. J. v. RUIJVEN, J. P. A. WESTERVELD, H. BAKKER, and S. RADELAAR
Diffusion of Antimony (^{125}Sb) in Polycrystalline Silicon
(9; 10.2; 22.1.2)
(a) Vol. **89**, No. 1, p. 105—115 (1985).
Erratum
(a) Vol. **91**, No. 1, p. 302 (1985).

K. ATOBE, N. NISHIMOTO, and M. NAKAGAWA
*Irradiation-Induced Aggregate Centers in Single Crystal Al_2O_3
(10.2 and 20.1; 11; 22.6)
(a) Vol. **89**, No. 1, p. 155—162 (1985).

G. A. ANDREEV and B. I. SMIRNOV
*Effect of Plastic Deformation on the Density of Mg-Doped LiF Crystals
(10.2; 22.5.2)
(a) Vol. **89**, No. 1, p. 185—189 (1985).

V. N. CHEBOTIN and V. A. MEZRIN
*Ordering of Defects, Thermodynamic and Transport Properties of Solid Oxide Electrolytes with a Fluorite Structure
(10.2; 14.4; 22.6.1)
(a) Vol. **89**, No. 1, p. 199—211 (1985).

K. SCHMALZ, F.-G. KIRSCHT, S. NIESE, I. BABANSKAJA, M. KITTLER, H. RICHTER, J. SCHÖNEICH, and W. SEIFERT
*On the Intrinsic Gettering in Cu-Contaminated Cz-Si
(10.2 and 13.4; 13.3; 22.1.2)
(a) Vol. **89**, No. 1, p. 389—401 (1985).

C. H. LING, C. Y. KWOK, and K. PRASAD
*Relative Hydrogen Content in Plasma-Enhanced CVD Silicon Nitride Films: Substrate Temperature Dependence and Effect of Thermal Annealing
(1.5 and 10.2; 22)
(a) Vol. **89**, No. 1, p. K39—K43 (1985).

V. F. STELMAKH, YU. R. SUPRUN-BELEVICH, V. D. TKACHEV, and A. R. CHELYADINSKII
Diffusion of Boron Implanted into Silicon
(9; 10.2; 11; 22.1.2)
(a) Vol. **89**, No. 1, p. K45—K49 (1985).

A. V. IVANOV and A. I. MELKER
*Interatomic Bond Rupture in the System of Two Coupled Anharmonic Chains with a Weak Interaction
(2 and 10.2; 4; 22.9)
(a) Vol. **89**, No. 1, p. K51—K54 (1985).

P. LOŠŤÁK, R. NOVOTNÝ, J. HORÁK, and J. KLIKORKA
*Properties of Sb_2Te_3 Single Crystals Doped with Tl Atoms
(10.2 and 20.1; 14.3; 22.7)
(a) Vol. **89**, No. 1, p. K55—K59 (1985).

V. N. Salomatov, N. N. Kristoffel, and I. A. Parfianovich
The $F_H(H^-)$ Center Energy Parameters and the Nature of F Center Unrelaxed Ground State in KCl
(13.4 and 20.1; 10.2; 22.5.2)
(b) Vol. **130**, No. 1, p. 293—296 (1985).

H. Söthe, P. Studzinski, and J.-M. Spaeth
Pairs of F-Centers and F^- Impurities in KCl Studied by ENDOR
(13.4 and 19; 10.2; 22.5.2)
(b) Vol. **130** No. 1, p. 339—346 (1985).

E. Radzhabov
Photodissociation of $O^{--}V_a^+$ Centers in Lithium Fluoride
(13.4 and 20.3; 10.2; 22.5.2)
(b) Vol. **130**, No. 1, p. K55—K58 (1985).

S. Lay, P. Delavignette, and J. Vicens
Analysis of Subgrain Boundaries in Tungsten Carbide Deformed at High Temperature
(1.4; 10.2; 22)
(a) Vol. **90**, No. 1, p. 53—68 (1985).

V. V. Aristov, W. Graeff, V. I. Polovinkina, and A. A. Snigirev
Dynamical Diffraction of X-Ray Spherical Waves in Perfect Crystals Using a Narrow Slit as Radiation Source
(1.1; 10.2; 22.1.2)
(a) Vol. **90**, No. 1, p. 69—75 (1985).

B. I. Smirnov, T. N. Snezhkova, and O. L. Volkova
*Plastic Deformation and Fracture of LiF Bicrystals with Twist Boundary
(10.2; 22.5.2)
(a) Vol. **90**, No. 1, p. 207—213 (1985).

N. Ch. Mitra and A. L. Bhattacharyya
*Dielectric Relaxations in Dysprosia (Dy_2O_3) Containing Charged Defects
(10.2 and 14.4; 22.6)
(a) Vol. **90**, No. 1, p. 301—306 (1985).

A. A. Sitnikova, L. M. Sorokin, I. E. Talanin, E. S. Falkevich, and E. G. Sheikhet
*Vacancy Type Microdefects in Dislocation-Free Silicon Single Crystals
(10.2; 22.1.2)
(a) Vol. **90**, No. 1, p. K31—K33 (1985).

A. V. Egranov and A. I. Nepomnyachikh
Emission Spectra of NaCl:Mg Crystals
(13.4 and 20.3; 10.2; 22.5.2)
(a) Vol. **90**, No. 1, p. K65—K69 (1985).

P. Enders
Structure and Bonding in Cubic IV–VI Crystals. VII. On the Influence of Defects on the Stability of the Cubic Structure
(4 and 13; 1.2; 10.2; 22)
(b) Vol. **130**, No. 2, p. 511—516 (1985).

N. T. BAGRAEV, L. S. VLASENKO, and A. A. LEBEDEV
*Gold Solid Solution Decay in Silicon. I. Nuclear Spin–Lattice Relaxation Data
(10.2 and 19; 22.1.2)
(a) Vol. **90**, No. 2, p. 607–619 (1985).

A. KESSLER
Bulk Conductivity, Polarization, and Thermal Stimulated Currents of Insulated Polar Crystals
(14.3.3 and 14.3.4; 10.2; 22.5.3; 22.5.4)
(a) Vol. **90**, No. 2, p. 715–736 (1985).

M. LAZARIDOU and K. EFTAXIAS
*On a Connection between Vacancy Formation Parameters and Melting Process in Rare Gas Solids
(8 and 10.2; 23)
(a) Vol. **90**, No. 2, p. K147–K149 (1985).

P. GAWORZEWSKI, E. HILD, and K. SCHMALZ
*Oxygen Donor Formation and Oxygen Precipitation in Czochralski-Silicon Due to Heat Treatment at 600 to 800 °C
(10.2; 14.3; 20.1; 22.1.2)
(a) Vol. **90**, No. 2, p. K151–K156 (1985).

A. GLEITZ and H. W. HELBERG
*Microwave Conductivity of Dislocations in Deformed Silicon Single Crystals
(10.2 and 14.3; 22.1.2)
(a) Vol. **90**, No. 2, p. K209–K213 (1985).

C. N. AFONSO, C. ORTIZ, and G. J. CLARK
*Defect Formation in LiF by Low Energy Ion Implantation
(10.2 and 11; 20.1; 22.5.2)
(b) Vol. **131**, No. 1, p. 87–96 (1985).

S. IHARA and K. SUZUKI
*Molecular Dynamical Study of β-$AgI_{1-x}Cl_x$
(10.2 and 14.4; 1.2; 22.5.4)
(b) Vol. **131**, No. 1, p. 97–103 (1985).

R. C. BHUNIYA and S. E. RAO
Colour Centres and Luminescence in KCl:Zn Crystals
(20.1 and 20.3; 10.2; 13.4; 22.5.2)
(b) Vol. **131**, No. 1, p. 339–348 (1985).

L. N. ALEKSANDROV, B. P. SURIN, and M. I. ZOTOV
*Dynamic Characteristics of Radiation Defects in Silicon
(10.2 and 11; 22.1.2)
(a) Vol. **91**, No. 1, p. 57–61 (1985).

K. NAKANO, K. SUZUKI, and S. TAKEUCHI
*Dislocations in SmS Single Crystals
(10.2; 22)
(a) Vol. **91**, No. 1, p. 73–77 (1985).

L. Bosi and M. Nimis
*An Analysis of the Classical Arguments Concerning Forecasts for the Z_1 Centre Formation in Alkali Halides
(10.2 and 13.4; 22.5.2)
(b) Vol. **131**, No. 2, p. K111–K116 (1985).

B. N. Mukashev, M. F. Tamendarov, S. Zh. Tokmoldin, and V. V. Frolov
*Hydrogen Implantation into Silicon. Infra-Red Absorption Spectra and Electrical Properties
(10.2 and 11; 4; 14.3; 20.1; 22.1.2)
(a) Vol. **91**, No. 2, p. 509–522 (1985).

S. A. Shevchenko, I. I. Khodos, and I. I. Snighireva
*Dislocation Dissociation and Electrical Properties of Plastically Deformed Germanium Single Crystals
(10.2; 14.3; 22.1.1)
(a) Vol. **91**, No. 2, p. 523–531 (1985).

V. I. Alshits, E. V. Darinskaya, and A. A. Urusovskaya
*The Investigation of the Shock Deformation Kinetics and Dislocation Dynamics upon Pulse Loading of CsI, NaCl, and LiF Crystals
(10.2; 22.5.2)
(a) Vol. **91**, No. 2, p. 533–542 (1985).

P. B. Fitzsimons and J. Corish
*The Trapping of Vacancies by Impurity Vacancy Complexes in Alkali Halide Crystals
(10.2; 22.5.2)
(a) Vol. **91**, No. 2, p. 543–548 (1985).

J. Prakash, Rahul, and A. K. Nishad
*Heating Rate Dependence of Peak and Position of Maximum Depolarization Current in ITC Measurements
(10.2 and 14.4; 22.5.2)
(a) Vol. **91**, No. 2, p. 685–692 (1985).

Y. Nishio, Y. Ootuka, K. Kajita, T. Iwata, and W. Sasaki
*Reverse Annealing Effect in the Neutron-Irradiated Si:P System
(10.2 and 14.3; 22.1.2)
(a) Vol. **91**, No. 2, p. 725–728 (1985).

J. L. Lindström, B. Svensson, and J. W. Corbett
*On the Core of the Thermal Donors in Silicon
(10.2; 22.1.2)
(a) Vol. **91**, No. 2, p. K107–K111 (1985).

H. B. Gon, S. K. Gupta, and K. V. Rao
Optical Absorption and Thermoluminescence of AC Field-Treated LiF Single Crystals later Irradiated with X-Rays
(20.1 and 20.3; 10.2; 22.5.2)
(a) Vol. **91**, No. 2, p. K143–K146 (1985).

R. Cywiński, E. Mugeński, W. Nowy-Wiechuła, and J. Wiechuła
*Aggregation and Precipitation of Mn^{2+}–Cation Vacancy Dipoles in NaCl Matrix
(10.2 and 20.3; 22.5.2)
(b) Vol. **132**, No. 2, p. K91 – K94 (1985).

T. Tsuboi
*Formation of the $Tl^0(1)$ Center in KCl
(10.2 and 20.1; 22.5.2)
(b) Vol. **132**, No. 2, p. K95 – K98 (1985).

V. I. Vdovin, L. A. Matveeva, G. N. Semenova, M. Ya. Skorohod, Yu. A. Tkhorik, and L. S. Khazan
*Mechanism of Misfit Dislocation Network Formation in Heteroepitaxial System Ge–GaAs (001)
(1.5 and 10.2; 22.1.1; 22.2.1)
(a) Vol. **92**, No. 2, p. 379 – 390 (1985).

D. Tsoukalas and P. Chenevier
Boron Diffusion in Silicon by a Vacancy Mechanism
(9; 10.2; 22.1.2)
(a) Vol. **92**, No. 2, p. 495 – 501 (1985).

M. Georgieva, P. Mishev, and G. Tumbev
*Doppler Broadening Measurements in γ-Irradiated LiF
(10.2; 22.5.2)
(a) Vol. **92**, No. 2, p. 503 – 506 (1985).

C. Templier, B . Boubeker, H. Garem, E. L. Mathé, and J. C. Desoyer
*Precipitation and Implanted Xenon in Silicon
(1.4 and 10.2; 11; 22.1.2)
(a) Vol. **92**, No. 2, p. 511 – 516 (1985).

C. A. Londos
Room-Temperature Irradiation of p-Type Silicon
(13.4; 10.2; 14.3.3; 22.1.2)
(a) Vol. **92**, No. 2, p. 609 – 614 (1985).

R. Burkova, S. Takova, and G. Dimitrov
*S-Pits on (111) Silicon Surfaces
(1.6 and 10.2; 22.1.2)
(a) Vol. **92**, No. 2, p. K107 – K108 (1985).

C. Dauwe and K. Motoko
*Three-Quantum Annihilation in the Plastic Crystal Norbornane
(1.2 and 10.2; 22.9)
(a) Vol. **92**, No. 2, p. K117 – K121 (1985).

11. Irradiation Effects (X-Ray Diffraction Investigations See 1 and 10)

Original Papers and Short Notes

A. AKILBEKOV, A. DAULETBEKOVA, and A. ELANGO
Photo- and Thermochemical Reactions with Participation of Br_3^- Centres in X-Rayed KBr
(10.2; 11; 20.1; 22.5.2)
(b) Vol. **127**, No. 2, p. 493–501 (1985).

S. V. NISTOR, D. SCHOEMAKER, I. URSU, and B. R. YANG
The Study of $Tl_2^+\langle 110\rangle$ Centres in NaCl and RbCl
(10.2 and 19; 11; 22.5.2)
(b) Vol. **127**, No. 2, p. 503–508 (1985).

M. A. IVANOV and L. B. KVASHNINA
*Multiple Scattering of Axially Channeled Particles by Thermal Vibrations and Electrons
(11; 6; 13)
(b) Vol. **127**, No. 2, p. 587–597 (1985).

F. BELL
*On the Measurement of Solid State Compton Profiles by Inelastic Fast-Ion–Atom Collisions under Channeling Conditions. A Case Study
(11; 13.1; 17)
(b) Vol. **127**, No. 2, p. 599–603 (1985).

E. KÜHRT, R. WEDELL, D. SEMRAD, and P. BAUER
*Theoretical Description of the Stopping Power of Light Ions in the Intermediate Energy Range
(11; 13.5.2; 21.1; 22.1.2; 23)
(b) Vol. **127**, No. 2, p. 633–639 (1985).

A. REQUARDT and G. LEHMANN
A Hole Center Causing Rose Coloration in $AlPO_4$ and $GaPO_4$
(13.4; 11; 19; 20.1; 20.3; 22.8.1)
(b) Vol. **127**, No. 2, p. 695–701 (1985).

A. REQUARDT and G. LEHMANN
Impurity-Related Electron Centers in α-Berlinite ($AlPO_4$)
(10.2 and 19; 11; 22.8.1)
(b) Vol. **127**, No. 2, p. K117–K120 (1985).

E. A. LITTLE
Void Swelling of Ferritic Steel during 1 MeV Electron Irradiation
(1.4; 11; 21.1.1)
(a) Vol. **87**, No. 2, p. 441–449 (1985).

G. NEUBERT and S. ROGASCHEWSKI
*Measurements of the Backscattering of 20 to 60 keV Electrons from Double Layers at Various Angles of Incidence
(11; 21; 21.3; 21.6)
(a) Vol. **87**, No. 2, p. 451–457 (1985).

H. DÜSTERHÖFT, R. DAUGS, and R. PIPPIG
A New Method for the Interpretation of Energy Distributions Measured on Secondary Ions
(17; 1.6; 11; 21.1; 21.1.1; 22.1.2)
(a) Vol. **87**, No. 2, p. 499–506 (1985).

S. JETSCHKE and K. HEHL
*Diffusion and Recrystallization Processes during Annealing of N^+- and P^+-Implanted $LiNbO_3$
(11; 9; 20.1; 22.8.1)
(a) Vol. **88**, No. 1, p. 193–205 (1985).

H.-J. FITTING and J. REINHARDT
Monte-Carlo Simulation of keV-Electron Scattering in Solid Targets
(17; 11; 22.1.2)
(a) Vol. **88**, No. 1, p. 245–259 (1985).

A. MERTENS, H. KLOSE, and NG. TANG
*Ion Beam Contacting of Silicon. II. Antimony in n-Silicon
(11 and 14.3.4; 22.1.2)
(a) Vol. **88**, No. 1, p. 369–374 (1985).

K. YASUDA, M. TAKEDA, H. MATSUDA, and A. YOSHIDA
*Radiation Effects in a Photo-Coupler with High-Energy Electrons and γ-Rays and Recovery by Current Annealing
(11 and 14.3.3; 22.1.2; 22.2.1)
(a) Vol. **88**, No. 1, p. 389–396 (1985).

M. SARKAR
*An Approximate Approach to Absorption Correction in Electron-Probe Microanalysis
(11 and 13)
(a) Vol. **88**, No. 1, p. K51–K53 (1985).

V. G. BARYSHEVSKII and S. V. CHEREPITSA
*Neutron Spin Precession and Spin Dichroism of Nonmagnetic Unpolarized Single Crystals
(1.1 and 11; 22)
(b) Vol. **128**, No. 2, p. 379–387 (1985).

A. M. BASU, A. CHOUDHURY, S. CHAUDHURI, P. MUKHOPADHYAY, and B. B. BALIGA
*The Influence of Impurities on the Range of 28 MeV α-Particles in NaCl Single Crystals
(10.2 and 11; 20.1; 22.5.2)
(b) Vol. **128**, No. 2, p. 483–488 (1985).

K. IWAHANA, P. KNOLL, and H. KUZMANY
*Effect of Irradiation with Intense Laser Light on Trans-Polyacetylene
(11; 20.3; 22.9)
(b) Vol. **128**, No. 2, p. 735–741 (1985).

I. URSU, V. CRĂCIUN, L. NANU, D. CRĂCIUN, I. N. MIHĂILESCU, AL. POPA, A. PANTELICĂ, and M. ŠĂLĂGEAN
On the Behaviour of GaAs Single Crystals Irradiated with a Free-Running Ruby Laser
(1.6; 11; 22.2.1)
(a) Vol. **88**, No. 2, p. 415–419 (1985).

A. L. GOLOVIN, R. M. IMAMOV, and E. A. KONDRASHKINA
Potentialities of New X-Ray Diffraction Methods in Structural Studies of Ion-Implanted Silicon Layers
(1.6; 11; 22.1.2)
(a) Vol. **88**, No. 2, p. 505–514 (1985).

V. L. Litvinov and N. A. Ukhin
*Fast-Neutron Radiation Damages in Heavily Doped p-Silicon
(11; 14.3; 22.1.2)
(a) Vol. **89**, No. 1, p. 95–104 (1985).

E. A. Koptelov and A. A. Semenov
Instability Conditions for a Spatially Homogeneous Void Distribution in Irradiated Metals
(10.1; 1.4; 11; 21)
(a) Vol. **89**, No. 1, p. 117–122 (1985).

K. Atobe, N. Nishimoto, and M. Nakagawa
Irradiation-Induced Aggregate Centers in Single Crystal Al_2O_3
(10.2 and 20.1; 11; 22.6)
(a) Vol. **89**, No. 1, p. 155–162 (1985).

A. P. Tyutnev, V. S. Saenko, V. N. Abramov, V. P. Sichkar, A. I. Karpechin, E. D. Pozhidaev, and A. V. Vannikov
*Dose Effects in Transient Radiation-Induced Conductivity in Polymers
(11 and 14.3; 22.9)
(a) Vol. **89**, No. 1, p. 311–320 (1985).

W. R. Fahrner, J. R. Laschinski, D. Bräunig, M. Knoll, and A. Neidig
*Damage Profiles after 50 to 500 MeV Ion Implantation as Deduced from Thyristor Leakage Currents
(11 and 14.3.3; 13.4; 22.1.2)
(a) Vol. **89**, No. 1, p. 347–351 (1985).

J. Sitek and M. Miglierini
Mössbauer Spectroscopy on Amorphous $Fe_xNi_{80-x}B_{20}$ after Neutron Irradiation
(1.4; 5; 11; 21.1; 21.1.1)
(a) Vol. **89**, No. 1, p. K31–K33 (1985).

V. F. Stelmakh, Yu. R. Suprun-Belevich, V. D. Tkachev, and A. R. Chelyadinskii
Diffusion of Boron Implanted into Silicon
(9; 10.2; 11; 22.1.2)
(a) Vol. **89**, No. 1, p. K45–K49 (1985).

S. C. Sabharwal and B. Ghosh
*Electrical Conductivity of $NH_4H_2PO_4$ Single Crystal
(11 and 14.4; 22.8.1)
(a) Vol. **89**, No. 1, p. K83–K87 (1985).

L. N. Aleksandrov
Simulation of the Influence of Mechanical Stresses on the Kinetics of Crystallization of Ion-Implanted Silicon Layers under Pulse Heating
(2 and 3; 1.5; 11; 22.1.2)
(a) Vol. **89**, No. 2, p. 443–449 (1985).

B. T. Deshmukh, B. H. Ghiya, and R. B. Pode
On the Dose Dependence of the Formation of Z_1 Centres in KCl:Ca
(10.2 and 20.3; 11; 20.1; 22.5.2)
(a) Vol. **89**, No. 2, p. 517–520 (1985).

A. V. Voitsekhovskii, A. P. Kokhanenko, and A. S. Petrov
*Radiation Defect Annealing of Electron Irradiated $Hg_{1-x}Cd_xTe$ Crystals under Various Thermal Treatments
(11 and 14.3; 22.4.4)
(a) Vol. **90**, No. 1, p. 241–251 (1985).

M. Kobiyama and S. Takamura
Internal Friction and Electrical Resistivity Measurements in Cu and Ag Alloys after Neutron Irradiation at Low Temperature. I. Experiments
(10.1 and 14.1; 11; 21.1; 21.6)
(a) Vol. **90**, No. 1, p. 253–267 (1985).

S. Takamura and M. Kobiyama
Internal Friction and Electrical Resistivity Measurements in Cu and Ag Alloys after Neutron Irradiation at Low Temperature. II. Discussion
(10.1 and 14.1; 11; 21.1; 21.6)
(a) Vol. **90**, No. 1, p. 269–276 (1985).

G. A. Kachurin, N. P. Stepina, M. Voelskow, and E. Wieser
*Annealing Behaviour of Low-Energy-Implanted Silicon
(11; 14.3; 22.1.2)
(a) Vol. **90**, No. 1, p. 285–290 (1985).

V. G. Kapinos and P. A. Platonov
*The Molecular Dynamics Simulation of Some Mechanisms of Vacancy Spatial Rearrangements in the Cascade Region in the Stage of the Thermal Spike
(10.1 and 11; 21.1)
(a) Vol. **90**, No. 1, p. 291–300 (1985).

E. V. Kučys, A. B. Abišalas, R. V. Karpiejus, and G. B. Petrauskas
*GaAs Epilayer Conductivity Compensation by High-Energy Proton Bombardment
(11 and 14.3.1; 13.4; 22.2.1)
(a) Vol. **90**, No. 1, p. K45–K48 (1985).

S. A. Nepijko, H. Pagnia, and V. I. Styopkin
*Electron Induced Off-Switching in Au Island Film Diodes
(11 and 14.1; 21.6)
(a) Vol. **90**, No. 1, p. K113–K116 (1985).

A. I. Melker, S. N. Romanov, and V. S. Tokmakova
*Displacement Spike Annealing in α-Iron Irradiated by PKA in the Energy Range 20 to 100 keV
(11; 10.1; 21.1.1)
(b) Vol. **130**, No. 2, p. 531–537 (1985).

W. Wesch, E. Wendler, G. Götz, K. Unger, H. Röppischer, and Chr. Resagk
*Laser Induced Defects in GaAs Layers
(10.2 and 11; 20.1; 22.2.1)
(b) Vol. **130**, No. 2, p. 539–546 (1985).

G. V. Dedkov and Kh. Shhedov
*Planar Channeling Radiation of Positrons and Electrons at Moderate Energies in Simple Crystals
(11; 13; 22.1; 22.1.1; 22.1.2)
(b) Vol. **130**, No. 2, p. 719–726 (1985).

J. Osvald
*Influence of Electron Lithography on Electrical Parameters of MOS Structures
(11 and 14.3.3; 13.3; 22.1.2)
(a) Vol. **90**, No. 2, p. K225–K227 (1985).

H. Nitta
*Semiclassical Theory of Dechanneling and the Diffusion Coefficients
(9 and 11; 22.1.2)
(b) Vol. **131**, No. 1, p. 75–86 (1985).

C. N. Afonso, C. Ortiz, and G. J. Clark
*Defect Formation in LiF by Low Energy Ion Implantation
(10.2 and 11; 20.1; 22.5.2)
(b) Vol. **131**, No. 1, p. 87–96 (1985).

W. Brauer and M. Rösler
*On the Relation between Stopping Power and Proton-Induced Electron Emission for Intermediate Energies: Aluminum
(11 and 13.5.2; 17; 21)
(b) Vol. **131**, No. 1, p. 177–183 (1985).

L. N. Aleksandrov, B. P. Surin, and M. I. Zotov
*Dynamic Characteristics of Radiation Defects in Silicon
(10.2 and 11; 22.1.2)
(a) Vol. **91**, No. 1, p. 57–61 (1985).

A. K. Rai, R. S. Bhattacharya, and M. H. Rashid
*Ion Beam Mixing of Alternate Au–Ge Layer with GaAs
(11 and 14.3.4; 1.6; 21.6; 22.1.1; 22.2.1)
(a) Vol. **91**, No. 1, p. 287–293 (1985).

G. Balchaitis, J. Kameneckas, G. Petrauskas, and A. Sakalas
*Physical and Chemical Changes in the Surface Layer of GaAs Induced by Recoil Implantation of Al Atoms
(11 and 14.3.2; 22.2.1)
(a) Vol. **91**, No. 1, p. 295–300 (1985).

R. Pflueger, J. C. Corelli, and J. W. Corbett
*Radiation-Enhanced Oxygen-Related Thermal Donor Formation in Neutron-Transmutation-Doped Floating-Zone Silicon
(10.2 and 11; 22.1.2)
(a) Vol. **91**, No. 1, p. K49–K54 (1985).

P. Gehrmann, K. Lenkeit, and R. Stolle
*Measurements of Proton Channeling Energy Losses in Silicon in the Intermediate Energy Region
(11; 22.1.2)
(b) Vol. **131**, No. 2, p. 519–526 (1985).

B. N. Mukashev, M. F. Tamendarov, S. Zh. Tokmoldin, and V. V. Frolov
*Hydrogen Implantation into Silicon. Infra-Red Absorption Spectra and Electrical Properties
(10.2 and 11; 4; 14.3; 20.1; 22.1.2)
(a) Vol. **91**, No. 2, p. 509–522 (1985).

YU. G. CHUKALKIN, V. V. PETROV, V. R. SHTIRTS, and B. N. GOSHCHITSKII
*Effects of Radiation Disorder in Chromium Spinels
(1 and 11; 22.8.2)
(a) Vol. **92**, No. 2, p. 347–354 (1985).

T. N. YANG, G. P. YU, and L. J. CHEN
*The Characterization of Nitrogen- Ion-Irradiation Induced Dislocation Loops in Zirconium
(1.4 and 11; 21)
(a) Vol. **92**, No. 2, p. 399–406 (1985).

M. WAGNER, H.-D. GEILER, and G. GÖTZ
Time-Resolved Investigation of Large-Area Explosive Crystallization of Amorphous Silicon Layers
(2 and 3; 1.2; 11; 22.1.2)
(a) Vol. **92**, No. 2, p. 413–420 (1985).

C. TEMPLIER, B. BOUBEKER, H. GAREM, E. L. MATHÉ, and J. C. DESOYER
Precipitation of Implanted Xenon in Silicon
(1.4 and 10.2; 11; 22.1.2)
(a) Vol. **92**, No. 2, p. 511–516 (1985).

Z. CZYŻEWSKI
*Backscattering Factor of the ZAF-Correction Procedure in Quantitative Electron-Probe Microanalysis
(11; 13; 21; 21.1; 21.6)
(a) Vol. **92**, No. 2, p. 563–566 (1985).

K. D. GLINCHUK, V. I. GUROSHEV, A. V. PROKHOROVICH, and N. S. ZAYATS
*Photoluminescence in Electron-Irradiated Unannealed and Annealed GaAs:Zn – Irradiation-Induced 1.26 and 1.39 eV Emission Bands
(11 and 20.3; 13.4; 22.2.1)
(a) Vol. **92**, No. 2, p. K145–K149 (1985).

12. Mechanical Properties (Plastic Deformations See 10)

Original Papers

J. SMELA
*Application of the Self-Consistent Field Approximation to the Calculation of the Bulk Modulus of Regular Face-Centered Crystals
(12)
(b) Vol. **127**, No. 2, p. K99–K104 (1985).

R. KÜHNERT and B. MICHEL
*Moiré Techniques by Means of Scanning Electron Microscopy
(10 and 12)
(a) Vol. **89**, No. 2, p. K163–K165 (1985).

A. P. FILIPPOV, G. N. GAIDUKOV, and S. K. MAKSIMOV
*Elastic Fields of a Dislocation Inclined to a Free Surface
(10 and 12)
(a) Vol. **90**, No. 2, p. 215–224 (1985).

MASAHARU KATO and H.-R. PAK
Some Remarks on Thermodynamics of Stress-Induced Martensitic Transformations
(1.2; 8; 12)
(b) Vol. **130**, No. 2, p. 421–430 (1985).

V. A. STRELTSOV
*Defects in Crystals under Pressure
(10 and 12)
(a) Vol. **91**, No. 1, p. 89–98 (1985).

S. C. GOYAL and K. KUMAR
*Lattice Theory of Fourth-Order Elastic Constants of Primitive Lattices
(12)
(b) Vol. **131**, No. 2, p. 451–457 (1985).

A. K. HEAD
*The Sextic Polynomial of Crystal Elasticity
(12)
(b) Vol. **132**, No. 1, p. 117–123 (1985).

12.1 Metals

Original Papers and Short Notes

N. I. NOSKOVA, N. F. VILDANOVA, YU. I. FILIPPOV, and A. P. POTAPOV
*Strength, Plasticity, and Fracture of Ribbons of $Fe_5Co_{70}Si_{15}B_{10}$ Amorphous Alloy
(10.1 and 12.1; 2; 21.1; 21.1.1)
(a) Vol. **87**, No. 2, p. 549–557 (1985).

A. V. DOBROMYSLOV, G. V. DOLGIKH, and I. A. PERETURINA
Effect of Normal Stresses on Yield Stress of Molybdenum Single Crystals at Slip on the Plane $(\bar{2}11)$
(10.1; 12.1; 21)
(a) Vol. **87**, No. 2, p. 559–563 (1985).

R. KUŚKA
The Application of Landau's Theory for Studying First-Order Phase Transitions by the Internal Friction Method
(1.2; 10.1; 12.1; 21)
(a) Vol. **88**, No. 1, p. 151–156 (1985).

D. O. NORTHWOOD and I. O. SMITH
*Steady-State Creep and Strain Transients for Stress Change Tests in an Al–0.4 wt% Li Solid Solution Alloy
(10.1 and 12.1; 21)
(a) Vol. **88**, No. 1, p. 181–191 (1985).

A. NOHARA
Dislocation-Structure Changes during the Dip Tests and the Dynamical Internal Stress in Al–Mg Alloys at High Temperatures
(10.1; 12.1; 21; 21.3)
(a) Vol. **88**, No. 1, p. 213–222 (1985).

A. RAJPUT
Phonon Dispersion in Platinum and Palladium
(6; 8; 12.1; 21)
(b) Vol. **128**, No. 2, p. 411–417 (1985).

P. ALLIA, E. BONETTI, E. EVANGELISTA, L. TURRONI, and F. VINAI
Reversible and Irreversible Processes of Structural Relaxation and Dynamic Young Modulus Behaviour in the $Fe_{40}Ni_{38}Mo_4B_{18}$ Amorphous Alloy
(2; 12.1; 21.1; 21.1.1)
(a) Vol. **88**, No. 2, p. 521–527 (1985).

A. VILASSAKDANONT and K. N. SUBRAMANIAN
*Effect of Temperature on Some of the Mechanical Properties of Cu–10Ni–6Sn Spinodal Alloy
(10.1 and 12.1; 21.1)
(a) Vol. **88**, No. 2, p. 553–560 (1985).

L. T. MIYADA-NABORIKAWA and R. DE. BATIST
*Low-Temperature Internal Friction Peaks in Pure Zirconium Deformed at 300 K
(10.1 and 12.1; 21)
(a) Vol. **89**, No. 1, p. 191–198 (1985).

T. SOMA, K. TANAKA, and H.-MATSUO KAGAYA
*Pressure Derivative of Elastic Stiffness Constants for α-Phase Cu-Based Alloys
(12.1; 21.1)
(b) Vol. **129**, No. 2, p. K123–K125 (1985).

L. DELAEY, G. GUÉNIN, Y. MURAKAMI, P. F. GOBIN, G. VAN TENDELOO, and J. VAN LANDUYT
Strain Field Observation and Lattice Stability at the Tip of Thermoelastic Cu–Zn–Al Martensite Plates
(1.3; 1.4; 12.1; 21.1)
(a) Vol. **89**, No. 2, p. 457–465 (1985).

L. T. MIYADA-NABORIKAWA, R. DE BATIST, and P. DELAVIGNETTE
*Low-Temperature Relaxation Peaks in Zirconium Deformed at 8 K
(10.1 and 12.1; 21)
(a) Vol. **90**, No. 1, p. 277–283 (1985).

K. IWASAKI
*An Additional Evidence for the Sliding Mechanism of the Grain Boundary Internal Friction Peak
(10.1 and 12.1; 21)
(a) Vol. **90**, No. 1, p. K35–K37 (1985).

A. PAWELEK
*Reconsideration of the Activation Energy for Dislocation Creep
(10.1 and 12.1; 21)
(a) Vol. **90**, No. 1, p. K39–K44 (1985).

A. PAWEŁEK
*Erratum to: A Simple Model of Small Yield Point Due to Strain Rate Change Based on the Dynamics of the Source Generated Dislocation Group
(10.1 and 12.1; 21)
(a) Vol. **90**, No. 1, p. K118 (1985).

L. KALVODA
*Computing Method of Elastic Moduli and Young's Modulus for Texturized Materials with Hexagonal Crystal Symmetry
(12.1; 21)
(b) Vol. **130**, No. 2, p. 451–456 (1985).

M. OMRI and A. VINCENT
Attenuation Stress Hysteresis Loops and Low-Frequency Internal Friction Associated with Thermomechanical Breakaway of Dislocations
(7 and 10.1; 12.1; 21)
(b) Vol. **130**, No. 2, p. 547–560 (1985).

D. O. NORTHWOOD and I. O. SMITH
*Work-Hardening Rates during the Steady-State Creep of Aluminium and Alloys Determined from Stress-Change Tests
(10.1 and 12.1; 21)
(a) Vol. **90**, No. 2, p. 563–572 (1985).

G. TURILLI, H. SZYMCZAK, and H. K. LACHOWICZ
*Elastic Constants of the $(RE)_2Fe_{14}B$ Tetragonal Compounds (RE = Y, Nd, and Pr)
(12.1; 21.1.1; 21.4)
(a) Vol. **90**, No. 2, p. K143–K146 (1985).

T. SOMA, M. TAKAHASHI, and M.-MATSUO KAGAYA
*Compression Effect on the α–β Phase Transition of Cu-Based Binary Alloys
(1.2 and 12.1; 1.3; 21.1)
(b) Vol. **131**, No. 1, p. K11–K14 (1985).

R. SCHÜLBE and U. SCHMIDT
On the Influence of the Pre-Ageing Conditions on the Reversion Behaviour of an Al–1.87 at% Cu Alloy
(1.3; 12.1; 21; 21.1)
(a) Vol. **91**, No. 2, p. 433–438 (1985).

T. C. LEE, S. SHEKHAR, and K. N. SUBRAMANIAN
*In-Situ Straining of Cu–10Ni–6Sn Spinodal Alloy in the High Voltage Electron Microscope
(10.1 and 12.1; 21.1)
(a) Vol. **91**, No. 2, p. 549–557 (1985).

A. NOHARA and T. IMURA
Dislocation Behaviour and the Anomalous Yield-Stress Peak in a β-Brass Single Crystal
(10.1; 12.1; 21.1)
(a) Vol. **91**, No. 2, p. 559–567 (1985).

B. K. DAS, A. K. TEWARI, and D. L. BHATTACHARYA
*Study of Pressure on Metallic Tin Foil by Mössbauer Spectroscopy
(5 and 12.1; 21.7)
(b) Vol. **132**, No. 7, p. K23–K25 (1985).

L. DÍAZ, R. PAREJA, M. A. PEDROSA, and R. GONZÁLEZ
Positron Annihilation and Electron Microscopy Study in the Early Stage of Fatigue of Polycrystalline Copper
(10.1; 12.1; 21.1)
(a) Vol. **92**, No. 1, p. 159–168 (1985).

B. DUTTA, S. K. SARKAR, and D. ROY
Three-Body Interaction in H.C.P. Metals
(4 and 6; 12.1; 21.3)
(b) Vol. **132**, No. 2, p. 345–354 (1985).

K. BÄRNER and E. GMELIN
Coupling of a Structural and a High-Spin–Low-Spin Transition in $MnAs_{1-x}P_x$ Compounds
(18; 8; 12.1; 21.1)
(b) Vol. **132**, No. 2, p. 431–436 (1985).

12.2 Non-Metals

Review Article

S. Porowski and W. Trzeciakowski
The Effect of Pressure on Deep Impurity States with Large Lattice Relaxation
(13.4; 12.2; 14.3; 16; 22.2.3; 22.2.4; 22.4.3; 22.5; 22.5.2)
(b) Vol. **128**, No. 1, p. 11–22 (1985).

Original Papers and Short Notes

N. Dutt, O. P. Sharma, and J. Shanker
*Analysis of Photoelastic Behaviour of Fluorite-Structure Crystals
(4 and 12.2; 14.4; 22.5.3)
(b) Vol. **127**, No. 1, p. 67–71 (1985).

H.-Matsuo Kagaya and T. Soma
Calculation of Phonon Dispersion Curves for AlP, AlAs, and AlSb
(6; 12.2; 22.2)
(b) Vol. **127**, No. 1, p. 89–93 (1985).

N. V. K. Prabhakar and R. K. Singh
*Anharmonic Properties of Sodium Halides
(12.2; 22.5.2)
(b) Vol. **127**, No. 1, p. 95–101 (1985).

N. Dass and M. Kumari
Derivation of Some Equations of State for Solids. A New Approach
(8; 12.2; 22.5.2)
(b) Vol. **127**, No. 1, p. 103–108 (1985).

Yu. S. Boyarskaya, D. Z. Grabko, and D. S. Pishkova
*Correlation between Yield Stress and Microhardness in Crystals Such as NaCl
(10.2 and 12.2; 22.5.2)
(a) Vol. **87**, No. 1, p. 175–179 (1985).

R. K. Gupta, A. Shiromany, P. S. Bakhshi, and J. Shanker
Crystalline Properties of Alkali Halides Using the Generalized Huggins-Mayer Potentials
(4 and 8; 12.2; 22.5.2)
(b) Vol. **127**, No. 2, p. 473–479 (1985).

A. Strigazzi
*A New Method for Measuring the Mixed Splay–Bend Elasticity and the Anchoring Strength in a Nematic Film
(1.5 and 12.2; 22)
(a) Vol. **87**, No. 2, p. 537–541 (1985).

A. P. Zdebskii, Yu. A. Olikh, and A. U. Savchuk
Saturation and Oscillation of Current in Semiconductors Subjected to Uniaxial Deformation
(14.3.4; 12.2; 22.4.1)
(a) Vol. **87**, No. 2, p. K209—K212 (1985).

F. S. Khan and P. B. Allen
*Temperature Dependence of the Elastic Constants of p^+ Silicon
(12.2; 13.1; 13.4; 22.1.2)
(b) Vol. **128**, No. 1, p. 31—38 (1985).

W. N. Bhende, P. S. Bakhshi, and J. Shanker
*Thermoelastic Behaviour and Higher-Order Elastic Constants of Ionic Crystals
(4 and 12.2; 6; 8; 22.5.2; 22.6)
(b) Vol. **128**, No. 1, p. 105—112 (1985).

S. R. Murthy and T. S. Rao
*Elastic Behaviour of Mixed Li–Ni Ferrites
(12.2 and 18.3; 22.8.2)
(a) Vol. **88**, No. 1, p. 239—243 (1985).

H. Bahadur and R. Parshad
*On Obtaining *f–T* Characteristics without Hysteresis in Quartz
(10.2 and 12.2; 22.6)
(a) Vol. **88**, No. 1, p. K35—K38 (1985).

Z. Šimša, Y. Kawai, and V. A. M. Brabers
*Anelasticity and Relaxation Effects in Single Crystal Mn Ferrites
(12.2 and 18.3; 7; 22.8.2)
(a) Vol. **88**, No. 2, p. 611—617 (1985).

R. K. Varshney and J. Shanker
*Temperature Dependence of Fourth-Order Elastic Constants of Alkali Halide Crystals
(12.2; 22.5.2)
(b) Vol. **129**, No. 1, p. 61—72 (1985).

N. Dutt, G. G. Agrawal, and J. Shanker
Investigation of the Equation of State and the Grüneisen-Anderson Parameters for Fluorite-Structure Crystals
(4; 6; 8; 12.2; 22.5.3; 22.8.1)
(b) Vol. **129**, No. 1, p. 73—79 (1985).

E. Anastassakis and M. Cardona
*Effects of Strains on the Dynamic Effective Charge of III–V Semiconductors
(6 and 12.2; 4; 22.2)
(b) Vol. **129**, No. 1, p. 101—116 (1985).

A. Casajus and A. R. Weisheit
*A New Method for Measurement of Stress in the Neighbourhood of Window Edges in Multiple Layers
(12.2)
(a) Vol. **89**, No. 1, p. K35—K36 (1985).

A. K. Sood, E. Anastassakis, and M. Cardona
*Raman Piezospectroscopy in GaAs Revisited
(6 and 12.2; 22.2.1)
(b) Vol. **129**, No. 2, p. 505–512 (1985).

T. Ito
A Pseudopotential Approach to the Structural and Thermodynamical Properties of III–V Ternary Semiconductor Alloys
(1.2 and 8; 12.2; 22.2.4)
(b) Vol. **129**, No. 2, p. 559–568 (1985).

C. Varotsos and M. Lazaridou
*Comment on the Elastic Constants of CaF_2–SrF_2 Mixed Crystals
(12.2; 22.5.4)
(b) Vol. **129**, No. 2, p. K95–K97 (1985).

J. E. MacDonald, G. A. Saunders, M. S. Omar, and B. R. Pamplin
*Comparison of the Elastic Behaviour of $CuGe_4P_3$ with that of $CuGe_2P_3$
(12.2; 22.8)
(a) Vol. **89**, No. 2, p. K137–K141 (1985).

A. Batana and J. Faour
*On the Calculation of the Pressure Dependence of the Transverse Optical Mode Frequency of Cubic Ionic Crystals
(4 and 12.2; 22.5.2)
(b) Vol. **130**, No. 1, p. 115–120 (1985).

T. Soma and H.-Matsuo Kagaya
Mode Grüneisen Parameters for Long-Wave Phonons of III–V Covalent Compounds under Pressure
(6; 4; 12.2; 22.2.1; 22.2.3)
(b) Vol. **130**, No. 2, p. 497–501 (1985).

J. J. Couderc, I. Dudouit, Ch. Hennig-Michaeli, and C. Levade
*The Interaction between Slip and Twinning Systems in Natural Sphalerite Experimentally Deformed
(10.2 and 12.2; 22.4.1)
(a) Vol. **90**, No. 2, p. 581–593 (1985).

W. Scheibner, R. Häszler, M. Stephan, and M. Jäckel
*On the Filler Induced Modification of the Glass Transition Temperature and Hardness of an Epoxy Resin Composite Material
(12.2; 2; 8; 22.9)
(a) Vol. **90**, No. 2, p. K139–K141 (1985).

E. G. Tsitsishvili
*Piezo-Optical Activity of Cubic Crystals
(12.2 and 20.1; 13.1; 13.5.1; 22.5)
(b) Vol. **131**, No. 1, p. 279–290 (1985).

B. Martin, J. M. Pastor, and J. A. de Saja
*Ferroelectric Behaviour of TGS and TGSe from Microhardness Measurements
(12.2; 1.2; 22.9)
(b) Vol. **131**, No. 1, p. K5–K9 (1985).

E. M. KOSTOVA and M. K. KOSTOV
*Experimental Study of Electrogyration in $Li_2Gd_4(MoO_4)_7$ Crystals
(12.2 and 14.4; 22.8)
(a) Vol. **92**, No. 2, p. K105 – K106 (1985).

13. Electron States

Review Articles

S. SUGAI
*Lattice Vibrations in the Charge-Density-Wave States of Layered Transition Metal Dichalcogenides
(6 and 13; 1.2; 20.1; 22)
(b) Vol. **129**, No. 1, p. 13–39 (1985).

K. H. FISCHER
Spin Glasses (II)
(18.2 and 18.4; 2; 5; 7; 8; 13; 14; 21; 21.1; 21.6; 22.6; 22.8)
(b) Vol. **130**, No. 1, p. 13–71 (1985).

V. CHRISTOPH and G. RÖPKE
*Theory of Inverse Linear Response Coefficients
(13 and 14; 15; 21; 22)
(b) Vol. **131**, No. 1, p. 11–42 (1985).

Original Papers and Short Notes

M. C. BÖHM
*Consistent Evaluation of Long-Range and Short-Range Contributions to the Quasi-Particle Energies of Transition Metal Solids
(13; 21.1)
(b) Vol. **127**, No. 1, p. 209–220 (1985).

E. HEINER
*A Dynamical Theory for a Disordered Hubbard Chain far from Equilibrium. I. Extended Master Equations
(13 and 14.3.4; 22)
(b) Vol. **127**, No. 1, p. 253–265 (1985).

J. ŠLECHTA
*On SCCF Theory of Falicov-Ramirez-Kimball Transitions in 1D Binary Alloys
(13 and 14)
(b) Vol. **127**, No. 1, p. 403–412 (1985).

A. BHATTACHARYA, A. GHOSAL, and D. CHATTOPADHYAY
*Small-Signal AC Response of Two-Dimensional Hot Electrons in Quantum-Well Structures of Polar Semiconductors
(13 and 14.3.4; 22.2.1)
(b) Vol. **127**, No. 1, p. 427–432 (1985).

P. ENDERS
*Structure and Bonding in Cubic IV–VI Crystals. IV. Evidence for Covalent Bonding in the IV–VI Family from the Extended Mulliken Scheme
(4 and 13; 1.2; 22)
(b) Vol. **127**, No. 2, p. 449–457 (1985).

J. A. Tuszyński and J. M. Dixon
*Tensorial Coupling and the Reduced Matrix Elements of One- and Two-Body Operators
(13)
(b) Vol. **127**, No. 2, p. 549—555 (1985).

L. Skála
*On the Electronic Structure of Clusters: Convergence to the Solid State Properties
(13)
(b) Vol. **127**, No. 2, p. 567—571 (1985).

G. Seifert and H. Eschrig
*LCAO-Xα Calculations of Transition Metal Clusters
(13; 21.1; 21.6)
(b) Vol. **127**, No. 2, p. 573—585 (1985).

M. A. Ivanov and L. B. Kvashnina
Multiple Scattering of Axially Channeled Particles by Thermal Vibrations and Electrons
(11; 6; 13)
(b) Vol. **127**, No. 2, p. 587—597 (1985).

V. Drchal and J. Kudrnovský
*The Theory of Auger Spectra from Alloys. The Effect of Electron Correlations in Partially Filled Narrow Bands
(13; 17; 21.1)
(b) Vol. **127**, No. 2, p. 611—619 (1985).

D. A. Fushman
Kinetic Properties of the Two-Level Systems in Glasses with Account of Tunneling
(2; 13)
(b) Vol. **127**, No. 2, p. 679—688 (1985).

K. Dieterich and M. Wagner
*On the Temperature Behaviour of DC Conductivity in One-Dimensional Conductors above T_c
(13 and 14; 6)
(b) Vol. **127**, No. 2, p. 715—726 (1985).

S. Guha and C. Basu
*Parametric Excitation of Electron-Acoustic Waves by Helicon Waves in Magnetised Semiconductor-Plasma
(13 and 20; 22)
(b) Vol. **127**, No. 2, p. K137—K142 (1985).

P. Enders
*Structure and Bonding in Cubic IV–VI Crystals. V. Lattice Repulsion. Comparison of Lattice and Vibronic Anharmonicity
(4 and 13; 22)
(b) Vol. **128**, No. 1, p. 75—82 (1985).

N. A. Nikolov, P. N. Malinov, K. G. Kostov, and G. V. Mihailova
*Influence of Collisions on the Resonances in Bounded Semiconductor Plasmas
(13 and 14.3.4)
(b) Vol. **128**, No. 1, p. 323—328 (1985).

L. Skála
*On the $e^{-\alpha t}$ Damping of the Coherent Memory Functions. One Property of the Equilibrium State
(13)
(b) Vol. **128**, No. 2, p. K149–K152 (1985).

P. Enders
*Structure and Bonding in Cubic IV–VI Crystals. VI. Acoustic and Optical Deformation Potentials
(4 and 13; 22)
(b) Vol. **129**, No. 1, p. 89–99 (1985).

S. Chakraborty, A. Manna, and A. K. Ghosh
*Atomic Scattering Factors of Solid Elements
(13; 21; 21.1; 21.1.1; 21.3; 22.1.1; 22.1.2)
(b) Vol. **129**, No. 1, p. 211–220 (1985).

D. A. Fushman
Saturation of Tunneling States in Glasses in the Presence of the Phonon Bottleneck
(2; 6; 13)
(b) Vol. **129**, No. 1, p. K21–K25 (1985).

N. J. Castellani, D. B. Leroy, and W. Lambrecht
*Core-Hole Screening in Small Metallic Particles as a Function of Core-Hole Position
(13; 21; 21.2)
(b) Vol. **129**, No. 1, p. K69–K73 (1985).

D. J. González, C. M. Zapata, and J. A. Alonso
*Density Functional Pseudopotential Calculation of the Cohesive Properties of Disordered Solid Alloys of Alkaline-Earth Metals. Influence of the Ionic Pseudopotential
(1.3 and 13; 4; 21.3)
(b) Vol. **129**, No. 2, p. 483–488 (1985).

L. Wendler
*Electron–Phonon Interaction in Dielectric Bilayer Systems. Effect of the Electronic Polarizability
(6 and 13; 13.5.3; 14.4; 22.2.1; 22.2.2)
(b) Vol. **129**, No. 2, p. 513–530 (1985).

P. Blaudeck
*Shell Model for Clusters of Dense-Packed s-Scatterers and Calculation of the Density of States
(2 and 13)
(b) Vol. **129**, No. 2, p. 729–734 (1985).

S. Adam, Gh. Adam, and A. Corciovei
*One-Particle Crystal Field Hamiltonians
(1 and 13)
(b) Vol. **129**, No. 2, p. 763–774 (1985).

S. Yu. Potapenko and A. M. Satanin
*The Resistance of a One-Dimensional System with a Stationary Random Potential which is a Markov Process
(13 and 14; 2)
(b) Vol. **129**, No. 2, p. 805–811 (1985).

P. Enders
*Structure and Bonding in Cubic IV–VI Crystals. VII. On the Influence of Defects on the Stability of the Cubic Structure
(4 and 13; 1.2; 10.2; 22)
(b) Vol. **130**, No. 2, p. 511–516 (1985).

W. Nolting and U. Dubil
*$T = 0$ Magnetic Polaron in f-Systems with Antiferromagnetic s–f Exchange
(13 and 18).
(b) Vol. **130**, No. 2, p. 561–573 (1985).

K. Kułakowski and A. Z. Maksymowicz
*Charge Transfer in Chromium–Transition Metal Alloys
(13; 14; 21.1)
(b) Vol. **130**, No. 2, p. 629–635 (1985).

G. V. Dedkov and E. Kh. Shidov
Planar Channeling Radiation of Positrons and Electrons at Moderate Energies in Simple Crystals
(11; 13; 22.1; 22.1.1; 22.1.2)
(b) Vol. **130**, No. 2, p. 719–726 (1985).

H. Fehske
*On a Simple Functional Moment Approach to Itinerant Magnetism. Application to Ni
(13 and 18.1; 18.2; 21.1)
(b) Vol. **130**, No. 2, p. K121–K125 (1985).

G. D. Boulakis
*The Compton Profile of Polycrystalline $TiH_{1.98}$
(13; 21)
(b) Vol. **130**, No. 2, p. K127–K130 (1985).

U. Pietsch, J. Mahlberg, and K. Unger
Investigation of Dynamical Bond Charge Transfer in GaAs by Changing X-Ray Reflection Power under High Electric Field
(4; 13; 22.2.1)
(b) Vol. **131**, No. 1, p. 67–73 (1985).

S. Ligenza
*Dynamical Nature of Superexchange Interaction in $Y_3Fe_5O_{12}$ Studied by Neutron Spectroscopy
(13 and 18.3; 22.8.2)
(b) Vol. **131**, No. 1, p. 105–112 (1985).

E. Heiner
*A Dynamical Theory for a Disordered Hubbard Chain far from Equilibrium. II. The Low-Field Case
(13 and 14.3.4; 22)
(b) Vol. **131**, No. 1, p. 213–224 (1985).

B. Schuh
*Hybrid Pairing and Heavy Fermions
(13 and 14.2)
(b) Vol. **131**, No. 1, p. 243–248 (1985).

A. Gras-Marti, V. E. Anderson, R. H. Ritchie, and J. C. Ashley
*Influence of Damping on the Energy Loss Straggling of Low Velocity Protons in an Electron Gas
(13 and 20)
(b) Vol. **132**, No. 2, p. 509—513 (1985).

K. K. Danilevich-Tovsiyuk and B. A. Lukiyanets
On the Electron–Phonon Interaction in Layered Crystals
(6; 12.2; 13)
(b) Vol. **132**, No. 2, p. K81—K85 (1985).

Z. Czyżewski
Backscattering Factor of the ZAF-Correction Procedure in Quantitative Electron-Probe Microanalysis
(11; 13; 21; 21.1; 21.6)
(a) Vol. **92**, No. 2, p. 563—566 (1985).

13.1 Band Structure

Review Articles

Original Papers and Short Notes

B. I. Reser
*Mixed Basis Method with Overlapping Atomic Orbitals
(13.1; 22.1.2)
(b) Vol. **127**, No. 1, p. 267–272 (1985).

K. M. Paraskevopoulos
*Electron Effective Mass Dependence on Carrier Concentration in $TlBiTe_2$ Monocrystals
(13.1 and 20.1; 14.3; 22.8)
(b) Vol. **127**, No. 1, p. K45–K50 (1985).

P. Enders
*Test of Two Relations between Tight-Binding Parameters for Tetrahedral Solids
(13.1; 22)
(b) Vol. **127**, No. 1, p. K51–K53 (1985).

K. R. Allakhverdiev, T. G. Mamedov, V. E. Shteinshreiber, and I. K. Efendieva
*Fundamental Absorption Edge of TlSe and $TlInTe_2$ Crystals
(13.1 and 20.1; 13.5.1; 22; 22.8)
(b) Vol. **127**, No. 1, p. K55–K60 (1985).

G. P. Peka, O. A. Tokalin, and N. K. Dryapiko
*Cathodoluminescence of Graded Band-Gap Semiconductor Structures
(13.1 and 20.3; 13.3; 22.2.4)
(a) Vol. **87**, No. 1, p. 293–303 (1985).

Ç. Tuncay and M. Tomak
*Empirical Tight-Binding Band Structure of Wurtzite Semiconductors SiC, ZnSe, and ZnTe
(13.1; 22.3; 22.4.2; 22.4.3)
(b) Vol. **127**, No. 2, p. 543–547 (1985).

F. Bell
On the Measurement of Solid State Compton Profiles by Inelastic Fast-Ion–Atom Collisions under Channeling Conditions. A Case Study
(11; 13.1; 17)
(b) Vol. **127**, No. 2, p. 599–603 (1985).

S. De, I. Bose, and S. Chatterjee
*Electronic Energy Bands of Pr under Pressure
(13.1; 21.4)
(b) Vol. **127**, No. 2, p. 605–610 (1985).

M. Pugnet, T. Amand, A. Cornet et M. Brousseau
*Absorption interbande induite dans le silicium sous forte excitation picoseconde à 1,06 μm
(Interband Absorption Induced in Silicon by Strong Picosecond Excitation at 1.06 μm)
(13.1 and 20.1; 22.1.2)
(b) Vol. **127**, No. 2, p. 621–631 (1985).

Qui-Yi Ye and Zhi-Cheng Wang
*Transition Region Effect in the SiO_2–SiO_x–Si System on the Si Two-Dimensional Electron Gas
(13.1 and 13.3; 22.1.2)
(b) Vol. **127**, No. 2, p. 641–649 (1985).

M. DE DIOS LEYVA and J. LÓPEZ GONDAR
*Zero Energy Gap Conditions and Band Inversion in Superlattices
(13.1; 22.2.4)
(b) Vol. **128**, No. 2, p. 575–581 (1985).

E. K. ARUSHANOV, L. L. KULYUK, L. N. LUKYANOVA, A. N. NATEPROV, S. I. RADAUTSAN, and A. A. SHTANOV
*Stimulated Emission Spectra and Valence Band Structure in the $Cd_3(As_xP_{1-x})_2$ System
(13.1 and 20.3; 22.3.1)
(b) Vol. **128**, No. 2, p. 583–589 (1985).

YU. S. GRUSHKO, YU. B. PADERNO, K. YA. MISHIN, L. I. MOLKANOV, G. A. SHADRINA, E. S. KONOVALOVA and E. M. DUDNIK
*A Study of the Electronic Structure of Rare Earth Hexaborides
(13.1; 22)
(b) Vol. **128**, No. 2, p. 591–597 (1985).

G. A. JONES and P. T. LANDSBERG
*Brillouin Zones and Fundamental Regions
(13.1)
(b) Vol. **128**, No. 2, p. 619–625 (1985).

M. MONDAL and K. P. GHATAK
*Effect of Size Quantization on the Effective Electron Mass in Ultrathin Bi Films
(13.1; 21.7)
(b) Vol. **128**, No. 2, p. K133–K137 (1985).

S. A. ZEINALOV
*Conduction Band Structure of $In_{0.5}Ga_{0.5}Sb$:Te
(13.1 and 14.3; 22.2.4)
(b) Vol. **128**, No. 2, p. K139–K142 (1985).

V. V. SOBOLEV, V. E. GRACHEV, and A. D. SMIRNOVA
*The Band Structure and Reflectance Spectra of $CdSnP_2$ Ternary Compound
(13.1 and 20.1; 22.8)
(b) Vol. **128**, No. 2, p. K143–K147 (1985).

P. K. BASU and K. BHATTACHARYYA
Modification of the Theory of Two-Dimensional Mobility Due to Scattering by Alloy Clusters
(14.3; 13.1; 22.2.4)
(b) Vol. **128**, No. 2, p. K175–K177 (1985).

P. J. POMONIS
Electron Density and Work Function Dependence of the Enthalpy of Formation of Metal Oxides
(4; 8; 13.1; 13.3; 22.6)
(a) Vol. **88**, No. 2, p. 515–519 (1985).

J. MERTSCHING
*Inverse Spectral Theory and Incommensurate Peierls Phases
(13.1; 13.5.3)
(b) Vol. **129**, No. 1, p. 235–246 (1985).

U. NEITZEL and K. BÄRNER
Optical and Magnetic Investigations of Some NiAs-Type Transition Metal Compounds
(18 and 20.1; 13.1; 21.1)
(b) Vol. **129**, No. 2, p. 707—715 (1985).

R. STEARNS, R. BRAUNSTEIN, and L. MULDAWER
*Optical Properties of β'-AuZn by Wavelength-Modulated Derivative Spectroscopy
(13.1 and 20.1; 21.1; 21.6)
(b) Vol. **129**, No. 2, p. 735—743 (1985).

M. MONDAL and K. P. GHATAK
*Influence of Magnetic Quantization on the Effective Electron Mass in Ternary Semiconductors
(13.1; 18; 22.4.4)
(b) Vol. **129**, No. 2, p. 745—753 (1985).

C. SUGIURA and S. MURAMATSU
*Sulfur K X-Ray Absorption-Edge Structures and Electronic States of Transition-Metal Sulfides MS and MS_2 (M = Fe, Co, Ni)
(13.1 and 20; 22)
(b) Vol. **129**, No. 2, p. K157—K161 (1985).

G. GOBSCH, G. PAASCH, H. ÜBENSEE, R. HERRMANN, and W. KRAAK
*Calculation of the Subband Structure of Inversion Layers in p-InSb Bicrystals
(13.1; 13.3; 14.3.2; 22.2.3)
(b) Vol. **129**, No. 2, p. K167—K171 (1985).

A. L. PUSHKARCHUK, P. P. MARDILOVICH, A. I. TROKHIMETS, S. G. GAGARIN, and G. M. ZHIDOMIROV
*Band Structure Variation of the (0001) Face for Corundum during Formation and Hydration
(13.1 and 13.3; 22.6)
(b) Vol. **129**, No. 2, p. K181—K185 (1985).

A. SLAOUI and P. SIFFERT
*Determination of the Electron Effective Mass and Relaxation Time in Heavily Doped Silicon
(13.1; 20.1; 22.1.2)
(a) Vol. **89**, No. 2, p. 617—622 (1985).

S. A. AZIMOV and T. M. RAZYKOV
*Energy Band Model of Ultrathin Solar Cells with Thickness up to 5 μm
(13.1 and 14.3.3; 16; 22; 22.4.4)
(a) Vol. **89**, No. 2, p. K223—K226 (1985).

NGUYEN NHU DAT and NGUYEN VAN HIEU
*On the Microscopic Theory of the Resonant Nonlinear Refractive Index of InSb
(13.1 and 20.1; 22.2.3)
(b) Vol. **130**, No. 1, p. 235—241 (1985).

P. T. LANDSBERG, A. NEUGROSCHEL, F. A. LINDHOLM, and C. T. SAH
*A Model Band-Gap Shrinkage in Semiconductors with Application to Silicon
(13.1 and 13.4; 22.1.2)
(b) Vol. **130**, No. 1, p. 255—266 (1985).

F. Thuselt and M. Rösler
*Universal Approximation Formulas for the Gap Shift in Doped Semiconductors
(13.1; 22.1.2; 22.2.1)
(b) Vol. **130**, No. 2, p. K139—K144 (1985).

H. J. Fischbeck and J. Mertsching
*Commensurate–Incommensurate Transition in a Quasi-One-Dimensional Conductor with a Nearly 1/3-Filled Band
(13.1)
(b) Vol. **131**, No. 1, p. 165—175 (1985).

G. Dräger, W. Czolbe, E. Schulz, A. Simunek, J. Drahokoupil, Yu. N. Kucherenko, and V. V. Nemoshkalenko
*Valence Band Investigations of GaS by Means of Polarized X-Ray Emission Spectra
(13.1 and 20; 22.3)
(b) Vol. **131**, No. 1, p. 193—200 (1985).

F. M. Gashimzade and D. G. Guliev
*Continuity Chords of Bands and Band Structure of Semiconducting Compounds with TlSe-Type Structure
(13.1; 22)
(b) Vol. **131**, No. 1, p. 201—206 (1985).

E. G. Tsitsishvili
Piezo-Optical Activity of Cubic Crystals
(12.2 and 20.1; 13.1; 13.5.1; 22.5)
(b) Vol. **131**, No. 1, p. 279—290 (1985).

K. R. Allakhverdiev, T. G. Mamedov, V. V. Panfilov, M. M. Shukyurov, and S. I. Subbotin
Influence of Hydrostatic Pressure on the Fundamental Absorption Edge of $TlGaSe_2$, $TlGaS_2$, and $TlInS_2$ Crystals
(12.2 and 20.1; 1.2; 13.1; 22.8)
(b) Vol. **131**, No. 1, p. K23—K28 (1985).

W. Henrion, G. Beddies, and H. Lange
Optical Characterization of Amorphous and Crystalline Molybdenum Silicide Films
(20.1; 2; 13.1; 21)
(b) Vol. **131**, No. 1, p. K55—K58 (1985).

A. K. Abass, H. A. Jassim, K. J. Majeid, and R. H. Misho
*Optical Parameters of Chemically Deposited Tin Disulphide
(13.1 and 20.1; 22)
(a) Vol. **91**, No. 1, p. 129—133 (1985).

T. M. Razykov, B. Kh. Kadyrov, and M. A. Khodyaeva
*Energy Band Models of n-$Zn_xCd_{1-x}S$–p-CdTe and n-$Zn_xCd_{1-x}S$–p-Si ($0 \leqq x \leqq 1$) Heterojunctions
(13.1 and 14.3.3; 22.4.4)
(a) Vol. **91**, No. 1, p. K87—K91 (1985).

T. CHASSÉ, U. BERG, and O. BRÜMMER
*Photoelectron Diffraction and Band Structure Effects in ARXPS from the Valence Bands of GeS
(13.1 and 17; 22)
(b) Vol. **132**, No. 1, p. 141–144 (1985).

E. DE CASTRO and P. OLIVO
Quantum Effects in Accumulation Layers of Si–SiO_2 Interfaces in the WKB Effective Mass Approximation
(13.4 and 14.3.2; 13.1; 22.1.2; 22.6)
(b) Vol. **132**, No. 1, p. 153–163 (1985).

P. ENDERS
*Acoustic and Optical Deformation Potentials in Cubic IV–VI Compounds. Strong Dependence upon the Lattice Distortion
(13.1; 6; 22)
(b) Vol. **132**, No. 1, p. 165–172 (1985).

E. ERBARUT and M. TOMAK
*A Localized Orbital Description of Ge–Si Alloy
(13.1; 22.1.1; 22.1.2)
(b) Vol. **132**, No. 1, p. 173–177 (1985).

T. KATTERLE and W. LORENZ
Quantumchemical Study of Imperfect Semiconductor Surfaces Including Step and Kink Sites. Silicon (111) and Chemisorption of H, OH, and H_2O
(13.3; 13.1; 22.1.2)
(b) Vol. **132**, No. 1, p. 225–238 (1985).

A. STERN and K. HÜBNER
*The One-Band Tight-Binding Hamiltonian: Fivefold Rings and Band Edges
(13.1; 22)
(b) Vol. **132**, No. 1, p. K51–K56 (1985).

L. A. BUGAEV and R. V. VEDRINSKII
*Hartree-Fock Approximation in the Scattering Wave Method. K-Absorption of Ionic Crystals in the Single-Scattering Formalism (XANES and EXAFS)
(13.1; 20; 22.5.2)
(b) Vol. **132**, No. 2, p. 459–464 (1985).

F. LOPEZ-AGUILAR and J. COSTA-QUINTANA
*Electronic Structure of the Cubic Laves Phase of γ-$CeAl_2$
(13.1; 21.4)
(b) Vol. **132**, No. 2, p. 485–494 (1985).

K. HEMACHANDRAN and A. R. CHETAL
*X-Ray Spectroscopic Studies of Azurite Mineral
(13.1 and 20; 22.8.1)
(b) Vol. **132**, No. 2, p. 503–507 (1985).

13.2 Fermi Surfaces

Original Paper

G. Urushadze
Acoustic Phenomena in Tunnel Junctions
(7 and 14.1; 13.2; 21)
(b) Vol. **132**, No. 1, p. 51–56 (1985).

13.3 Surface and Interface States

Original Papers and Short Notes

G. Paasch
*Alternative Quasi-Classical Transport Equations. I. General Formalism for the Inclusion of Wave Functions
(13.3 and 14.3.2)
(b) Vol. **127**, No. 1, p. 385–394 (1985).

G. P. Peka, O. A. Tokalin, and N. K. Dryapiko
Cathodoluminescence of Graded Band-Gap Semiconductor Structures
(13.1 and 20.3; 13.3; 22.2.4)
(a) Vol. **87**, No. 1, p. 293–303 (1985).

H. Toutah and G. Pananakakis
*Study of Electronic States Located at the Si–SiO_2 Interface and within the Oxide Volume of VLSI MOS Capacitors Using Capacitance and Conductance Measurements
(13.3 and 14.3.3; 14.3.2; 22.1.2)
(a) Vol. **87**, No. 1, p. 311–317 (1985).

Qui-Yi Ye and Zhi-Cheng Wang
*Transition Region Effect in the SiO_2–SiO_x–Si System on the Si Two-Dimensional Electron Gas
(13.1 and 13.3; 22.1.2)
(b) Vol. **127**, No. 2, p. 641–649 (1985).

D. Weaire and J. P. Kermode
*Electron States in the GaAs–$Ga_xAl_{1-x}As$ Superlattice
(13.3; 22.2.4)
(b) Vol. **127**, No. 2, p. K143–K145 (1985).

Th. Schurig and R. Herrmann
*Some Properties of n-Inversion Layers at Grain Boundaries in InSb Bicrystals
(13.3 and 14.3.4; 22.2.3)
(b) Vol. **127**, No. 2, p. K159–K163 (1985).

H. Pagnia, J. Schnellbächer, and N. Sotnik
*A Regeneration Model for Conducting Filaments in MIM Diodes
(13.3 and 14.3.3; 22.6)
(a) Vol. **87**, No. 2, p. 709–717 (1985).

V. S. Lysenko, V. I. Zimenko, I. P. Tyagulskii, I. N. Osiyuk, O. V. Snitko, and T. N. Sytenko
*Flash-Lamp Annealing of Si–SiO_2 Transition Layer Defects
(11 and 13.3; 14.3.2; 22.1.2)
(a) Vol. **87**, No. 2, p. K175–K180 (1985).

R. Herrmann, W. Kraak, and G. Nachtwei
Electrical Properties of Grain Boundaries in InSb Bicrystals
(14.3; 13.3; 22.2.3)
(b) Vol. **128**, No. 1, p. 337–344 (1985).

M. Mondal and K. P. Ghatak
Theoretical Analysis of Oscillatory Heat Capacity in *n*-Channel Inversion Layers on Ternary Semiconductors
(8; 13.3; 22.4.4)
(b) Vol. **128**, No. 1, p. K21–K26 (1985).

Tran Thoai Duy Bao
*Static Polarizability and Dipolar Surface Plasmon of Small Metal Particles
(13.3; 13.5.2)
(b) Vol. **128**, No. 2, p. 731–734 (1985).

Y. H. Ohtsuki and R. Kawai
*Surface Wake Potential (Comments to the Paper by Kononets and Filippov [1])
(13.3)
(b) Vol. **128**, No. 2, p. K169–K170 (1985).

P. J. Pomonis
Electron Density and Work Function Dependence of the Enthalpy of Formation of Metal Oxides
(4; 8; 13.1; 13.3; 22.6)
(a) Vol. **88**, No. 2, p. 515–519 (1985).

S. I. Kirillova, V. E. Primachenko, and O. V. Snitko
Nonequilibrium Depletion Relaxation in Strong Electric Fields under Various Conditions at the Silicon Surface
(14.3.2 and 14.3.4; 13.3; 22.1.2)
(a) Vol. **88**, No. 2, p. 647–654 (1985).

X. B. Liao, J. Y. Lian, and G. L. Kong
Analysis for Steady State Field Effect Measurements on a-Si:H Films
(14.3.1; 2; 13.3; 22.1.2)
(a) Vol. **88**, No. 2, p. 697–704 (1985).

S. Nara
*On the Shape Dependence of the Kubo Effect and Polyhedral Shapes
(13.3)
(b) Vol. **129**, No. 1, p. 287–291 (1985).

F. Bechstedt and R. Enderlein
*Electron–LO-Phonon Interaction near Interfaces. Application to Scattering in Inversion Layer Electrons on SiO_2–Si Interfaces
(6 and 13.3; 22.1.2; 22.6)
(b) Vol. **129**, No. 1, p. 349–362 (1985).

A. Stern and K. Hübner
*Electronic Density of States of a Valence-Alternation-Pair Defect at the Si–SiO_2 Interface
(13.3; 22.1.2; 22.6)
(b) Vol. **129**, No. 1, p. K89–K93 (1985).

A. Jakubowski and K. Iniewski
*Technical Method of Determination of the Interface Trap Density
(13.3 and 14.3.3)
(a) Vol. **89**, No. 1, p. 383–388 (1985).

L. Skala
*Electron Transfer in Chemisorption
(13.3; 4)
(b) Vol. **130**, No. 2, p. K157–K162 (1985).

J. Osvald
Influence of Electron Lithography on Electrical Parameters of MOS Structures
(11 and 14.3.3; 13.3; 22.1.2)
(a) Vol. **90**, No. 2, p. K225–K227 (1985).

F. Bechstedt and R. Enderlein
*Dielectric Screening, Polar Phonons, and Longitudinal Electronic Excitations of Quantum Well Double Heterostructures. Application to Light Scattering from Charge Density Fluctuations
(6 and 13.3; 14.4; 20; 22.2.4)
(b) Vol. **131**, No. 1, p. 53–66 (1985).

A. Beckmann
*Electronic Structure of Perfect Chemisorbed Monolayers and of Disordered Surfaces
(13.3)
(b) Vol. **131**, No. 1, p. 319–327 (1985).

H. Flietner
*U-Shaped Distributions at Semiconductor–Insulator Interfaces and the Nature of the Related Defect Centres
(13.3; 4; 10.2; 22.1.2)
(a) Vol. **91**, No. 1, p. 153–164 (1985).

E. Klier
*A Study of Semiinsulating GaAs Surfaces by Thermostimulated Currents
(13.3 and 14.3.2; 22.2.1)
(a) Vol. **91**, No. 2, p. 631–635 (1985).

V. A. Budyanu, S. N. Chechuy, I. A. Damaskin, S. A. Fedoseev, A. A. Nasakin, S. L. Pyshkin, M. I. Valkovskaya, and V. P. Zenchenko
Investigation of $A^{III}B^{V}$–Si Heterojunctions Grown by Laser Deposition
(14.3.1 and 14.3.3; 13.3; 22.1.2; 22.2.1; 22.2.2)
(a) Vol. **91**, No. 2, p. 737–744 (1985).

V. M. Fomin and E. P. Pokatilov
*Phonons and the Electron–Phonon Interaction in Multi-Layer Systems
(6 and 13.3; 22)
(b) Vol. **132**, No. 1, p. 69–82 (1985).

T. Katterle and W. Lorenz
*Quantumchemical Study of Imperfect Semiconductor Surfaces Including Step and Kink Sites. Silicon (111) and Chemisorption of H, OH, and H_2O
(13.3; 13.1; 22.1.2)
(b) Vol. **132**, No. 1, p. 225–238 (1985).

13.4 Impurity and Defect States

Review Articles

S. Porowski and W. Trzeciakowski
*The Effect of Pressure on Deep Impurity States with Large Lattice Relaxation
(13.4; 12.2; 14.3; 16; 22.2.3; 22.2.4; 22.4.3; 22.5; 22.5.2)
(b) Vol. **128**, No. 1, p. 11–22 (1985).

Yu. A. Katulevskii
Injection of Degenerate Electron–Hole Plasma
(14.3.4; 13.1; 13.4; 14.3.3; 22)
(a) Vol. **88**, No. 1, p. 11–42 (1985).

M. Godlewski
*On the Application of the Photo-EPR Technique to the Studies of Photoionization, DAP Recombination, and Non-Radiative Recombination Processes
(13.4 and 19; 16; 22.2; 22.4.1; 22.5)
(a) Vol. **90**, No. 1, p. 11–52 (1985).

B. L. Gelmont, B. T. Kolomiets, and K. D. Tsendin
Impurity Conduction of Chalcogenide Vitreous Semiconductors
(2 and 14.3; 13.1; 13.4; 22.3; 22.8)
(a) Vol. **91**, No. 2, p. 319–337 (1985).

Yu. E. Perlin and A. A. Kaminskii
*Nonradiative Transitions of the Trivalent Lanthanides in Insulating Laser Crystals
(13.4 and 20.3; 6; 20.2; 22.8.1)
(b) Vol. **132**, No. 1, p. 11–40 (1985).

Original Papers and Short Notes

A. A. Andreev, A. V. Zherzdev, A. I. Kosarev, K. V. Kougia, and I. S. Shlimak
*Photoluminescence and Recombination of Excess Carriers in Amorphous Hydrogenated Silicon
(13.4 and 20.3; 16; 22.1.2)
(b) Vol. **127**, No. 1, p. 273–278 (1985).

Prem Chand and M. Umar
*Electron Paramagnetic Resonance Study of Copper(II) Ions in Single Crystal of Strontium Formate Dihydrate
(13.4 and 19; 22.9)
(b) Vol. **127**, No. 1, p. 279–286 (1985).

Y. Nakaoka
Crystal Structure Dependence of Hole Stability in Ionic Crystals
(13.5.3; 6; 13.4; 22.5.2; 23)
(b) Vol. **127**, No. 1, p. 327–338 (1985).

M. J. Otero de la Gandara, J. C. Conesa, and J. Soria
*Paramagnetic Centers in UV-Irradiated $Zn_4O(BO_2)_6$
(13.4 and 19; 11; 22.8.1)
(a) Vol. **87**, No. 1, p. K81 – K84 (1985).

Th. Müssig and F. Granzer
Long Living Conduction Electrons in Decomposed Na_xAg_{1-x} Cl Crystals, Detected by EPR Spectroscopy
(19; 13.4; 22.5.4)
(a) Vol. **87**, No. 1, p. K85—K88 (1985)

P. Pipinys, A. Rimeika, and A. Kiveris
*On a Trap Ionization Mechanism by Electric Field in CdF_2
(13.4 and 20.3; 14.3.4; 22.5.3)
(a) Vol. **87**, No. 1, p. K89 – K91 (1985).

Zhen-ya Li and Jui-teng Lin
*A Continuum Model for the Impurity State of Ferromagnets
(13.4 and 18.2)
(b) Vol. **127**, No. 2, p. 517 – 521 (1985).

B. R. Yang, A. Bouwen, and D. Schoemaker
*Electron Spin Resonance Study of Fe^+ Centers in NaCl : Fe^{++} and Motion of the Associated Vacancies
(13.4 and 19; 10.2; 22.5.2)
(b) Vol. **127**, No. 2, p. 657—662 (1985).

M. J. J. Lammers and G. Blasse
*Luminescence Properties of Rare-Earth-Activated Gadolinium Fluoride (GdF_3) and Oxyfluorides (GdOF)
(13.4 and 20.3; 22.8)
(b) Vol. **127**, No. 2, p. 663 – 671 (1985).

A. Maduemezia and F. J. Vega-Catalan
*A Refinement of the Mollwo-Ivey Rule for F-Centres in Alkali Halides
(13.4 and 20.1; 22.5.2)
(b) Vol. **127**, No. 2, p. 689 – 694 (1985).

A. Requardt and G. Lehmann
*A Hole Center Causing Rose Coloration in $AlPO_4$ and $GaPO_4$
(13.4; 11; 19; 20.1; 20.3; 22.8.1)
(b) Vol. **127**, No. 2, p. 695 – 701 (1985).

Yu. P. Gnatenko, Yu. I. Zhirko, N. P. Gavaleshko, and P. I. Babii
On the Possibility of Electron–Hole Drop Formation on Neutral Acceptors in CdTe Crystal
(13.5 and 20.1; 13.4; 22.4.3)
(b) Vol. **127**, No. 2, p. K125 – K128 (1985).

Ľ. Hrivnák
Band-Gap Narrowing in Semi-Insulating GsAs
(13.1 and 14.3; 13.4; 22.2.1)
(b) Vol. **127**, No. 2, p. K165 – K170 (1985).

B. G. TAGIEV, G. M. NIFTIEV, and F. SH. AIDAEV
Photoluminescence of GaSe:Tm Single Crystals
(20.3; 13.4; 22.3)
(b) Vol. **128**, No. 1, p. K65—K68 (1985).

K. STAVREV, G. ST. NIKOLOV, and K. KYNEV
*Orbital and Term Energies for Manganese(II) in Trigonal Fields of C_{3v} Symmetry
(13.4; 20.1; 22.4.1)
(b) Vol. **128**, No. 1, p. K69—K73 (1985).

A. BREZINI
*Localization in One-Dimensional Random Potential
(13.4)
(b) Vol. **128**, No. 1, p. K81—K85 (1985).

S. D. LETYUCHENKO, E. I. SLINKO, K. D. TOVSTYUK, A. G. KHANDOZHKO, and N. K. TOVSTYUK
*Jahn-Teller Effect in Li-Doped SnTe
(13.4 and 19; 10.2; 22)
(b) Vol. **128**, No. 1, p. K87—K92 (1985).

A. A. ALYBAKOV, O. M. ARBOTOEV, V. A. GUBANOVA, and K. KUDABAEV
*Cr^{5+} Paramagnetic Centers in X-Irradiated $KH_2PO_4:Cr^{3+}$ Single Crystals
(13.4; 11; 19; 20.1; 22.8.1)
(b) Vol. **128**, No. 1, p. K93—K96 (1985).
Erratum
(b) Vol. **130**, No. 1, p. K89 (1985).

B. RAY and P. SEPHTON
Reflection and Fluorescence Emission Spectra in the $Ag_{1-x}Cu_xI$ System
(20.1 and 20.3; 13.4; 22.5.4)
(a) Vol. **88**, No. 1, p. 261—265 (1985).

W. HOHENAU
*On the Kinetics of X-Ray Induced Luminescence in Natural Quartz
(13.4 and 20.3; 22.6)
(a) Vol. **88**, No. 1, p. 267—276 (1985).

J. A. GIACOMETTI, G. F. LEAL FERREIRA, and B. GROSS
*Negative Charge Transport in Fluorethylenepropylene by the Constant Current Method
(13.4 and 14.3.4; 22.9)
(a) Vol. **88**, No. 1, p. 297—307 (1985).

V. A. KALUKHOV and S. I. CHIKICHEV
*The Influence of Isoelectronic Impurities on Intrinsic Deep Levels in Liquid Phase Epitaxial Gallium Arsenide
(13.4; 10.2; 22.2.1)
(a) Vol. **88**, No. 1, p. K59—K61 (1985).
Erratum
(a) Vol. **89**, No. 1, p. K115 (1985).

V. I. Kirillov, D. I. Materikin, E. I. Pribylova, Yu. A. Kapustin, and S. I. Rembeza
*On the Nature of Impurity Optical Absorption Bands in Silicon Doped by Noble Metals
(13.4 and 20.1; 22.1.2)
(b) Vol. **128**, No. 2, p. K163–K167 (1985).

K. Yasuda, T. Wada, H. Masuda, M. Takeda, and S. Ikuta
*Defect Distribution in Crystalline Silicon Irradiated with High-Energy Electrons
(11 and 13.4; 14.3; 22.1.2)
(a) Vol. **88**, No. 2, p. 543–551 (1985).

G. Yu. Rudko, V. I. Sidorenko, and V. P. Svirgun
*Self-Induced Polarization Plane Rotation in the Degenerated Absorption Band of F_A Centers in KCl Crystals
(13.4 and 20.1; 10.2; 22.5.2)
(a) Vol. **88**, No. 2, p. 625–630 (1985).

L. A. Nefediev and V. V. Samartsev
*The Dynamic Echo-Hologram Transformation in Three-Level Systems
(13.4 and 20)
(a) Vol. **88**, No. 2, p. 631–635 (1985).

V. S. Lysenko, M. M. Lokshin, A. N. Nazarov, and T. E. Rudenko
RF Plasma Annealing of Implanted MIS Structures
(14.3.3; 11; 13.4; 22.1.2)
(a) Vol. **88**, No. 2, p. 705–712 (1985).

R. Pickenhain, V. Geist, W. Seifert, and K. Jacobs
Green and Yellow Light Emitting Diodes Produced from Vapour Phase Epitaxial GaP :N. II. Defects Produced by 1 MeV Proton Irradiation
(14.3.3 and 20.3; 13.4; 22.2.2)
(a) Vol. **88**, No. 2, p. 721–731 (1985).

V. V. Petrov, V. S. Prosolovich, V. D. Tkachev, Yu. A. Karpov, and M. G. Millvidskii
Donor Centers in Irradiated Si Doped with Rare-Earth Elements
(10.2; 11; 13.4; 14.3; 22.1.2)
(a) Vol. **88**, No. 2, p. K141–K144 (1985).

S. P. Chernov, L. I. Devyatkova, O. N. Ivanova, A. A. Kaminskii, V. V. Mikhailin, S. N. Rudnev, and T. V. Uvarova
$5d^1 4f^{N-1}$–$4f^N$ Absorption and Luminescence of Ce^{3+}, Pr^{3+}, and Nd^{3+} Ions in BaY_2F_8 Single Crystals
(20.2 and 20.3; 13.4; 22.8)
(a) Vol. **88**, No. 2, p. K169–K173 (1985).

S. B. Iyer and V. Kumar
*Investigation of the $E_c - 0.11$ eV Level of Iron in Silicon
(13.4; 22.1.2)
(a) Vol. **88**, No. 2, p. K179–K182 (1985).

V. D. Tkachev, A. M. Zaitsev, and V. V. Tkachev
*Chemical Activity of Noble Gases in Diamond
(10.2 and 13.4; 20.3; 22.1)
(b) Vol. **129**, No. 1, p. 129–133 (1985).

I. MERTIG, E. MROSAN, R. ZELLER, P. H. DEDERICHS, and P. ZIESCHE
*Calculation of the Thermoelectric Power of Ga, Ni, Rh, and Pd Impurities in Copper
(13.4 and 15; 14.1; 21.1)
(b) Vol. **129**, No. 1, p. 407–414 (1985).

T. R. GAWRON and J. MYCIELSKI
*Erratum to: Shallow Impurity States in Strong Magnetic Field
(13.4; 20.1; 22)
(b) Vol. **129**, No. 1, p. 448 (1985).

A. V. MUDRYI, A. L. PUSHKARCHUK, and A. G. ULYASHIN
*Noble Gas Atoms in Si: Energy Spectrum Distortion of Vacancy Complexes
(10.2 and 13.4; 22.1.2)
(b) Vol. **129**, No. 1, p. K83–K87 (1985).

A. B. AHMED and R. K. GARTIA
*Application of the Fractional Glow Technique in the Analysis of a Complex Thermoluminescence Pattern
(13.4 and 20.3; 22.8.1)
(a) Vol. **89**, No. 1, p. 231–235 (1985).

D. LAPRAZ, F. GAUME, and M. BARLAND
*On the Thermoluminescent Mechanism of a Calcium Fluorapatite Single Crystal Doped with Mn^{2+}
(13.4 and 20.3; 14.3; 20.1; 22.8)
(a) Vol. **89**, No. 1, p. 249–253 (1985).

U. K. REDDY
Electrical and Optical Properties of Non-Crystalline As–Se–Cd Thin Films
(2 and 14.3.1; 13.4; 20.1; 22.8)
(a) Vol. **89**, No. 1, p. 255–261 (1985).

R. OBERSCHMID
*Absorption Centers of $Bi_{12}GeO_{20}$ and $Bi_{12}SiO_{20}$ Crystals
(13.4 and 20.1; 1.1; 16; 22.6)
(a) Vol. **89**, No. 1, p. 263–270 (1985).

W. R. FAHRNER, J. R. LASCHINSKI, D. BRÄUNIG, M. KNOLL, and A. NEIDIG
Damage Profiles after 50 to 500 MeV Ion Implantation as Deduced from Thyristor Leakage Currents
(11 and 14.3.3; 13.4; 22.1.2)
(a) Vol. **89**, No. 1, p. 347–351 (1985).

H. KRAUSE
Trap Induction and Breakdown Mechanism in SiO_2 Films
(14.3.4; 13.4; 22.6)
(a) Vol. **89**, No. 1, p. 353–362 (1985).

B. POUMELLEC, J. F. MARUCCO, and F. LAGNEL
*Electron Transport in TiO_{2-x} at Intermediate Temperatures $300\ K < T < 1500\ K$
(13.4 and 14.3; 15; 22.6)
(a) Vol. **89**, No. 1, p. 375–382 (1985).

K. Schmalz, F.-G. Kirscht, S. Niese, I. Babanskaja, M. Kittler, H. Richter, J. Schöneich, and W. Seifert
*On the Intrinsic Gettering in Cu-Contaminated Cz-Si
(10.2 and 13.4; 13.3; 22.1.2)
(a) Vol. **89**, No. 1, p. 389–401 (1985).

Yu. V. Zhilyaev, V. V. Krivolapchuk, A. V. Rodionov, V. V. Rossin, T. V. Rossina, and Yu. N. Sveshnikov
The Investigation of a Transition Layer in Epitaxial GaAs by the Low Temperature Photoluminescence Technique
(20.3; 13.4; 22.2.1)
(a) Vol. **89**, No. 1, p. K61–K64 (1985).

Z. Januškevičius, V. Karalkevičiutė, and A. Sakalas
*The Nature of Acceptor Centres in Zinc and Cadmium Diphosphide
(13.4; 22.3)
(a) Vol. **89**, No. 1, p. K65–K69 (1985).

G. Balčaitis, Z. Januškevičius, and A. Sodeika
*On the Nature of Energy Levels in $ZnGeP_2$
(13.4; 22.8)
(a) Vol. **89**, No. 1, p. K71–K74 (1985).

S. Shigetomi and T. Matsumori
Photoluminescence Spectra of Si-Implanted GaAs
(20.3; 10.2; 13.4; 22.2.1)
(a) Vol. **89**, No. 1, p. K79–K82 (1985).

V. Z. Polinger, L. F. Chibotaru, and I. B. Bersuker
*The Effect of Spin–Vibrational Interaction in Magnetic Properties of Exchange Tetraclusters
(13.4 and 18; 22.8)
(b) Vol. **129**, No. 2, p. 615–624 (1985).

G. Jacobs
Transient I_e Centres in Alkali Halides
(6 and 20.1; 13.4; 22.5.2)
(b) Vol. **129**, No. 2, p. 755–762 (1985).

R. Capelletti, M. Manfredi, and M. Solzi
*Energy Transfer Mechanisms in the NaCl:Pb^{++}, Mn^{++} System along Aggregation Processes
(13.4 and 20.1; 10.2; 22.5.2)
(b) Vol. **129**, No. 2, p. 789–798 (1985).

K. E. Tarkpea and A. Ots
*Thermal Stability of Fe^+ Centres and Thermoluminescence of ZnS Single Crystals
(13.4 and 20.3; 19; 22.4.1)
(b) Vol. **129**, No. 2, p. 799–803 (1985).

C. C. Becerra, A. Paduan-Filho, and N. F. Oliveira, Jr.
*Energy Level Crossing of the Ni^{++} Ion Induced by Magnetic Field in $NiSnCl_6 \cdot 6\,H_2O$ and α-$NiSO_4 \cdot 6\,H_2O$
(13.4; 22.8)
(b) Vol. **129**, No. 2, p. K173–K175 (1985).

N. Sielaff, H. Nelkowski, H. Pradella, J. Sahm, and K. Zink
EPR Measurements on Fe–Cu Complexes in ZnSe
(19; 13.4; 22.4.2)
(b) Vol. **129**, No. 2, p. K177—K179 (1985).

A. Tomita and N. Takeuchi
Thermally Stimulated Exoelectron Emission and Spectra of Thermoluminescence in NaCl:Ca, Cu and NaCl:Cu Single Crystals
(17 and 20.3; 13.4; 22.5.2)
(a) Vol. **89**, No. 2, p. 609–615 (1985).

K. Žďánský
*Analysis of Non-Exponential Filling in DLTS on p–n Junctions. Application to Dominant Traps in AlGaAs Lasers
(13.4 and 14.3.3; 22.2.4)
(a) Vol. **89**, No. 2, p. 629–637 (1985).

B. G. Tagiev, G. M. Niftiev, and F. Sh. Aidaev
*Injection and Thermodepolarization Currents in GaS:Er Single Crystals
(13.4 and 14.3.4; 22.3)
(a) Vol. **89**, No. 2, p. 639–645 (1985).

K. Mochizuki, H. Onodera, and K. Igaki
Photoconductivity of CdS_xSe_{1-x} with Controlled Deviation from Stoichiometry
(16; 13.4; 20.3; 22.4.4)
(a) Vol. **89**, No. 2, p. 673–678 (1985).

G.-H. Chen, A. Morimoto, M. Kumeda, and T. Shimizu
ESR Studies on Doped Hydrogenated Amorphous Silicon–Carbon
(2 and 19; 13.4; 22.1.2)
(a) Vol. **89**, No. 2, p. K181–K183 (1985).

P. I. Baranskii, V. V. Savyak, and Yu. V. Simonenko
Influence of Hot Annealing and Cooling Rate on the Hall Effect and Piezoresistance in Transmutationally Doped and Ordinary Silicon Crystals
(14.3; 13.4; 22.1.2)
(a) Vol. **89**, No. 2, p. K185–K189 (1985).

B. G. Tagiev, S. A. Abushov, G. M. Niftiev, Ch. M. Briskina, V. F. Zolin, V. M. Markushev, and F. Sh. Aidaev
*Photo- and Electroluminescence of Neodymium in GaS
(13.4 and 20.3; 22.3)
(a) Vol. **89**, No. 2, p. K191–K196 (1985).

B. Olejníková
*A Method of Determination of the Trap Depth of Deep Centres by Capacitance Measurements
(13.4 and 14.3.4; 22.1.2)
(a) Vol. **89**, No. 2, p. K197–K200 (1985).

A. E. BELYAEV, O. P. GORODNICHII, V. S. VIKHNIN, and R. M. VINETSKII
*The Nonlinear Effect of Uniaxial Stress for Shallow Donors in Silicon
(13.4; 22.1.2)
(b) Vol. **130**, No. 1, p. K47–K50 (1985).

E. RADZHABOV
*Photodissociation of $O^{--}V_a^+$ Centers in Lithium Fluoride
(13.4 and 20.3; 10.2; 22.5.2)
(b) Vol. **130**, No. 1, p. K55–K58 (1985).

S. W. BIERNACKI
*The Splitting of Impurity d-States in Zincblende Crystals
(13.4; 20.1; 22.4.2)
(b) Vol. **130**, No. 1, p. K59–K63 (1985).

V. A. ZASUKHA and S. V. VOLKOV
*Selection Rules for Electron Transfer between Impurities in Crystals
(13.4)
(b) Vol. **130**, No. 1, p. K65–K68 (1985).

D. SACHELARIE
Analysis of Gold-Doped MOS Structures
(14.3.3; 13.3; 13.4; 22.1.2)
(a) Vol. **90**, No. 1, p. 401–407 (1985).

E. V. KUČYS, A. B. ABIŠALAS, R. V. KARPIEJUS, and G. B. PETRAUSKAS
GaAs Epilayer Conductivity Compensation by High-Energy Proton Bombardment
(11 and 14.3.1; 13.4; 22.2.1)
(a) Vol. **90**, No. 1, p. K45–K48 (1985).

A. A. KAMINSKII, K. KURBANOV, S. E. SARKISOV, M. M. SATTAROVA, T. V. UVAROVA, and P. P. FEDOROV
Stimulated Emission of Nd^{3+} Ions in Non-Stoichiometric $Cd_{1-x}Ce_xFe_{2+x}$ and $Cd_{1-x}Nd_xFe_{2+x}$ Fluorides with Fluorite Structure
(20.2 and 20.3; 13.4; 22.8)
(a) Vol. **90**, No. 1, p. K55–K60 (1985).

S. SHIGETOMI, T. IKARI, Y. KOGA, and S. SHIGETOMI
*Annealing Behaviour of Photoacoustic Spectra of Undoped InSe
(7 and 13.4; 16; 22.3)
(a) Vol. **90**, No. 1, p. K61–K64 (1985).

A. V. EGRANOV and A. I. NEPOMNYACHIKH
*Emission Spectra of NaCl:Mg Crystals
(13.4 and 20.3; 10.2; 22.5.2)
(a) Vol. **90**, No. 1, p. K65–K69 (1985).

E. CHRYSSOU and B. IATRIDIS
*Thermally Stimulated Exoelectron Emission (TSEE) of Heat Treated $CaSO_4 \cdot 2\,H_2O$
(13.4 and 17; 22.8.1)
(a) Vol. **90**, No. 1, p. K75–K79 (1985).

B. P. Tyagi and K. Sen
*Electrical Properties of Polycrystalline Silicon in the Dark and under Illumination
(13.4 and 14.3; 16; 22.1.2)
(a) Vol. **90**, No. 2, p. 709–713 (1985).

V. Ghiordănescu and L. Dollingher
Some Observations on the Absorption of Copper in Photochromic Glass
(20.1; 13.4; 22.8)
(a) Vol. **90**, No. 2, p. K199–K204 (1985).

J. T. Lin and Z. Y. Li
Microscopic Model for the Relaxation and Excitation of Localized Spin Waves
(13.5.4 and 18; 13.4)
(b) Vol. **131**, No. 1, p. 141–149 (1985).

M. C. Böhm
*Evaluation of Suitable Mean-Field (Hartree-Fock) Representations for Perturbational Studies of Disordered Solids
(13.4; 4)
(b) Vol. **131**, No. 1, p. 309–318 (1985).

D. Neogy and T. Purohit
*The Behavior of Active Centers in a Laser Host. A Crystal Field Investigation on Nd^{3+} and Er^{3+} in LaF_3 Single Crystals
(13.4; 8; 18; 19; 22.5)
(b) Vol. **131**, No. 1, p. 329–338 (1985).

R. C. Bhuniya and S. E. Rao
Colour Centres and Luminescence in KCl:Zn Crystals
(20.1 and 20.3; 10.2; 13.4; 22.5.2)
(b) Vol. **131**, No. 1, p. 339–348 (1985).

S. W. Biernacki
*Effects of the Crystal Field in the Spectra of Iron Group Impurities in ZnSe
(13.4; 22.4.2)
(b) Vol. **131**, No. 1, p. 349–356 (1985).

Prem Chand, G. C. Upreti, M. Umar, and R. J. Singh
*EPR of Cu^{2+} in Barium Formate Single Crystal
(13.4 and 19; 22.9)
(b) Vol. **131**, No. 1, p. 357–367 (1985).

H. Taguchi and M. Shimada
*X-Ray Photoelectron Spectroscopy Study of $CaMnO_{3-\delta}$
(4 and 13.4; 17; 22.8)
(b) Vol. **131**, No. 1, p. K59–K62 (1985).

S. K. Lilov
*Calculation of the Fermi Level, Fermi Distribution, and Activity Coefficient for p-SiC:Sc
(13.4; 22.3)
(b) Vol. **131**, No. 1, p. K81–K85 (1985).

V. A. Kovarskii, L. V. Chernysh, and M. K. Sheinkman
*Shake-Effect in Radiative Auger Recombination of Carriers on Two-Electron Impurity Centres in Semiconductors
(13.4; 13.5.1; 20.3; 22)
(b) Vol. **131**, No. 2, p. 677–682 (1985).

W. Ulrici
*Photoinduced Optical Absorption of Cr^{4+} in Semi-Insulating GaAs:Cr
(13.4; 16; 19; 20.1; 22.2.1)
(b) Vol. **131**, No. 2, p. 707–717 (1985).

W. Ulrici, K. Friedland, L. Eaves, and D. P. Halliday
*Optical and Electrical Properties of Vanadium-Doped GaAs
(13.4; 14.3; 20.1; 22.2.1)
(b) Vol. **131**, No. 2, p. 719–728 (1985).

A. Schenk, R. Enderlein, and D. Suisky
*Electroabsorption for Deep Level to Band Transitions Accompanied by Multiphonon Processes
(13.4 and 20.1; 6; 14.3.4; 22)
(b) Vol. **131**, No. 2, p. 729–739 (1985).

T. Schröder, O. Dössel, H. Nahme, and N. Schwentner
*Transient Absorption of Xe and Kr Guest Atoms in Ar Crystals
(13.4 and 20.1; 23)
(b) Vol. **131**, No. 2, p. 741–746 (1985).

L. Bosi and M. Nimis
*An Analysis of the Classical Arguments Concerning Forecasts for the Z_1 Centre Formation in Alkali Halides
(10.2 and 13.4; 22.5.2)
(b) Vol. **131**, No. 2, p. K111–K116 (1985).

B. Mengel, H.-U. Müller, F. Ludwig, V. L. Konstantinov, and R. Herrmann
Magneto-Optical Investigations in the Submillimeter Wavelength Range on p-InSb Bicrystals
(18 and 20.1; 13.4; 22.2.3)
(b) Vol. **131**, No. 2, p. K145–K147 (1985).

F. Ludwig, H.-U. Müller, B. Mengel, and R. Herrmann
Magneto-Optical Investigations on n-InSb in the Far Infrared
(18 and 20.1; 13.4; 22.2.3)
(b) Vol. **131**, No. 2, p. K149–K152 (1985).

T. Sh. Abesadze and M. Punkkinen
*Discrete Saturation of the Optical Absorption Line of Pr^{3+} in CaF_2 under the Condition of Pumping
(13.4 and 20.1; 19; 22.5.3)
(b) Vol. **131**, No. 2, p. K153–K155 (1985).

E. Redjai and G. Massé
*Donor–Acceptor Pair-Transitions in $AgInS_2$
(13.4 and 20.3; 22.8)
(b) Vol. **131**, No. 2, p. K157–K159 (1985).

H.-J. MÜLLER
*Influence of Local Environment on Bound Excitons and Bound Excitonic Molecules in II–V Alloys. I. Theory and Numerical Results for $GaAs_{1-x}P_x$:N
(13.4 and 13.5.1; 22.2.4)
(b) Vol. **132**, No. 1, p. 239–254 (1985).

B. ČABRIĆ, LJ. RISTOVSKI, and G. DAVIDOVIĆ
*Impurity Induced Energy Transfer in a Linear Polymer Chain
(13.4; 6; 13.5.1; 22.9)
(b) Vol. **132**, No. 1, p. 261–270 (1985).

M. INOUE, M. KOYANO, H. NEGISHI, Y. UEDA, and H. SATO
Localized Impurity Level and Carrier Concentration in Self-Intercalated TiS_2 Crystals
(14.3; 13.4; 22)
(b) Vol. **132**, No. 1, p. 295–303 (1985).

M. J. SPRINGIS and and J. A. VALBIS
*Red Luminescence of Color Centres in Sapphire
(13.4 and 20.3; 10.2; 22.6)
(b) Vol. **132**, No. 1, p. K61–K65 (1985).

P. GAWORZEWSKI and E. HILD
The Effect of Pre-Heat-Treatments on the Formation Kinetics of Thermal Donors in Silicon
(10.2; 13.4; 22.1.2)
(a) Vol. **92**, No. 1, p. 129–135 (1985).

M. I. DARBY, G. R. EVANS, M. N. READ, and D. J. SHAUGHNESSY
*Interaction of Hydrogen Impurities with Edge Dislocations in α-Zirconium
(10.1 and 13.4; 21)
(a) Vol. **92**, No. 1, p. 149–158 (1985).

E. GRILLI, M. GUZZI, E. CAMERLENGHI, and F. PIO
*Photoluminescence of $Zn_xCd_{1-x}In_2S_4$ Layered Alloys ($0.7 < x < 1$)
(13.4 and 20.3; 22.8)
(a) Vol. **92**, No. 1, p. 213–220 (1985).

YE MING, E. KRÄTZIG, and R. ORLOWSKI
Photorefractive Effects in $LiNbO_3$:Cr Induced by Two-Step Excitation
(20.1; 13.4; 16; 22.8.1)
(a) Vol. **92**, No. 1, p. 221–229 (1985).

A. N. ANAGNOSTOPOULOS, C. MANOLIKAS, and A. KAROUTIS
*Electrical Effects Associated with the Ordering Process in $CdInGaS_4$ Crystals. II. Electron Traps Determined with Conductivity Measurements
(13.4 and 14.3.4; 22.8)
(a) Vol. **92**, No. 1, p. 231–235 (1985).

V. K. JAIN and M. S. JAHAN
*Low-Temperature Thermoluminescent Characteristics of CaF_2:Mn
(13.4 and 20.3; 22.5.3)
(a) Vol. **92**, No. 1, p. 237–248 (1985).

S. N. Lukin and O. P. Teslya
*The Structure of the Magnetic Centre in Cs_3CoCl_5
(13.4; 19; 22.5.4)
(b) Vol. **132**, No. 2, p. 601–606 (1985).

H.-J. Müller
*Influence of Local Environment on Bound Excitons and Bound Excitonic Molecules in III–V Alloys. II. Numerical Results for $Ga_xIn_{1-x}P$:N and the Role of Disorder Topology for the Optical Spectrum
(13.4 and 13.5.1; 20.3; 22.2.4)
(b) Vol. **132**, No. 2, p. 607–617 (1985).

J. P. Jouart, C. Bissieux, G. Mary, and M. Egée
*Study of the Sensitization Effect of Eu^{2+} Ions on the Trivalent Rare-Earth Fluorescence in CdF_2 Crystals
(13.4 and 20.3; 22.5.3)
(b) Vol. **132**, No. 2, p. 619–625 (1985).

R. R. Atabekyan, V. L. Vinetskii, V. A. Gevorkyan, L. P. Godenko, and R. K. Ezoyan
Photoconductivity of Corundum Irradiated with High Energy Electrons at Excitation of F and F^+ Centres
(16; 13.4; 22.6)
(b) Vol. **132**, No. 2, p. 635–640 (1985).

G. Abraham, H. Nelkowski, H. Pradella, J. Sahm, and K. Zink
Manganese Pairs in ZnSe Detected by EPR Measurements
(19; 13.4; 22.4.2)
(b) Vol. **132**, No. 2, p. K133–K138 (1985).

V. P. Seth, V. K. Jain, A. Yadav, and R. S. Bansal
ESR Study of Vanadyl Ion in Potassium Succinate
(19; 13.4; 22.8)
(b) Vol. **132**, No. 2, p. K139–K142 (1985).

J. K. Srivastava and S. J. Supe
*Thermoluminescence of $CaSO_4$:Dy. A Semiempirical Approach for a Qualitative Analysis of Non-linear Thermoluminescence Effects
(13.4; 20.3; 22.8.1)
(a) Vol. **92**, No. 2, p. 573–588 (1985).

L. Tarricone, E. Grilli, M. Guzzi, and L. Zanotti
*Optical Absorption and Recombination Processes in $CdGaInS_4$ Layered Crystals
(13.4 and 20.3; 16; 20.1; 22.8)
(a) Vol. **92**, No. 2, p. 595–602 (1985).

G. M. Mileva and L. I. Pavlov
*Third-Order Optical Polarizability of ZnS at Low Temperature
(13.4; 20.1; 22.4.1)
(a) Vol. **92**, No. 2, p. 603–607 (1985).

13.5 Elementary Excitations (Phonons See 6)

Review Article

H. W. Streitwolf
*Physical Properties of Polyacetylene
(13.5 and 13.5.3; 6; 18; 19; 20.1; 22.9)
(b) Vol. **127**, No. 1, p. 11–54 (1985).

Original Papers and Short Notes

A. A. Eremko, Yu. B. Gaididei, and A. A. Vakhnenko
*Dissociation-Accompanied Raman Scattering by Davydov Solitons
(13.5; 20.1)
(b) Vol. **127**, No. 2, p. 703–713 (1985).

Yu. P. Gnatenko, Yu. I. Zhirko, N. P. Gavaleshko, and P. I. Babii
*On the Possibility of Electron–Hole Drop Formation on Neutral Acceptors in CdTe Crystal
(13.5 and 20.1; 13.4; 22.4.3)
(b) Vol. **127**, No. 2, p. K125–K128 (1985).

S. Mochizuki
Radiative Optical Modes of Vibration in CeO_2 Thin Films
(6 and 20.1; 13.5; 22.6)
(b) Vol. **128**, No. 1, p. 23–30 (1985).

B. Orel, M. Klanjšek, V. Moiseenko, M. Volnyanskii, N. Le Clavé, B. Pasquier, and A. Novak
*Isotopic Impurity Modes and Polariton Dispersion in Li_2GeO_3 Crystals
(6 and 13.5; 20.1; 22.8.1)
(b) Vol. **128**, No. 1, p. 53–67 (1985).

F. Gashi, R. Gashi, Z. Shemsedini, and R. B. Žakula
*The Energy Transport from One Molecular Chain to the Other which are in Mutual Contact by a Small Number of Crystal Sites
(13.5; 22.9)
(b) Vol. **128**, No. 1, p. 259–273 (1985).

L. Wendler
Phonon-Polaritons in Bilayer Systems. s-Polarized Guided Wave Phonon-Polaritons
(6 and 20.1; 13.5; 22.2.1; 22.2.2)
(b) Vol. **128**, No. 2, p. 425–437 (1985).

M. Yu. Nikolaev, N. V. Ryzhanova, A. V. Vedyaev, and S. M. Zubritskii
*Ferromagnetism and Elementary Excitations in the Hubbard Model
(13.5 and 18.2)
(b) Vol. **128**, No. 2, p. 513–523 (1985).

13.5.1 Excitons

Original Papers and Short Notes

C. TRALLERO GINER and O. SOTOLONGO COSTA
One-Phonon Resonance Raman Scattering
(6 and 20.1; 13.5.1; 22.4.1)
(b) Vol. **127**, No. 1, p. 121–130 (1985).

I. G. LANG, S. T. PAVLOV, A. V. PROKAZNIKOV, and A. V. GOLTSEV
General Theory of Secondary Radiation in Condensed Matter
(20; 6; 13.5.1)
(b) Vol. **127**, No. 1, p. 187–200 (1985).

P. REINEKER, J. KÖHLER, and U. SCHMID
*Triplet Excitons in Non-Resonant Molecular Pairs. Saturation Effects in Electron Spin Resonance
(13.5.1; 19; 22.9)
(b) Vol. **127**, No. 1, p. 237–244 (1985).

K. R. ALLAKHVERDIEV, T. G. MAMEDOV, V. E. SHTEINSHREIBER, and I. K. EFENDIEVA
Fundamental Absorption Edge of TlSe and $TlInTe_2$ Crystals
(13.1 and 20.1; 13.5.1; 22; 22.8)
(b) Vol. **127**, No. 1, p. K55–K60 (1985).

N. A. DROZDOV, N. I. ZYATKOVA, A. A. PATRIN, and V. D. TKACHEV
Photoluminescence of Silicon Annealed at Low Temperatures (420 to 550 °C)
(10.2 and 20.3; 13.4; 13.5.1; 22.1.2)
(a) Vol. **87**, No. 1, p. K45–K48 (1985).

P. V. VODOLAZSKII, B. R. KIYAK, M. G. MATSKO, V. B. NOSOV, G. T. PETROVSKII, and A. V. SHATILOV
*Reversible and Irreversible Changes in Exciton Spectra of ZnSe Luminescence
(13.5.1 and 20.3; 22.4.2)
(a) Vol. **87**, No. 1, p. K69–K71 (1985).

V. M. ASNIN, V. I. STEPANOV, R. ZIMMERMANN, and M. RÖSLER
*Screening of the Spin–Orbit Exciton by the Electron–Hole Liquid in Germanium
(13.5.1 and 20.1; 13.5.2; 22.1.1)
(b) Vol. **127**, No. 2, p. 651–655 (1985).

B. J. SCHMID and P. REINEKER
*Erratum to: The Mixing of Frenkel and Charge-Transfer Excited States. II. Numerical Treatment of Non-Symmetric Dimers
(13.5.1; 22.9)
(b) Vol. **127**, No. 2, p. 735 (1985).

R. R. GUSEINOV
*Self-Energy of Charge in Superlattice
(13.5.1)
(b) Vol. **127**, No. 2, p. K133–K136 (1985).

R. Stegmann, G. Oelgart, and F. Thuselt
*The Influence of Coulomb Screening and of NN_i-Pairs on the Recombination of Bound Excitons in $GaAs_{1-x}P_x$:N Mixed Crystals
(13.5.1 and 20.3; 22.2.4)
(a) Vol. **87**, No. 2, p. 641–650 (1985).

A. Meisen and G. Heber
*Magnetic Excitons in a Spin-Cluster System
(13.5.1 and 18)
(b) Vol. **128**, No. 1, p. 209–218 (1985).

V. M. Fomin and E. P. Pokatilov
*Excitons in Periodic Structures with Homopolar Semiconductors and Anisotropic Dielectrics
(13.5.1 and 14.4; 22)
(b) Vol. **128**, No. 1, p. 251–257 (1985).

H. Geistlinger and W. Weller
*The Influence of Correlation and Band Structure Effects on an Exciton Weakly Bound to an Isoelectronic Impurity
(13.4 and 13.5.1)
(b) Vol. **128**, No. 1, p. 313–321 (1985).

D. V. Kapor, D. Lj. Mirjanić, S. D. Stojanović, J. P. Šetrajčić, and B. S. Tošić
*On the Possibility of the Realization of Superfluid Excitation Transfer in Molecular Crystals
(13.5.1; 6)
(b) Vol. **128**, No. 1, p. K49–K52 (1985).

U. F. Kozmidis-Luburić, D. Lj. Mirjanić, J. P. Šetrajčić, and B. S. Tošić
*The Influence of Phonons on Two-Particle Exciton States in Molecular Chains
(13.5.1; 6)
(b) Vol. **128**, No. 1, p. K53–K57 (1985).

N. I. Grigorchuk
*Energy Spectrum and Light Absorption Coefficient for Bound Electron–Exciton States in One- and Two-Dimensional Molecular Crystals
(13.5.1 and 20.1; 22.9)
(b) Vol. **128**, No. 2, p. 599–610 (1985).

I. Rückmann, K. Jarašiūnas, and E. Gaubas
*Transient Light-Induced Gratings Due to Excitons in CdSe
(13.5.1 and 20.1; 22.4.2)
(b) Vol. **128**, No. 2, p. 627–641 (1985).

H. Geistlinger and W. Weller
*A Variational Method in Wannier Representation for an Electron Weakly Bound to an Isoelectronic Impurity
(13.4 and 13.5.1)
(b) Vol. **128**, No. 2, p. 709–715 (1985).

TH. KLEINEFELD and W. VON DER OSTEN
*Erratum to: Localized and Free Biexcitons in Silver Bromide
(13.5.1 and 20.3; 22.5.1)
(b) Vol. **128**, No. 2, p. 808 (1985).

G. D. GUSEINOV, E. F. BAGIRZADE, S. S. ABDINBEKOV, and N. Z. GASANOV
*Modulation Spectra Peculiarities in $A^{III}B^{III}X_2^{VI}$ Anisotropic Semiconductors
(13.5.1 and 20.1; 14.3.4; 22.3.1)
(b) Vol. **128**, No. 2, p. K129–K131 (1985).

E. L. ALBUQUERQUE, N. S. ALMEIDA, and M. C. OLIVEROS
*Light Scattering by Surface Ripples in a Spatially Dispersive Medium
(13.5.1 and 20; 7)
(b) Vol. **129**, No. 1, p. 177–186 (1985).

M. ROSENZWEIG
*Magneto-Reflectance of the B-Exciton of CdS
(13.5.1 and 20.1; 22.4.1)
(b) Vol. **129**, No. 1, p. 187–202 (1985).

V. M. FOMIN and E. P. POKATILOV
*Excitons in Multi-Layer Systems
(13.5.1; 22)
(b) Vol. **129**, No. 1, p. 203–209 (1985).

V. CAPOZZI, S. CANEPPELE, M. MONTAGNA, and F. LÉVY
*Radiative Decay from Free and Bound Excitons in GaSe
(13.5.1 and 20.3; 22.3)
(b) Vol. **129**, No. 1, p. 247–257 (1985).

C. M. SOTOMAYOR-TORRES, M. D. LUMB, and G. WALKER
*Bound Exciton, Free-To-Bound and Electron–Hole Plasma Luminescence in $Ga_xIn_{1-x}P$ ($0.54 \leqq x \leqq 0.76$)
(13.5.1 and 20.3; 22.2.4)
(b) Vol. **129**, No. 1, p. 279–286 (1985).

A. S. LYUTOVICH, K. L. LYUTOVICH, V. P. POPOV, and L. N. SAFRONOV
*Radiative Recombination at Centres in Germanium–Silicon Solid Solutions
(13.4 and 13.5.1; 20.3; 22.1.1; 22.1.2)
(b) Vol. **129**, No. 1, p. 313–320 (1985).

S. A. MOSKALENKO, S. S. RUSSU, V. V. BALTAGA, A. I. BOBRYSHEVA, and A. V. LELYAKOV
*The Para- and Ortho-Phases of High-Density Excitons in Cu_2O
(13.5.1; 22.6)
(b) Vol. **129**, No. 2, p. 657–665 (1985).

A. A. O'CONNEL-BRONIN
Resonance LO Phonon-Assisted Luminescence and Light and Heavy Excitons in NaI Crystals
(6 and 20.3; 13.5.1; 22.5.2)
(b) Vol. **129**, No. 2, p. 675–681 (1985).

CH. NÖLDEKE, W. METZGER, R. P. HUEBENER, and H. SCHNEIDER
*Exciton Impact Ionization and Electric Conductivity in Germanium at Low Temperature
(13.5.1 and 20.3; 14.3; 22.1.1)
(b) Vol. **129**, No. 2, p. 687–695 (1985).

V. MAY, F. BOLDT, and K. HENNEBERGER
*Many-Body Theory for the Dense Exciton Gas of Direct Semiconductors. I. General Considerations
(13.5.1; 22)
(b) Vol. **129**, No. 2, p. 717–728 (1985).

M. SELG
*Theory of Vibrational Relaxation of R_2^* Centres in Rare Gas-Solids. Application to Ne_2^*
(13.5.1 and 20.3; 6; 20.1; 23)
(b) Vol. **129**, No. 2, p. 775–784 (1985).

R. R. GUSEINOV
*Energy Spectrum of the Cross Motion of Excitons in a Superlattice
(13.5.1)
(b) Vol. **130**, No. 1, p. 205–209 (1985).

V. I. SOKOLOV, T. P. SURKOVA, M. P. KULAKOV, and A. V. FADEEV
*New Experimental Evidence Concerning the Nature of L, M, and N Lines in ZnSe:Co
(13.4 and 13.5.1; 14.3.4; 20.1; 22.4.2)
(b) Vol. **130**, No. 1, p. 267–272 (1985).

G. I. PILIPENKO, V. G. PLEKHANOV, V. I. TYUTYUNNIK, O. I. TYUTYUNNIK, and F. F. GAVRILOV
*Direct Observation of Bound Excitons in the Spectra of Resonance Secondary Emission of LiH:Mg Crystals
(13.5.1 and 20.1; 20.3; 22.5)
(b) Vol. **130**, No. 1, p. 273--280 (1985).

V. A. BIBIK, I. V. BLONSKII, M. S. BRODIN, and N. A. DAVYDOVA
*Structural Phase Transition in the Layer Semiconductor PbI_2 Induced by Laser Irradiation
(1.2 and 13.5.1; 20.1; 22.5)
(a) Vol. **90**, No. 1, p. K11–K14 (1985).

E. P. POKATILOV, S. I. BERIL, V. M. FOMIN, and G. A. POGORILKO
*Wannier-Mott Exciton States in Two-Layer Periodic Structures
(13.5.1; 22.2.1; 22.2.4)
(b) Vol. **130**, No. 2, p. 619–628 (1985).

F. BOLDT, K. HENNEBERGER, and V. MAY
*Many-Body Theory for the Dense Exciton Gas of Direct Semiconductors. II. Calculation of Exciton Level Shift and Damping in Dependence on Exciton Density
(13.5.1; 22)
(b) Vol. **130**, No. 2, p. 675–687 (1985).

V. N. VIVCHARUK, YU. V. GUDYMA, V. N. ERMAKOV, and V. V. NITSOVICH
*Exciton Mechanism of Exchange Interaction in Semiconductors with Magnetic Impurities
(13.4 and 13.5.1; 18; 22.4.3)
(b) Vol. **130**, No. 2, p. K131–K134 (1985).

R. ZIMMERMANN and H. STOLZ
*The Mass Action Law in Two-Component Fermi Systems Revisited. Excitons and Electron–Hole Pairs
(13.5.1; 22)
(b) Vol. **131**, No. 1, p. 151–164 (1985).

H. ROSSMANN and F. HENNEBERGER
*Resonatorless Optical Bistability of CdS. Reconsideration of Transient Effects Due to Lattice Heating
(13.5.1 and 20; 22.4.1)
(b) Vol. **131**, No. 1, p. 185–192 (1985).

S. F. TEREKHOVA, N. A. ONISHCHENKO, O. N. TALENSKII, and S. A. PENDUR
*The Refractive Index and the Exciton Transition Oscillator Strength of CdSe Single Crystals
(13.5.1 and 20.1; 22.4.2)
(b) Vol. **131**, No. 1, p. 207–212 (1985).

A. V. SHERMAN
*Exciton Absorption Spectrum
(13.5.1 and 20.1; 6)
(b) Vol. **131**, No. 1, p. 225–233 (1985).

P. I. BARDETSKII, M. I. SHMIGLYUK, and SH. D. TIRON
*Non-Linear Optical Nutation on Interexciton Transitions in Cu_2O
(13.5.1; 22.6)
(b) Vol. **131**, No. 1, p. 235–241 (1985).

E. G. TSITSISHVILI
Piezo-Optical Activity of Cubic Crystals
(12.2 and 20.1; 13.1; 13.5.1; 22.5)
(b) Vol. **131**, No. 1, p. 279–290 (1985).

V. MAY, F. BOLDT, and K. HENNEBERGER
*Many-Body Theory for the Dense Exciton Gas of Direct Semiconductors. III. Response of the Many-Exciton System on an Externally Driven Light Field
(13.5.1; 22.4.1)
(b) Vol. **131**, No. 2, p. 605–614 (1985).

M. S. BRODIN, I. V. BLONSKII, A. O. GUSHCHA, and V. V. TISHENKO
*Secondary Emission Spectra and Energy Relaxation of Polaritons in Layer Polar Semiconductors
(6 and 13.5.1; 20.1; 20.3; 22.5)
(b) Vol. **131**, No. 2, p. 615–620 (1985).

F. V. KUSMARTSEV
*The Narrow Bands of Exciton Absorption in Polar Solids and Alloys. Urbach's Rule
(13.5.1 and 20.1; 22.5; 22.5.4)
(b) Vol. **131**, No. 2, p. 621–629 (1985).

V. A. KOVARSKII, L. V. CHERNYSH, and M. K. SHEINKMAN
Shake-Effect in Radiative Auger Recombination of Carriers on Two-Electron Impurity Centres in Semiconductors
(13.4; 13.5.1; 20.3; 22)
(b) Vol. **131**, No. 2, p. 677–682 (1985).

B. M. Nitsovich, G. M. Pestryakov, and V. D. Falenchuk
*Temperature Anomalies of Exciton Absorption in GeO_2
(13.5.1 and 20.1; 22.6)
(b) Vol. **131**, No. 2, p. K135–K138 (1985).

F. F. Charfi and M. Zouaghi
*Hot-Exciton Orientation in Indirect-Gap $GaAs_{1-x}P_x$ Alloys
(6 and 13.5.1; 20.3; 22.2.4)
(b) Vol. **132**, No. 1, p. 109–115 (1985).

V. May
*Light Transmission at the 1 s-Exciton Resonance of CdS in the Presence of a Density Profile of Excitons
(13.5.1 and 20.1; 22.4.1)
(b) Vol. **132**, No. 1, p. 145–151 (1985).

H.-J. Müller
*Influence of Local Environment on Bound Excitons and Bound Excitonic Molecules in III–V Alloys. I. Theory and Numerical Results for $GaAs_{1-x}P_x$:N
(13.4 and 13.5.1; 22.2.4)
(b) Vol. **132**, No. 1, p. 239–254 (1985).

B. Čabrić, Lj. Ristovski, and G. Davidović
Impurity Induced Energy Transfer in a Linear Polymer Chain
(13.4; 6; 13.5.1; 22.9)
(b) Vol. **132**, No. 1, p. 261–270 (1985).

Yu. A. Bychkov
*The Joint Effect of Coulomb Interaction and Spin–Orbit Coupling on the Spectrum of 2D Magnetic Excitons
(13.5.1)
(b) Vol. **132**, No. 1, p. K57–K60 (1985).

E. A. Popov, M. M. Kotlyarskii, and I. S. Edelman
*Temperature and Field Dependence of Exciton-Magnon Absorption of $NaMnCl_3$
(13.5.1 and 13.5.4; 18.2; 20.1; 22.5.4)
(b) Vol. **132**, No. 2, p. 465–470 (1985).

R. Levy, J. Y. Bigot, M. Frindi, and B. Hönerlage
*Optical Switching Behaviour in Two-Beam Experiments with CuCl
(13.5.1; 20.1; 22.5)
(b) Vol. **132**, No. 2, p. 495–502 (1985).

H.-J. Müller
*Influence of Local Environment on Bound Excitons and Bound Excitonic Molecules in III–V Alloys. II. Numerical Results for $Ga_xIn_{1-x}P$:N and the Role of Disorder Topology for the Optical Spectrum
(13.4 and 13.5.1; 20.3; 22.2.4)
(b) Vol. **132**, No. 2, p. 607–617 (1985).

A. B. Henriques and J. P. von der Weid
*Decay Kinetics of Triplet Molecular Excitons in Alkali Cyanides
(13.5.1 and 20.3; 22.9)
(b) Vol. **132**, No. 2, p. 627–633 (1985).

A. I. Bobrysheva and Yu. M. Shvera
*The Influence of the Excitonic Longitudinal–Transverse Splitting on the Biexciton Binding Energy
(13.5.1; 22.4.1; 22.5.2)
(b) Vol. **132**, No. 2, p. K107–K110 (1985).

I. Broser, K.-H. Pantke, and M. Rosenzweig
*Transmission and Damping of Excitonic Polaritons in CdS
(13.5.1 and 20.1; 22.4.1)
(b) Vol. **132**, No. 2, p. K117–K120 (1985).

13.5.2 Plasmons

Original Papers and Short Notes

S. GHOSH and S. DIXIT
*Effect of Relativistic Mass Variation of the Electron on the Parametric Instability of Acoustic Waves in Transversely Magnetised Piezoelectric Semiconducting Plasmas
(13.5.2; 7; 22.2.3)
(b) Vol. **127**, No. 1, p. 245–252 (1985).

S. GHOSH and R. B. SAXENA
*Stimulated Brillouin Scattering of High-Power Helicon Waves in Longitudinally Magnetized Cubic Piezoelectric Semiconducting Plasmas
(13.5.2; 7; 22.2.3)
(b) Vol. **127**, No. 2, p. 557–566 (1985).

E. KÜHRT, R. WEDELL, D. SEMRAD, and P. BAUER
Theoretical Description of the Stopping Power of Light Ions in the Intermediate Energy Range
(11; 13.5.2; 21.1; 22.1.2; 23)
(b) Vol. **127**, No. 2, p. 633–639 (1985).

V. M. ASNIN, V. I. STEPANOV, R. ZIMMERMANN, and M. RÖSLER
Screening of the Spin–Orbit Exciton by the Electron–Hole Liquid in Germanium
(13.5.1 and 20.1; 13.5.2; 22.1.1)
(b) Vol. **127**, No. 2, p. 651–655 (1985).

A. I. KASIYAN and P. I. RUSSU
*The Effect of Collective Electron–Hole Plasma Modes on Charge Carrier Mobility in Polar Semiconductors
(6 and 13.5.2; 14.3.4; 22; 22.2.1)
(b) Vol. **128**, No. 1, p. 243–249 (1985).

TRAN THOAI DUY BAO
Static Polarizability and Dipolar Surface Plasmon of Small Metal Particles
(13.3; 13.5.2)
(b) Vol. **128**, No. 2, p. 731–734 (1985).

S. GHOSH and M. S. PATHAK
*Transverse Phonon–Helicon Interactions in Semiconducting Plasmas
(6 and 13.5.2; 7; 22.2.3)
(b) Vol. **129**, No. 1, p. 81–87 (1985).

A. I. KASIYAN and P. I. RUSSU
*Charge Carrier Mobility Caused by Scattering from Hybrid Plasmon–Phonon Modes
(6 and 13.5.2; 14.3.4; 22; 22.2.1)
(b) Vol. **129**, No. 1, p. 439–447 (1985).

S. SEN, D. N. BOSE, and M. S. HEDGE
*Plasmon Energies and Dielectric Constants of the III–VI Semiconductors In_2Te_3 and Ga_2Te_3
(4 and 13.5.2; 22.3)
(b) Vol. **129**, No. 1, p. K65–K67 (1985).

S. GHOSH and S. DIXIT
*Effect of Relativistic Mass Variation of the Electron on the Modulational Instability of Laser Beams in Transversely Magnetised Piezoelectric Semiconducting Plasmas
(7 and 13.5.2; 22.2.3)
(b) Vol. **130**, No. 1, p. 219–224 (1985).

R.-L. MA and P.Y. YU
A New Technique for Measuring Temporal Profiles of Picosecond and Femtosecond Laser Pulses Using Self-Induced Deflection in a Semiconductor Wedge
(20.2; 13.5.2; 22.1.2)
(a) Vol. **90**, No. 1, p. 343–348 (1985).

W. BRAUER and M. RÖSLER
*On the Relation between Stopping Power and Proton-Induced Electron Emission for Intermediate Energies: Aluminum
(11 and 13.5.2; 17; 21)
(b) Vol. **131**, No. 1, p. 177–183 (1985).

S. GHOSH and S. DIXIT
*Stimulated Raman Scattering and Raman Instability of an Intense Helicon Wave in Longitudinally Magnetized n-Type Piezoelectric Semiconducting Plasma
(13.5.2; 6; 22.2.3)
(b) Vol. **131**, No. 1, p. 255–265 (1985).

S. L. DRECHSLER and M. BOBETH
Dielectric Properties of Trans-Polyacethylene. I. The Macroscopic Longitudinal Dielectric Function of a Peiers-Fröhlich Semiconductor of Commensurability 2 within RPA
(14.3; 13.5.2; 22.9)
(b) Vol. **131**, No. 1, p. 267–278 (1985).

D. R. MUSIN and V. N. NISHANOV
*Plasma Oscillations in Non-Uniform Quasi-Two-Dimensional Systems
(13.5.2)
(b) Vol. **132**, No. 1, p. K45–K49 (1985).

13.5.3 Polarons

Review Article

H. W. Streitwolf
*Physical Properties of Polyacetylene
(13.5 and 13.5.3; 6; 18; 19; 20.1; 22.9)
(b) Vol. **127**, No. 1, p. 11–54 (1985).

Original Papers and Short Notes

Pan Jin-Sheng
*The Surface or Interface Polaron in Polar Crystals. I. Weak-Coupling Limit
(13.5.3)
(b) Vol. **127**, No. 1, p. 307–318 (1985).

Y. Nakaoka
*Crystal Structure Dependence of Hole Stability in Ionic Crystals
(13.5.3; 6; 13.4; 22.5.2; 23)
(b) Vol. **127**, No. 1, p. 327–338 (1985).

N. A. Mironova and V. N. Skvortsova
*Fine Structure of the NiO:Co Infrared Absorption Spectrum
(13.5.3 and 20.1; 22.6)
(b) Vol. **127**, No. 1, p. K81–K82 (1985).

Pan Jin-Sheng
*The Surface or Interface Polaron in Polar Crystals. II. Strong Coupling Limit
(13.5.3)
(b) Vol. **128**, No. 1, p. 287–295 (1985).

Pan Jin-Sheng
*The Surface or Interface Polaron in Polar Crystals. III. The Phase-Transition-Like Transition for the Surface Polaron
(13.5.3)
(b) Vol. **128**, No. 2, p. 663–672 (1985).

J. Loos
*On the Frequency-Dependent Small-Polaron Conductivity in the High-Temperature Regime
(13.5.3 and 14)
(b) Vol. **128**, No. 2, p. 797–805 (1985).

J. Mertsching
Inverse Spectral Theory and Incommensurate Peierls Phases
(13.1; 13.5.3)
(b) Vol. **129**, No. 1, p. 235–246 (1985).

L. WENDLER
Electron-Phonon Interaction in Dielectric Bilayer Systems. Effect of the Electronic Polarizability
(6 and 13; 13.5.3; **14.4**; 22.2.1; 22.2.2)
(b) Vol. **129**, No. 2, p. 513–530 (1985).

J. RÖSELER
*Application of the Density-Functional Formalism to Bound Polarons
(13.5.3; 22)
(b) Vol. **130**, No. 1, p. 303–314 (1985).

G. P. TRIBERIS
*Thermoelectric Power Due to Small Polaron Hopping Motion in Disordered Systems
(13.5.3 and 15; 2)
(b) Vol. **132**, No. 2, p. 641–652 (1985).

13.5.4 Magnons

Original Papers and Short Notes

K. G. Fischer and G. Heber
*Magnons in an Anisotropic Ferromagnetic Heisenberg Chain with One Static π-Kink
(13.5.4 and 18.2)
(b) Vol. **131**, No. 1, p. 117–128 (1985).

A. L. Tronconi and O. A. C. Nunes
*Spin-Flip Intraband Absorption of Intense Laser Field in the Europium Compounds
(13.5.4 and 20.1; 22.6)
(b) Vol. **131**, No. 1, p. 137–140 (1985).

J. T. Lin and Z. Y. Li
*Microscopic Model for the Relaxation and Excitation of Localized Spin Waves
(13.5.4 and 18; 13.4)
(b) Vol. **131**, No. 1, p. 141–149 (1985).

Yu. M. Seidov and A. M. Suleimanov
*Theory of Absorption and Scattering of Light by Singlet–Triplet Magnetic Systems
(13.5.4 and 20.1; 18.2; 18.4)
(b) Vol. **131**, No. 2, p. 545–554 (1985).

R. Blinc, P. Cevc, J. Seliger, and M. Koren
*EPR of Tl^{2+} Doped Incommensurate K_2SeO_4
(13.5.4 and 19; 22.8.1)
(b) Vol. **131**, No. 2, p. K167–K172 (1985).
Erratum
(b) Vol. **132**, No. 2, p. K143 (1985).

P. Pawlicki
*Calculation of the Magnon Spectrum for Mn_3GaC Metamagnets
(13.5.4; 18.2; 18.4; 22.8)
(b) Vol. **132**, No. 1, p. K37–K40 (1985).

V. B. Kokshenev
*Spin–Phonon Excitations in Paramagnetic Crystals. Heat Transfer at High Temperatures
(6 and 13.5.4; 8)
(b) Vol. **132**, No. 2, p. 375–382 (1985).

E. A. Popov, M. M. Kotlyarskii, and I. S. Edelman
*Temperature and Field Dependence of Exciton-Magnon Absorption of $NaMnCl_3$
(13.5.1 and 13.5.4; 18.2; 20.1; 22.5.4)
(b) Vol. **132**, No. 2, p. 465–470 (1985).

14. Electrical Properties. Transport Phenomena

Review Articles

Original Papers and Short Notes

S. Yu. Potapenko and A. M. Satanin
*The Resistance of a One-Dimensional System with a Stationary Random Potential which is a Markov Process
(13 and 14; 2)
(b) Vol. **129**, No. 2, p. 805–811 (1985).

K. Kułakowski and A. Z. Maksymowicz
Charge Transfer in Chromium–Transition Metal Alloys
(13; 14; 21.1)
(b) Vol. **130**, No. 2, p. 629–635 (1985).

Z. A. Kasamanyan, V. M. Gasparyan, and E. S. Yuzbashyan
*Oscillatory Current along a Tunnel-Thin Potential Barrier in a Magnetic Field
(13.3 and 14)
(b) Vol. **130**, No. 2, p. K149–K152 (1985).

C. S. Sunandana and K. S. Rao
ESR and Other Properties of Copper-Substituted Tellurium Vanadate Glasses
(2; 14; 19; 20.1; 22.6.1)
(a) Vol. **90**, No. 2, p. 681–690 (1985).

A. S. Davydov and Yu. B. Gaididei
Facilitated Electron Tunneling through Polymer Molecules
(13.4; 14; 22.9)
(b) Vol. **132**, No. 1, p. 189–201 (1985).

14.1 Metals. Semi-Metals

Review Article

S. A. SHABALOVSKAYA
Phase Transitions in the Intermetallic Compound TiNi with Charge-Density Wave Formation
(1.2 and 13.1; 6; 14.1; 21.1)
(b) Vol. **132**, No. 2, p. 327–344 (1985).

Original Papers and Short Notes

D. SEN and S. K. SEN
*Transport in Simple Metals in the Framework of a Unified Pseudopotential Study
(8 and 14.1; 15; 21; 21.2)
(b) Vol. **127**, No. 1, p. 377–383 (1985).

G. SCHUBERT and V. CHRISTOPH
*Calculation of Thermopower and Thermal Conductivity Using Correlation Functions
(14.1 and 15; 21)
(b) Vol. **127**, No. 1, p. K89–K93 (1985).

J. A. CUSIDÓ, A. ISALGUÉ, and J. TEJADA
On the Amorphous to Crystalline Transformation of $Fe_{80}B_{20}$ by Means of Electrical and Thermal Conductivity, X-Ray Diffraction, and Mössbauer Measurements
(1.2 and 2; 5; 8; 14.1; 21.1.1)
(a) Vol. **87**, No. 1, p. 169–174 (1985).

J. F. DINHUT, P. BOUILLAUD et J. P. RIVIÈRE
*Irradiation aux ions d'un film mince d'alliaga Fe 65 at%–Co 35 at%. Production d'un verre métallique Fe 50 at%–Co 30 at%–P 20 at%
(Ion Irradiation of Thin Films of Fe (65 at%)–Co (35 at%) Alloy. Production of a Metallic Fe (50 at%)–Co (30 at%)–P (20 at%) Glass)
(11 and 14.1; 1.3; 21.1; 21.1.1)
(a) Vol. **87**, No. 1, p. 199–205 (1985).

H. KIDO, T. HOSHIKAWA, M. SHIMADA, and M. KOIZUMI
*Crystal Structure and Magnetic and Electrical Properties of Pseudo-Ternary $HoCu_{1-x}Zn_xSi$ and $NdCu_{0.6}Zn_{0.4}Si$ Compounds
(14.1 and 18.1; 1.3; 18.2; 22.8)
(a) Vol. **87**, No. 1, p. 273–277 (1985).

S. M. J. AKHTAR and E. E. KHAWAJA
*A Study of the Resistivity and the Thermoelectric Power of Thin Films of Sb and Bi
(14.1 and 15; 21.7)
(a) Vol. **87**, No. 1, p. 335–340 (1985).

P. MUTH
*A Common Maximum of the Minority Spin-Band Resistivities of Some Nickel Alloys
(14.1; 13.1; 21.1)
(b) Vol. **127**, No. 2, p. K151–K153 (1985).

R. AIDUN, S. ARAJS, and C. A. MOYER
*Formation of Localized Magnetic Moments in Chromium–Manganese Alloys
(14.1 and 18.2; 21.1)
(b) Vol. **128**, No. 1, p. 133 – 139 (1985).

A. BATH, J. HUGEL and R. KLEIM
Thermoelectric Power of Liquid Germanium and Tin
(2 and 15; 14.1; 22.1; 22.1.1)
(b) Vol. **128**, No. 2, p. 761 – 767 (1985).

J. G. GASSER, J. L. BRETONNET, and A. BRUSON
*Temperature Dependence of the Liquid Tin Resistivity
(2 and 14.1; 21.7)
(b) Vol. **128**, No. 2, p. 789 – 796 (1985).

G. SURYANARAYANA REDDY, H. C. GUPTA, and B. B. TRIPATHI
*Non-Linear Dependence of the Temperature Coefficient of Electrical Resistivity on Electron–Phonon Interaction Constant in the Case of Transition Metal Amorphous Alloys
(14.1; 2; 21.1)
(b) Vol. **128**, No. 2, p. K183 – K186 (1985).

A. H. PHILLIPS, N. K. GOBRAN, and R. KAMEL
Structure Analysis of the Metallic Glass $Fe_{40}Ni_{40}B_{20}$ Using the Tangent Lattice Model
(1.2; 14.1; 21.1; 21.1.1)
(a) Vol. **88**, No. 2, p. 463 – 467 (1985).

P. KWAPULIŃSKI, J. RASEK, and J. W. MOROŃ
The Kinetics of Structural Defect Recovery in Pure Iron in the Temperature Interval 300 to 800 K
(10.1; 14.1; 21.1.1)
(a) Vol. **88**, No. 2, p. 561 – 573 (1985).

I. MERTIG, E. MROSAN, R. ZELLER, P. H. DEDERICHS, and P. ZIESCHE
Calculation of the Thermoelectric Power of Ga, Ni, Rh, and Pd Impurities in Copper
(13.4 and 15; 14.1; 21.1)
(b) Vol. **129**, No. 1, p. 407 – 414 (1985).

M. I. KATSNELSON and A. S. SHCHERBAKOV
*Temperature Dependence of Conductivity at Small Electron Mean Free Path
(2 and 14.1; 21.1)
(b) Vol. **129**, No. 2, p. 813 – 822 (1985).

A. P. ZHERNOV, KH. M. PASHAEV, and A. M. ALIEV
*The Non-Equilibrium Electron Distribution Function in the Electrical Resistance Problem for Potassium Metal. Influence of N- and U-Processes
(14.1; 21.2)
(b) Vol. **129**, No. 2, p. 823 – 833 (1985).

J. OLEJNICZAK, A. ROJEK, and C. SUŁKOWSKI
*Superconductivity and Resistivity of $LaRh_{1.1}Sn_4$
(14.1 and 14.2; 21.4; 21.7)
(b) Vol. **130**, No. 1, p. K73 – K76 (1985).

B. R. Sood
*Magnetoconductance in Metallic Granular Gallium Films
(14.1 and 18; 2; 21)
(b) Vol. **130**, No. 2, p. 801–806 (1985).

P. Kopčanský
*Resonance Maximum of the Electrical Resistivity in Amorphous Ferromagnets and Dilute Alloys
(14.1 and 18.2; 21)
(b) Vol. **130**, No. 2, p. K175–K178 (1985).

P. Urwank, E. Wieser, A. Hässner, Ch. Kaufmann, H. Lippmann, and I. Melzer
Formation of $MoSi_2$ by Light Pulse Irradiation
(1.2 and 11; 14.1; 21)
(a) Vol. **90**, No. 2, p. 463–468 (1985).

Z. Gierak, J. W. Moroń, and J. Rasek
Phenomena of Structural Defect Recovery in Molybdenum
(1.4 and 10.1; 14.1; 21)
(a) Vol. **90**, No. 2, p. 515–521 (1985).

D. S. Haroutunian, A. G. Karagyozyan, and G. N. Karajian
*Anomalous Paramagnetic State in AgMn Spin Glass and the Kondo Effect
(1.3 and 14.1; 18.1; 21)
(b) Vol. **131**, No. 1, p. K87–K90 (1985).

A. R. Ali
On the Behaviour of Interaction of Copper Atoms with Lattice Defects in Quenched and Cold-Worked Ni–Cu Alloys
(10.1; 14.1; 21.1)
(a) Vol. **91**, No. 1, p. 51–56 (1985).

Y. Kreislerová, T. Zemčík, and H. Hemschik
Mössbauer, Electrical, and Magnetic Studies of the Sn-Doped Ni_3Fe Alloy
(5 and 18.2; 14.1; 21.1; 21.1.1)
(a) Vol. **91**, No. 1, p. 115–122 (1985).

B. B. Dasgupta
*Anomalous Skin Effect in a Cylindrical Conductor
(14.1)
(a) Vol. **91**, No. 1, p. 235–241 (1985).

R. Fromageau, A. Magnouche, and N. Gerard
*The Resistivity of Dissolved Hydrogen in Palladium and Amorphous $Pd_{80}Si_{20}$
(2 and 14.1; 21.6)
(a) Vol. **91**, No. 1, p. K75–K80 (1985).

P. Muth and V. Christoph
*The Spontaneous Resistivity Anisotropy of Iron-Based Alloys
(13.1 and 14.1; 21.1.1)
(b) Vol. **131**, No. 2, p. 769–773 (1985).

14.2 Superconductivity. Superconducting Materials and Devices

Original Papers and Short Notes

P. C. Maithani and M. D. Tiwari
Lattice Specific Heat and Superconducting Transition Temperature of a V:W Alloy
(6 and 8; 14.2; 21.1)
(b) Vol. **127**, No. 1, p. 79–88 (1985).

M. Mück, H. Rogalla, and B. David
*DC-SQUID Operation above 20 K
(14.2)
(a) Vol. **87**, No. 1, p. K105–K107 (1985).

G. Giaquinta, C. Di. Mauro, and R. Fazio
*On the Universality of the Resistive Transition Due to Vortex Depairing in 2D Superconductors in the High k Limit
(14.2)
(b) Vol. **127**, No. 2, p. K155–K157 (1985).

M. Takeda, H. Yoshida, and H. Hashimoto
*Local Tetragonality and Atomic Structure in Nb_3Sn Superconductor Studied by High Resolution Electron Microscopy
(1.4 and 14.2; 21)
(a) Vol. **87**, No. 2, p. 473–482 (1985).

K. Bortel
*Magnetic Properties of the Transition Metals in a Superconductor
(14.2 and 18; 21.1)
(b) Vol. **128**, No. 1, p. 177–185 (1985).

G. Giaquinta, C. Di Mauro, and R. Fazio
*Vortex Unbinding, Fluctuations and Enhanced Resistivity in 2D Superconductors
(14.2)
(b) Vol. **128**, No. 1, p. K97–K99 (1985).

E. Kolley and W. Kolley
*Electron Localization Effects on T_c in Strong-Coupling Superconductivity
(14.2)
(b) Vol. **128**, No. 1, p. K101–K104 (1985).

Z. Ivanov, B. Todorov, and V. Gancheva
Thin ε-Phase Films Prepared by Alloy Evaporation
(1.5; 14.2)
(a) Vol. **88**, No. 1, p. 89–92 (1985).

M. Arita, H.-U. Nissen, and W. Schauer
Structure Images of the Hexagonal Nb_5Ge_3 ($D8_8$) Structure
(1.4; 14.2; 21)
(a) Vol. **88**, No. 1, p. K1–K4 (1985).

S.-L. Drechsler and A. P. Zhernov
*Electron–Electron Interaction and Isotope Effects in Superconducting Metal–Hydrogen (Deuterium) Systems Like $PdH_x(D_x)$
(14.2; 21)
(b) Vol. **130**, No. 1, p. 381 – 386 (1985).

G. Giaquinta, C. Di Mauro, and R. Fazio
*Universal Flux Flow Resistance in Dirty 2D Superconductors below the Vortex Unbinding Temperature
(14.2)
(b) Vol. **130**, No. 1, p. K69 – K71 (1985).

J. Olejniczak, A. Rojek, and C. Sułkowski
*Superconductivity and Resistivity of $LaRh_{1.1}Sn_4$
(14.1 and 14.2; 21.4; 21.7)
(b) Vol. **130**, No. 1, p. K73 – K76 (1985).

J. Schumann and P. Otschik
*Phonon Structure ($\alpha^2 F(\omega)$) of Superconducting V_3Si Monocrystals by Tunneling Measurements
(6 and 14.2; 21.1)
(b) Vol. **130**, No. 2, p. 437 – 450 (1985).

I. M. Chapnik
*Superconducting Transition Temperature and the Formation of Closed Electron Shells in the Atoms of Superconducting Compounds
(14.2)
(b) Vol. **130**, No. 2, p. K179 – K182 (1985).

A. Gilabert, F. de Schutter, and Y. Bruynseraede
*Magnetic Properties of Superconducting Oxygen-Doped Vanadium Foils
(14.2 and 18)
(a) Vol. **90**, No. 2, p. 651 – 656 (1985).

Wang Rong-Yao, Luo Qi-Guang, and Zhang Xiao
*A Formulation for the Critical Temperature T_c of $L1_2$-Type Superconductors
(14.2; 21)
(a) Vol. **90**, No. 2, p. 763 – 769 (1985).

B. Schuh
*Hybrid Pairing and Heavy Fermions
(13 and 14.2)
(b) Vol. **131**, No. 1, p. 243 – 248 (1985).

G. S. Ameta and U. N. Upadhyaya
*Electronic Properties of Ternary Superconductors
(14.2; 18.2; 21.4)
(b) Vol. **131**, No. 1, p. 369 – 386 (1985).

Z. Przełożny
*Critical Currents in Type-II Superconductors
(14.2)
(b) Vol. **131**, No. 1, p. 387 – 400 (1985).

14.3 Semiconductors

Review Articles

S. Porowski and W. Trzeciakowski
The Effect of Pressure on Deep Impurity States with Large Lattice Relaxation
(13.4; 12.2; 14.3; 16; 22.2.3; 22.2.4; 22.4.3; 22.5; 22.5.2)
(b) Vol. **128**, No. 1, p. 11–22 (1985).

B. L. Gelmont, B. T. Kolomiets, and K. D. Tsendin
*Impurity Conduction of Chalcogenide Vitreous Semiconductors
(2 and 14.3; 13.1; 13.4; 22.3; 22.8)
(a) Vol. **91**, No. 2, p. 319—337 (1985).

Original Papers and Short Notes

G. Ch. Christakudis, S. K. Plachkova, E. S. Avilov, L. E. Shelimova, and N. Kh. Abrikosov
*Transport Phenomena of Ferroelectric Solid Solutions
$(GeTe)_{1-x}((Ag_2Te)_{1-y}(Sb_2Te_3)_y)_x$ with $y = 0.60$, $0 \leqq x \leqq 1$ at 300 K
(14.3; 8; 14.4.1; 15; 22.7.1)
(b) Vol. **127**, No. 1, p. 419–426 (1985).

K. M. Paraskevopoulos
Electron Effective Mass Dependence on Carrier Concentration in $TlBiTe_2$ Monocrystals
(13.1 and 20.1; 14.3; 22.8)
(b) Vol. **127**, No. 1, p. K45–K50 (1985).

R. Fremunt, Y. Hirayama, F. Arai, and T. Sugano
*Activation of S^+-Implanted InP by Scanned CW Laser Irradiation
(11 and 14.3; 22.2.2)
(a) Vol. **87**, No. 1, p. K49–K52 (1985).

M. Averous, J. P. Desfours, and J. R. Vellas
*Long- and Short-Range Magnetic Interactions in EuO
(14.3 and 18.2; 22.6)
(b) Vol. **127**, No. 2, p. 523–529 (1985).

E. Gerlach
*Overshoot in Low-Conduction Semiconductors
(14.3; 22)
(b) Vol. **127**, No. 2, p. 727–733 (1985).

Ľ. Hrivnák
*Band-Gap Narrowing in Semi-Insulating GaAs
(13.1 and 14.3; 13.4; 22.2.1)
(b) Vol. **127**, No. 2, p. K165–K170 (1985).

A. KUCZKOWSKI and T. SŁUPKOWSKI
*The Electronic Properties of Macromolecular Structures with Conjugated Bonds Obtained by Chemical Modification of Certain Aliphatic Polymers
(14.3 and 15; 22.9)
(a) Vol. **88**, No. 1, p. 291—295 (1985).

S. A. MAZEN, A. A. GHANI, and A. H. ASHOUR
*Charge Transport in Cu–Cd Ferrospinels
(14.3 and 15; 22.8.2)
(a) Vol. **88**, No. 1, p. 343—346 (1985).

R. ASSENOV, V. A. MOSHNIKOV, and D. A. YASKOV
On the Behaviour of Iodine in PbTe and SnTe
(10.2; 14.3; 22)
(a) Vol. **88**, No. 1, p. K27—K30 (1985).

S. SHIGETOMI, T. IKARI, Y. KOGA, and S. SHIGETOMI
*Drift Mobility of Electrons Parallel to the *c*-Axis in InSe
(14.3; 13.4; 16; 22.3)
(a) Vol. **88**, No. 1, p. K71—K73 (1985).

H. TAGUCHI
*Electrical Properties of $CaMnO_{3-\delta}$
(14.3; 22.6)
(a) Vol. **88**, No. 1, p. K79—K82 (1985).

I. V. GUTSUL, I. G. KIRNAS, P. G. LITOVCHENKO, V. I. MARUSYAK, V. M. NITSOVICH, S. E. OSTAPOV, and E. E. PETROSYAN
*Study of the Hall Coefficient in p-Type Germanium Taking into Account Anisotropic Charge Carrier Scattering
(14.3; 22.1.1)
(b) Vol. **128**, No. 2, p. 753—759 (1985).

S. A. ZEINALOV
*Conduction Band Structure of $In_{0.5}Ga_{0.5}Sb:Te$
(13.1 and 14.3; 22.2.4)
(b) Vol. **128**, No. 2, p. K139—K142 (1985).

P. I. BARANSKII, A. E. BELYAEV, and A. V. ZHIDKOV
*Extreme Disordering of Electron Scattering in a Gapless Semiconductor
(14.3 and 18; 22.4.4)
(b) Vol. **128**, No. 2, p. K171—K174 (1985).

P. K. BASU and K. BHATTACHARYYA
*Modification of the Theory of Two-Dimensional Mobility Due to Scattering by Alloy Clusters
(14.3; 13.1; 22.2.4)
(b) Vol. **128**, No. 2, p. K175—K177 (1985).

H. BÖTTGER, P. SZYLER, and D. WEGENER
*Numerical Investigation of Non-Ohmic Hopping Conduction at Low Temperatures
(2 and 14.3)
(b) Vol. **128**, No. 2, p. K179—K182 (1985).

U. Henninger and R. Keiper
*On Theoretical Modelling of Continuously Inhomogeneous Semiconductor Structures
(14.3; 22)
(b) Vol. **129**, No. 1, p. 399–406 (1985).

R. Herrmann, W. Kraak, G. Nachtwei, and Th. Schurig
*Two-Dimensional Electron Gas in the n-Inversion Layer of p-Type InSb Bicrystals
(13.1 and 14.3; 22.2.3)
(b) Vol. **129**, No. 1, p. 415–426 (1985).

H. Ishikawa, M. Aoyama, O. Uemura, and T. Satow
*Electronic Properties of the Liquid Tl–GeTl_2Te_4 and Tl–Sn Tl_2Te_4 Systems
(2 and 14.3; 15; 18; 22.7.1)
(b) Vol. **129**, No. 1, p. 427–434 (1985).

V. Narasimhan, H. V. Keer, and D. K. Chakrabarty
*Structural and Electrical Properties of the $LaCo_{1-x}Mn_xO_3$ and $LaCo_{1-x}Fe_xO_3$ Systems
(1.2 and 14.3; 15; 22.8.2)
(a) Vol. **89**, No. 1, p. 65–71 (1985).

V. L. Litvinov and N. A. Ukhin
Fast-Neutron Radiation Damages in Heavily Doped p-Silicon
(11; 14.3; 22.1.2)
(a) Vol. **89**, No. 1, p. 95–104 (1985).

D. Lapraz, F. Gaume, and M. Barland
On the Thermoluminescent Mechanism of a Calcium Fluoraptite Single Crystal Doped with Mn^{2+}
(13.4 and 20.3; 14.3; 20.1; 22.8)
(a) Vol. **89**, No. 1, p. 249–253 (1985).

L. G. Khvostantsev, A. I. Orlov, N. Kh. Abrikosov, and L. D. Ivanova
Kinetic Properties and Phase Transitions in Sb_2Te_3 under Hydrostatic Pressure up to 9 GPa
(1.2 and 15; 14.3; 22.7)
(a) Vol. **89**, No. 1, p. 301–309 (1985).

A. P. Tyutnev, V. S. Saenko, V. N. Abramov, V. P. Sichkar, A. I. Karpechin, E. D. Pozhidaev, and A. V. Vannikov
*Dose Effects in Transient Radiation-Induced Conductivity in Polymers
(11 and 14.3; 22.9)
(a) Vol. **89**, No. 1, p. 311–320 (1985).

J. Krištofik, J. J. Mareš, and V. Šmíd
*The Effect of Pressure on Conductivity and Permittivity of As_2Te_3-Based Glasses
(14.3; 2; 22)
(a) Vol. **89**, No. 1, p. 333–345 (1985).

B. Poumellec, J. F. Marucco, and F. Lagnel
*Electron Transport in TiO_{2-x} at Intermediate Temperatures 300 K $< T <$ 1500 K
(13.4 and 14.3; 15; 22.6)
(a) Vol. **89**, No. 1, p. 375–382 (1985).

W. Kraak, R. Herrmann, and G. Nachtwei
*Two-Dimensional Properties of n-Inversion Layers in InSb Grain Boundaries under High Hydrostatic Pressure
(13.1 and 14.3; 22.2.3)
(b) Vol. **130**, No. 2, p. 781–785 (1985).

H. K. Sy, D. K. Desai, and C. K. Ong
*Electron Screening and Mobility in Heavily Doped Silicon
(14.3; 22.1.2)
(b) Vol. **130**, No. 2, p. 787—792 (1985).

B. Mari, A. Segura, and A. Chevy
Free Carrier Absorption in n-Type Indium Selenide
(6 and 20.1; 14.3; 22.3)
(b) Vol. **130**, No. 2, p. 793—799 (1985).

O. P. Gorodnichii, I. N. Gorbatyuk, I. M. Rarenko, and R. M. Vinetskii
*Longitudinal Magnetoresistance in (Cd, Hg)Te Crystals in Quantizing Magnetic Fields
(14.3; 22.4.4)
(b) Vol. **130**, No. 2, p. K169–K173 (1985).

G. D. Guseinov, A. Kh. Matiev, A. U. Malsagov, S. Kh. Umarov, E. G. Abdullaev, and N. D. Yuraev
*Effect of Pressure on Electrical Conductivity of $TlInSe_2$ Single Crystals
(14.3; 22.8)
(a) Vol. **90**, No. 2, p. 703–707 (1985).

B. P. Tyagi and K. Sen
*Electrical Properties of Polycrystalline Silicon in the Dark and under Illumination
(13.4 and 14.3; 16; 22.1.2)
(a) Vol. **90**, No. 2, p. 709–713 (1985).

T. Słupkowski
*Electrical Conductivity of Polyester Polymer Containing Carbon Black
(14.3; 22.9)
(a) Vol. **90**, No. 2, p. 737–741 (1985).

M. Moldovanova and R. Assenov
*Peculiarities in the Temperature Dependence of the Hall Coefficient in Tin Telluride. Mobility
(14.3; 22)
(a) Vol. **90**, No. 2, p. 743—747 (1985).

P. Gaworzewski, E. Hild, and K. Schmalz
Oxygen Donor Formation and Oxygen Precipitation in Czochralski-Silicon
Due to Heat Treatment at 600 to 800 °C
(10.2; 14.3; 20.1; 22.1.2)
(a) Vol. **90**, No. 2, p. K151–K156 (1985).

Lu Zhiheng and Zhang Xiojie
The Observation on the Atom Displacements of Silicon Highly Doped by Arsenic during Post-Annealing
(11; 14.3; 22.1.2)
(a) Vol. **90**, No. 2, p. K157–K160 (1985).

D. K. CHAKRABARTY, M. K. JHA, and S. UNNIKRISHNAN
Preparation and Properties of Barium Substituted Gadolinium Cobaltite
(18.2; 1.1; 14.3; 22.8.1)
(a) Vol. **90**, No. 2, p. K175–K177 (1985).

R. Z. SADYKHOV, L. M. VALIEV, and A. D. NAMAZOV
*Magnetic Susceptibility and the Hall Effect in the $FeCr_{2-x}In_xS_4$ $(0 \leqq x \leqq 1)$ System
(14.3 and 18.1; 22.8)
(a) Vol. **90**, No. 2, p. K189–K191 (1985).

A. GLEITZ and H. W. HELBERG
*Microwave Conductivity of Dislocations in Deformed Silicon Single Crystals
(10.2 and 14.3; 22.1.2)
(a) Vol. **90**, No. 2, p. K209–K213 (1985).

S. L. DRECHSLER and M. BOBETH
*Dielectric Properties of Trans-Polyacetylene. I. The Macroscopic Longitudinal Dielectric Function of a Peiers-Fröhlich Semiconductor of Commensurability 2 within RPA
(14.3; 13.5.2; 22.9)
(b) Vol. **131**, No. 1, p. 267–278 (1985).

H. LEMKE
Doping Properties of Rhodium and Iridium in Silicon
(13.4; 14.3; 22.1.2)
(a) Vol. **91**, No. 1, p. 143–152 (1985).

N. A. KORNYUSHKIN, N. I. BRYZGALOVA, and A. A. FRANTSUZOV
*Relaxation of Tunnel Injection Limited Current in Case of Low Capture Probability
(13.4 and 14.3; 22.1.2)
(a) Vol. **91**, No. 1, p. 211–218 (1985).

A. I. LEBEDEV and KH. A. ABDULLIN
*Impurity Photoconductivity and Electrical Properties of $Pb_{1-x-y}Ge_xSn_yTe$ Doped with Indium
(14.3 and 16; 13.4; 22)
(a) Vol. **91**, No. 1, p. 225–234 (1985).

M. KITAO, K. YOSHII, and S. YAMADA
*Thermoelectric Power of Glassy $As_{40}Se_{60-x}Te_x$
(14.3 and 15; 2; 22)
(a) Vol. **91**, No. 1, p. 271–277 (1985).

M. SCHENK, H. BERGER, C. KLEINT, N. S. GOLOVANOVA, O. I. TANANAEVA, and V. P. ZLOMANOV
The Influence of Bi on Lattice Parameter and Carrier Concentration of PbTe
(10.2; 1.1; 14.3; 22.7)
(a) Vol. **91**, No. 1, p. K35–K39 (1985).

T. SATO, N. MORITA, and M. SHIMADA
*Synthesis and Electical Properties of $Li_xMg_{1-x}M_2O_4$ (M = Ti, V) $(0 \leqq x \leqq 1.0)$
(1.3 and 14.3; 22.8)
(a) Vol. **91**, No. 1, p. K81–K85 (1985).

G. Contreras, L. Tapfer, A. K. Sood, and M. Cardona
Physical Properties of Ion-Implanted Laser Annealed n-Type Germanium
(6 and 13.4; 14.3; 20.1; 22.1.1)
(b) Vol. **131**, No. 2, p. 475—487 (1985).

W. Ulrici, K. Friedland, L. Eaves, and D. P. Halliday
Optical and Electrical Properties of Vanadium-Doped GaAs
(13.4; 14.3; 20.1; 22.2.1)
(b) Vol. **131**, No. 2, p. 719—728 (1985).

M. Averous, B. A. Lombos, C. Fau, E. Ilbnouelghazi, J. C. Tedenac, G. Brun, and M. A. Bartkowski
*$Pb_{1-x}Gd_xTe$ a New Semimagnetic Semiconductor with Rare Earths
(14.3 and 18.4; 19; 22.8)
(b) Vol. **131**, No. 2, p. 759—768 (1985).

B. N. Mukashev, M. F. Tamendarov, S. Zh. Tokmoldin, and V. V. Frolov
Hydrogen Implantation into Silicon. Infra-Red Absorption Spectra and Electrical Properties
(10.2 and 11; 4; 14.3; 20.1; 22.1.2)
(a) Vol. **91**, No. 2, p. 509—522 (1985).

S. A. Shevchenko, I. I. Khodos, and I. I. Snighireva
Dislocation Dissociation and Electrical Properties of Plastically Deformed Germanium Single Crystals
(10.2; 14.3; 22.1.1)
(a) Vol. **91**, No. 2, p. 523—531 (1985).

H. Lemke
Doping Properties of Cobalt in Silicon
(13.4; 14.3; 22.1.2)
(a) Vol. **91**, No. 2, p. 649—659 (1985).

N. A. Kornyushkin, N. I. Bryzgalova, and A. A. Frantsuzov
*Relaxation of Current Flowing through Oxide Thermally Grown on Polycrystalline Silicon
(14.3; 22.6)
(a) Vol. **91**, No. 2, p. 677—684 (1985).

A. P. Tyuntnev, V. N. Abramov, P. I. Dubenskov, V. S. Saenko, E. D. Pozhidaev, and A. V. Vannikov
*Doping Effects in Transit Radiation-Induced Conductivity of Polymers
(14.3; 11; 22.9)
(a) Vol. **91**, No. 2, p. 715—724 (1985).

Y. Nishio, Y. Ootuka, K. Kajita, T. Iwata, and W. Sasaki
*Reverse Annealing Effect in the Neutron-Irradiated Si:P System
(10.2 and 14.3; 22.1.2)
(a) Vol. **91**, No. 2, p. 725—728 (1985).

I. V. Gutsul, I. G. Kirnas, P. G. Litovchenko, V. I. Marusyak, V. M. Nitsovich, S. E. Ostapov, and E. E. Petrosyan
*Hall Factor in p-Ge with Anisotropic Energy Spectrum
(14.3; 22.1.1)
(b) Vol. **132**, No. 1, p. 271—276 (1985).

V. D. Iskra
On the Theory of Scattering by Short-Range Potentials
(13.4; 14.3; 22)
(b) Vol. **132**, No. 2, p. 581–590 (1985).

14.3.1 Films

Original Papers and Short Notes

X. B. LIAO, J. Y. LIAN, and G. L. KONG
*Analysis for Steady State Field Effect Measurements on a-Si:H Films
(14.3.1; 2; 13.3; 22.1.2)
(a) Vol. **88**, No. 2, p. 697–704 (1985).

H. KOBAYASHI, S. TANAKA, V. SHANKER, M. SHIIKI, T. KUNOU, J. MITA, and H. SASAKURA
Multicolor Electroluminescent ZnS Thin Films Doped with Rare Earth Fluorides
(20.3; 14.3.1; 22.4.1)
(a) Vol. **88**, No. 2, p. 713–720 (1985).

U. K. REDDY
*Electrical and Optical Properties of Non-Crystalline As–Se–Cd Thin Films
(2 and 14.3.1; 13.4; 20.1; 22.8)
(a) Vol. **89**, No. 1, p. 255–261 (1985).

S. A. Y. AL-ISMAIL, K. ARSHAK, and C. A. HOGARTH
*Electron Spin Resonance Measurements and Electrical Characteristics, before and after Electroforming, of Thin Films of SiO/Nb_2O_5 and Nb_2O_5
(14.3.1 and 14.3.4; 2; 19; 22.6; 22.6.1)
(a) Vol. **89**, No. 1, p. 363–374 (1985).

M. MÜLLER, TH. FRAUENHEIM, and C. HAMANN
*Peierls Transition and Fluctuation Conductivity in Thin Lead-Phthalocyanine (PbPc) Films
(1.2 and 14.3.1; 22.9)
(a) Vol. **89**, No. 1, p. K89–K94 (1985).

YU. G. FARIVER, A. I. BAIRAMOV, T. D. DZHAFAROV, and M. S. SADIGOV
*The Effect of Substrate on the Electrical Properties of As_2S_3 Films
(14.3.1 and 14.3.3; 22)
(a) Vol. **89**, No. 1, p. K95–K98 (1985).

U. K. REDDY and P. J. REDDY
**I–U* Characteristics of Non-Crystalline As–Se–Cd Thin Films
(14.3.1; 2; 14.3.4; 22.8)
(a) Vol. **89**, No. 2, p. 679–682 (1985).

V. A. PILIPOVICH, B. A. BUDKEVICH, I. M. ROMANOV, G. D. IVLEV, I. A. GES, and S. P. ZHVAVYI
*Laser Annealing Effects on Electrochromic Properties of Amorphous Evaporated WO_3 Films
(14.3.1 and 20.1; 1.5; 22.6)
(a) Vol. **89**, No. 2, p. 709–717 (1985).

A. M. ANDRIESH, V. V. BIVOL, V. I. VERLAN, B. V. EFREMUSHKIN, M. S. IOVU, and E. G. KHANCHEVSKAYA
Influence of Electrode Material on the Electrical and Photoelectrical Properties of $p\text{-InP}/As_2Se_3$ Heterostructures
(14.3.3 and 16; 2; 14.3.1; 22; 22.2.2)
(a) Vol. **89**, No. 2, p. K231–K234 (1985).

U. RENSCH and G. EICHHORN
*Investigation of the State of Strain and of the Electrical Properties of AlN Layers on Monocrystalline Si
(1.5 and 14.3.1; 22.1.2; 22.2)
(a) Vol. **90**, No. 1, p. 135–141 (1985).

E. L. Frankevich, D. I. Kadyrov, I. A. Sokolik, A. I. Pristupa, V. M. Kobryanskii, and N. Y. Zurabyan
*On the Conductivity Mechanism of Weakly Doped Polyacetylene
(14.3.1; 19; 22.9)
(b) Vol. **132**, No. 1, p. 283–294 (1985).

I. G. Oganova and I. M. Aliev
On the Exponential Shape of the Absorption Edge in Heteroepitaxial n-InSb Layers
(20.1; 14.3.1; 22.2.3)
(b) Vol. **132**, No. 1, p. K41–K44 (1985).

Y. Q. Sheng, P. Munz, and R. Schultheiss
The Dtermination of the Thickness of Ultrathin Metal and Semiconductor Films on Conventional Semiconductor Substrates
(1.5; 14.3.1; 21; 21.1; 21.4; 21.6; 22.1.1)
(a) Vol. **92**, No. 1, p. 121–128 (1985).

P. R. Vaya
*Magnetoresistance of PbTe Films
(14.3.1; 13.1; 22)
(a) Vol. **92**, No. 2, p. 555–561 (1985).

P. Danesh
*Annealing Behaviour of a-Si:H:Cl Films
(14.3.1; 22.1.2)
(a) Vol. **92**, No. 2, p. 631–635 (1985).

M. Abd-Lefdil, M. Rolland, and M. Cadene
*n^+–n Silicon–Trans$(CH)_x$ Schottky Diode
(14.3.1 and 14.3.3; 22.9)
(a) Vol. **92**, No. 2, p. K171–K174 (1985).

14.3.2 Surfaces and Interfaces

Original Papers and Short Notes

G. PAASCH
*Alternative Quasi-Classical Transport Equations. I. General Formalism for the Inclusion of Wave Functions
(13.3 and 14.3.2)
(b) Vol. **127**, No. 1, p. 385–394 (1985).

H. TOUTAH and G. PANANAKAKIS
Study of Electronic States Located at the Si–SiO_2 Interface and within the Oxide Volume of VLSI MOS Capacitors Using Capacitance and Conductance Measurements
(13.3 and 14.3.3; 14.3.2; 22.1.2)
(a) Vol. **87**, No. 1, p. 311–317 (1985).

V. S. LYSENKO, V. I. ZIMENKO, I. P. TYAGULSKII, I. N. OSIYUK, O. V. SNITKO, and T. N. SYTENKO
Flash-Lamp Annealing of Si–SiO_2 Transition Layer Defects
(11 and 13.3; 14.3.2; 22.1.2)
(a) Vol. **87**, No. 2, p. K175–K180 (1985).

A. MORALES-ACEVEDO
*A Simple Model of MOS Transistors on Large-Grain Polycrystalline Silicon
(14.3.2 and 14.3.3; 22.1.2)
(a) Vol. **88**, No. 1, p. 381–387 (1985).

S. I. KIRILLOVA, V. E. PRIMACHENKO, and O. V. SNITKO
*Nonequilibrium Depletion Relaxation in Strong Electric Fields under Various Conditions at the Silicon Surface
(14.3.2 and 14.3.4; 13.3; 22.1.2)
(a) Vol. **88**, No. 2, p. 647–654 (1985).

V. I. ARKHIPOV, V. M. LOGIN, and S. YA. SHEVCHENKO
*Effect of Multiple Trapping on Photoreceptor Discharge Characteristics under the Condition of Surface Carrier Generation
(14.3.2 and 16; 2; 22.9)
(a) Vol. **89**, No. 1, p. 293–299 (1985).

S. Chandra and S. N. SAHU
*Electrodeposited Tungsten Selenide Films. II. Optical, Electrical, Electrochemical Characterization and Photoelectrochemical Solar Cell Studies
(14.3.2; 16; 20.1; 22)
(a) Vol. **89**, No. 1, p. 321–331 (1985).

G. GOBSCH, G. PAASCH, H. ÜBENSEE, R. HERRMANN, and W. KRAAK
Calculation of the Subband Structure of Inversion Layers in p-InSb Bicrystals
(13.1; 13.3; 14.3.2; 22.2.3)
(b) Vol. **129**, No. 2, p. K167–K171 (1985).

S. Kuźmiński and A. T. Szaynok
*Shallow Energy States in CdTe
(13.3 and 14.3.2; 16; 22.4.3)
(a) Vol. **89**, No. 2, p. 623–628 (1985).

I. Petr and O. Gilar
*Irradiation of MOS Transistors and Resulting Transport Processes in SiO_2
(14.3.2 and 14.3.3; 11; 22.1.2)
(a) Vol. **89**, No. 2, p. 703–708 (1985).

H. Übensee, G. Paasch, J.-P. Zöllner, Th. Fiedler, and G. Gobsch
Self-Consistent Theory of the Electronic Structure of Inversion Layers. I. A New Method Using the Modified Local Density Approximation
(13.3; 13.1; 14.3.2; 22.2.1; 22.2.3)
(b) Vol. **130**, No. 1, p. 387–402 (1985).

A. J. Muhammad, D. V. Morgan, and A. E. Guile
*Electronic Conduction in $Nb/Nb_2O_5/In$ Structures and the Effect of Space-Charge Overlap
(14.3.2 and 14.3.4; 22.6)
(a) Vol. **90**, No. 1, p. 371–381 (1985).

N'Guyen Duc Tien
*Characteristics with Negative Differential Resistance of a Device Using $Si–SiO_2$ Structure
(14.3.2 and 14.3.4; 22.1.2)
(a) Vol. **90**, No. 1, p. K103–K105 (1985)

S. Nojima, M. Oishi, H. Asahi, and H. Nagai
*An Interfacial Property of the InGaAs/Semi-Insulating InP Structure Prepared by Metal Organic Vapor Phase Epitaxy
(14.3.2 and 14.3.4; 22.2.2; 22.2.4)
(a) Vol. **90**, No. 2, p. K215–K218 (1985).

G. Balchaitis, J. Kameneckas, G. Petrauskas, and A. Sakalas
*Physical and Chemical Changes in the Surface Layer of GaAs Induced by Recoil Implantation of Al Atoms
(11 and 14.3.2; 22.2.1)
(a) Vol. **91**, No. 1, p. 295–300 (1985).

W. Braune, M. Schnürer, N. Kubicki, and R. Herrmann
*Carrier Mobility Determination in Inversion Layers of Anodized InSb MOS Structures
(14.3.2; 22.2.3)
(b) Vol. **131**, No. 2, p. K181–K183 (1985).

E. Klier
*A Study of Semiinsulating GaAs Surfaces by Thermostimulated Currents
(13.3 and 14.3.2; 22.2.1)
(a) Vol. **91**, No. 2, p. 631–635 (1985).

E. De Castro and P. Olivo
*Quantum Effects in Accumulation Layers of $Si–SiO_2$ Interfaces in the WKB Effective Mass Approximation
(13.4 and 14.3.2; 13.1; 22.1.2; 22.6)
(b) Vol. **132**, No. 1, p. 153–163 (1985).

14.3.3 Devices, Junctions (Contact Problems See 14.3.4)

Review Artikles

YU. A. KATULEVSKII
Injection of Degenerate Electron–Hole Plasma
(14.3.4; 13.1; 13.4; 14.3.3; 22)
(a) Vol. **88**, No. 1, p. 11–42 (1985).

J. S. KANG and D. K. SCHRODER
*The Pulsed MIS Capacitor. A Critical Review
(14.3.3; 22)
(a) Vol. **89**, No. 1, p. 13–43 (1985).

Original Papers and Short Notes

A. ARMIGLIATO, S. SOLMI, C. DONOLATO, P. NEGRINI, E. GABILLI, A. GARULLI, and M. KITTLER
Influence of Chlorine Implantation on Phosphorus Diffusivity and Oxidation-Induced Defects in Silicon
(10.2 and 11; 14.3.3; 22.1.2)
(a) Vol. **87**, No. 1, p. 207–216 (1985).

H. TOUTAH and G. PANANAKAKIS
*Study of Electronis States Located at the Si–SiO_2 Interface and within the Oxide Volume of VLSI MOS Capacitors Using Capacitance and Conductance Measurements
(13.3 and 14.3.3; 14.3.2; 22.1.2)
(a) Vol. **87**, No. 1, p. 311–317 (1985).

A. J. MUHAMMAD, D. V. MORGAN, and A. E. GUILE
*Contact Overlap in MIM Structures with Distributed Traps
(14.3.3 and 14.3.4; 13.4)
(a) Vol. **87**, No. 1, p. 355–362 (1985).

A. YU. LEIDERMAN
*On the Generation–Recombination Currents in p–n Junctions of Semiconductors with Continuous Gap-State Spectrum
(2 and 14.3.3; 14.3.4; 22.1.2)
(a) Vol. **87**, No. 1, p. 363–372 (1985).

J. FIDDICKE and G. OELGART
*The Importance of the Excitation Volume for the Determination of the Minority Carrier Diffusion Length
(14.3.3; 22.2.2)
(a) Vol. **87**, No. 1, p. 383–389 (1985).

M. M. CHANDRA and M. PRASAD
*On the Ideality Factor of Schottky Diodes at Low Temperatures
(14.3.3; 22.1.2)
(a) Vol. **87**, No. 1, p. K97–K100 (1985).

RAVIENDRA D. and J. K. SHARMA
*n-CdS/p-$AgInSe_2$ Solar Cells by Electrodeposition
(14.3.3 and 20.1; 1.5; 22.4.1; 22.8)
(a) Vol. **88**, No. 1, p. 365—368 (1985).

O. GILAR and I. PETR
*Combination of a Silicon Photodiode and a Scintillator as a Dose Rate Detector
(14.3.3 and 16; 22.1.2)
(a) Vol. **88**, No. 1, p. 375—380 (1985).

A. MORALES-ACEVEDO
*A Simple Model of MOS Transistors on Large-Grain Polycrystalline Silicon
(14.3.2 and 14.3.3; 22.1.2)
(a) Vol. **88**, No. 1, p. 381—387 (1985).

K. YASUDA, M. TAKEDA, H. MASUDA, and A. YOSHIDA
*Radiation Effects in a Photo-Coupler with High-Energy Electrons and γ-Rays and Recovery by Current Annealing
(11 and 14.3.3; 22.1.2; 22.2.1)
(a) Vol. **88**, No. 1, p. 389—396 (1985).

P. I. STOIANOVA and R. E. ANDREICHIN
*Coordinate Sensitive Photoelement on the Basis of the Photoelectret State in Vitreous Semiconductors
(14.3.3 and 16; 2; 22)
(a) Vol. **88**, No. 1, p. K91—K93 (1985).

R. GULATI, S. B. KAUSHIK, P. L. TRIVEDI, H. S. SHARMA, I. CHANDRA, and B. L. SHARMA
*Optical Effect in Normally-Off GaAs MESFET
(14.3.3; 16; 22.2.1)
(a) Vol. **88**, No. 1, p. K99—K103 (1985).

A. ROMANOWSKI
*The Impact of the Heat Generation of the Primary Scanning Beam on the SEM-EBIC Characteristics
(14.3.3; 22.1.2; 22.2.1)
(a) Vol. **88**, No. 2, p. 663—672 (1985).

V. S. LYSENKO, M. M. LOKSHIN, A. N. NAZAROV, and T. E. RUDENKO
*RF Plasma Annealing of Implanted MIS Structures
(14.3.3; 11; 13.4; 22.1.2)
(a) Vol. **88**, No. 2, p. 705—713 (1985).

R. PICKENHAIN, V. GEIST, W. SEIFERT, and K. JACOBS
*Green and Yellow Light Emitting Diodes Produced from Vapour Phase Epitaxial GaP:N. II. Defects Produced by 1 MeV Proton Irradiation
(14.3.3 and 20.3; 13.4; 22.2.2)
(a) Vol. **88**, No. 2, p. 721—731 (1985).

H. PAGNIA, J. SCHNELLBÄCHER, and N. SOTNIK
Correlation between Current and Induced Emission Processes in MIM Diodes
(17 and 20.3; 14.3.3)
(a) Vol. **88**, No. 2, p. K209—K212 (1985).

M. Duszak and A. Jakubowski
*Influence of Humidity on Electrical Properties of a MOS Structure with an Ultrathin Dielectric Film
(14.3.3 and 14.4; 22.1.2; 22.6)
(a) Vol. **89**, No. 2, p. 699–702 (1985).

I. Petr and O. Gilar
*Irradiation of MOS Transistors and Resulting Transport Processes in SiO_2
(14.3.2 and 14.3.3; 11; 22.1.2)
(a) Vol. **89**, No. 2, p. 703–708 (1985).

S. A. Azimov and T. M. Razykov
*Energy Band Model of Ultrathin Solar Cells with Thickness up to 5 μm
(13.1 and 14.3.3; 16; 22; 22.4.4)
(a) Vol. **89**, No. 2, p. K223–K226 (1985).

F. P. Korshunov, I. G. Marchenko, and V. T. Troshchinskii
*Variation of Switching Time of Silicon p–n–p–n Structures Irradiated by Fast Electrons
(11 and 14.3.3; 22.1.2)
(a) Vol. **89**, No. 2, p. K227–K230 (1985).

A. M. Andriesh, V. V. Bivol, V. I. Verlan, B. V. Efremushkin, M. S. Iovu, and E. G. Khanchevskaya
*Influence of Electrode Material on the Electrical and Photoelectrical Properties of p-InP/As_2Se_3 Heterostructures
(14.3.3 and 16; 2; 14.3.1; 22; 22.2.2)
(a) Vol. **89**, No. 2, p. K231–K234 (1985).

N. A. Avdeev, Yu. E. Gardin, V. A. Gurtov, and S. N. Kuznetsov
*Study of Charge Accumulation Kinetics in MOS Structures with Intrinsic Anodic Oxide on InSb by Measuring Transient Photocurrents
(13.3 and 14.3.3; 16; 22.2.3)
(a) Vol. **90**, No. 1, p. 393–400 (1985).

D. Sachelarie
*Analysis of Gold-Doped MOS Structures
(14.3.3; 13.3; 13.4; 22.1.2)
(a) Vol. **90**, No. 1, p. 401–407 (1985).

M. K. Agarwal, P. D. Patel, L. T. Talele, and D. Laxminarayana
*Optical Band Gaps of Molybdenum Sulphoselenide (MoS_xSe_{2-x}, $0 \leqq x \leqq 2$) Single Crystals from Spectral Response
(14.3.3 and 16; 13.1; 22)
(a) Vol. **90**, No. 1, p. K107–K111 (1985).

G. I. Ibaev, S. I. Kulieva, and T. G. Magerramov
*Erratum to: Frequency Dependences of the Dynamic Current–Voltage Characteristic of the Se–CdSe Heterostructure Based on a Single-Crystal Selenium Film
(14.3.1 and 14.3.3; 22.1.3; 22.4.2)
(a) Vol. **90**, No. 1, p. K117 (1985).
Erratum
(a) Vol. **92**, No. 2, p. K175 (1985).

V. A. Budyanu, S. N. Chechuy, I. A. Damaskin, S. A. Fedoseev, A. A. Nasakin, S. L. Pyshkin, M. I. Valkovskaya. and V. P. Zenchenko
*Investigation of $A^{III}B^{V}$–Si Heterojunctions Grown by Laser Deposition
(14.3.1 and 14.3.3; 13.3; 22.1.2; 22.2.1; 22.2.2)
(a) Vol. **91**, No. 2, p. 737–744 (1985).

P. Seidel, H.-J. Köhler, B. Mikolajczak, F. Müller, and H.-G. Meyer
*Explantation of the AC Josephson Effect in Nb–NbO_x–Pb Tunnel Junctions with the Stewart-McCumber Model
(14.2 and 14.3.3; 21; 22.6)
(a) Vol. **91**, No. 2, p. K165–K168 (1985).

A. Ramam, R. Gulati, and B. L. Sharma
*Variable Pinch-Off GaAs MESFET
(14.3.3; 22.2.1)
(a) Vol. **91**, No. 2, p. K169–K172 (1985).

V. S. Lysenko, A. N. Nazarov, and T. E. Rudenko
*Simple Method for the Determination of the Doping Profile in MIS Structures with Implanted Shallow p–n Junction
(11 and 14.3.3; 22.1.2)
(a) Vol. **91**, No. 2, p. K173–K178 (1985).

V. Brückner and F. Kerstan
*Fast Responce Time Measurements in Transistors Using Picosecond Optoelectronic Switches
(14.3.3; 22.2.1)
(a) Vol. **91**, No. 2, p. K179–K183 (1985).

F. Campabadal and X. Aymerich-Humet
On the Effects of Localized States and Tunnel Charge in Thin Insulating Films
(13.3 and 14.3.2; 14.3.3; 22.6)
(b) Vol. **132**, No. 1, p. 305–315 (1985).

D. Braunig, K. H. Yang, T. Y. Tan, and C. P. Schneider
*In Depth Generation Lifetime Profiling of Heat-Treated Czochralski Silicon
(14.3.3; 22.1.2)
(a) Vol. **92**, No. 1, p. 327–335 (1985).

T. Rang
*Erratum to: Simple Relationship between the Breakdown Voltage, Concentration, and Junction Depth for Diffused pn Junctions
(14.3.3; 22.1.2)
(a) Vol. **92**, No. 1, p. K88 (1985).

C. A. Londos
Room-Temperature Irradiation of p-Type Silicon
(13.4; 10.2; 14.3.3; 22.1.2)
(a) Vol. **92**, No. 2, p. 609–614 (1985).

W. A. Badaway and M. S. El-Dessouki
*A Study of the Stability of n-Si/SnO_2 Solar Cells under Normal Working Conditions
(14.3.3 and 16; 22.1.2)
(a) Vol. **92**, No. 2, p. K167–K170 (1985).

14.3.4 High-Field Phenomena, Space-Charge Effects, Inhomogeneitis, Injected Carriers (Electroluminescence See 20.3; Junctions See 14.3.3)

Review Article

YU. A. KATULEVSKI
*Injection of Degenerate Electron–Hole Plasma
(14.3.4; 13.1; 13.4; 14.3.3; 22)
(a) Vol. **88**, No. 1, p. 11–42 (1985).

Original Papers and Short Notes

E. HEINER
*A Dynamical Theory for a Disordered Hubbard Chain far from Equilibrium. I. Extended Master Equations
(13 and 14.3.4; 22)
(b) Vol. **127**, No. 1, p. 253–265 (1985).

H. KÖSTER, JR., O. V. KURNOSOVA, and I. N. YASSIEVICH
*Tunneling from Deep Levels of l–c Type in Electric Fields
(13.4 and 14.3.4; 22.2.1)
(b) Vol. **127**, No. 1, p. 339–350 (1985).

A. G. MIRONOV, V. A. MOROZOVA, and V. S. VAVILOV
Some Peculiarities of the Electrooptical Spectra of n-GaAs(Cr) at Liquid Helium Temperatures (Wannier Levels?)
(13.4 and 20.1; 14.3.4; 22.2.1)
(b) Vol. **127**, No. 1, p. 359–363 (1985).

A. I. AGAFONOV, A. F. PLOTNIKOV, and V. N. SELEZNEV
*Conductivity of Amorphous A_xB_{1-x} Semiconductors with Potential Fluctuations of the Mobility Edges
(13.4 and 14.3.4; 22)
(b) Vol. **127**, No. 1, p. 395–401 (1985).

A. BHATTACHARYA, A. GHOSAL, and D. CHATTOPADHYAY
*Small-Signal AC Response of Two-Dimensional Hot Electrons in Quantum-Well Structures of Polar Semiconductors
(13 and 14.3.4; 22.2.1)
(b) Vol. **127**, No. 1, p. 427–432 (1985).

V. K. ARORA
*Intracollisional Field Effect in Quasi-Hot Electron Transport
(14.3.4; 6; 22.1.1; 22.2.1)
(b) Vol. **127**, No. 1, p. K95–K97 (1985).

A. J. MUHAMMAD, D. V. MORGAN, and A. E. GUILE
*Contact Overlap in MIM Sturctures with Distributed Traps
(14.3.3 and 14.3.4; 13.4)
(a) Vol. **87**, No. 1, p. 355–362 (1985).

I. ZÓLOMY
Metal–(Tunnel) Insulator–Semiconductor Switch (MISS) with Negative Resistances both in Forward and Reverse Directions
(14.3.3; 14.3.4; 22.1.2)
(a) Vol. **87**, No. 2, p. K213—K216 (1985).

A. I. KASIYAN and P. I. RUSSU
The Effect of Collective Electron–Hole Plasma Modes on Charge Carrier Mobility in Polar Semiconductors
(6 and 13.5.2; 14.3.4; 22; 22.2.1)
(b) Vol. **128**, No. 1, p. 243—249 (1985).

N. A. NIKOLOV, P. N. MALINOV, K. G. KOSTOV, and G. V. MIHAILOVA
*Influence of Collisions on the Resonances in Bounded Semiconductor Plasmas
(13 and 14.3.4)
(b) Vol. **128**, No. 1, p. 323—328 (1985).

J. A. GIACOMETTI, G. F. LEAL FERREIRA, and B. GROSS
*Negative Charge Transport in Fluorethylenepropylene by the Constant Current Method
(13.4 and 14.3.4; 22.9)
(a) Vol. **88**, No. 1, p. 297—307 (1985).

I. L. BAGINSKII and E. G. KOSTSOV
*Theory of Transient Currents in Dielectrics at a Limited Level of Single Injection. I. The Case of Strong Injection
(14.3.4; 14.4; 22)
(a) Vol. **88**, No. 1, p. 331—341 (1985).

A. MERTENS, H. KLOSE, and NG. TANG
*Ion Beam Contacting of Silicon. II. Antimony in n-Silicon
(11 and 14.3.4; 22.1.2)
(a) Vol. **88**, No. 1, p. 369—374 (1985).

T. LALINSKÝ, P. KORDOS, B. KOVÁCS, I. MOJZES, B. SZENTPÁLI, and A. TICHY-RÁCZ
*Optimization of the Heat Treatment for Forming AuGe Based Contacts to n-GaAs
(14.3.4; 22.2.1)
(a) Vol. **88**, No. 1, p. K87—K90 (1985).

U. BRAATZ, F. FINK, and P. KLINGER
*Optoelectronic Switches on the Basis of Mercuric Iodide (Red Phase, α-HgI_2)
(14.3.4; 16; 20.2; 22.5)
(a) Vol. **88**, No. 1, p. K95—K97 (1985).

YU. K. POZHELA, E. V. STARIKOV, and P. N. SHIKTOROV
*Far Infrared Absorption by Hot Holes in p-Ge under $\boldsymbol{E} \perp \boldsymbol{B}$ Fields
(14.3.4 and 20.1; 22.1.1)
(b) Vol. **128**, No. 2, p. 653—661 (1985).

O. ZIEP and R. KEIPER
*On the Theory of the Electron–Phonom Scattering Rate Including the Intra-Collosional Field Effect
(13 and 14.3.4; 22)
(b) Vol. **128**, No. 2, p. 779—787 (1985).

U. K. Reddy and P. J. Reddy
I–U Characteristics of Non-Crystalline As–Se–Cd Thin Films
(14.3.1; 2; 14.3.4; 22.8)
(a) Vol. **89**, No. 2, p. 679—682 (1985).

D. A. Aronov, P. I. Knigin, and B. R. Mamatkulov
*The Enhancement of Exclusion Effect in Compensated Semiconductors under the Action of Impurity Illumation Generating Miniority Current Carriers
(14.3.3 and 14.3.4; 16; 22)
(a) Vol. **89**, No. 2, p. 683—691 (1985).

I. Thurzo and F. Dubecký
*On the Role of the Back Contact in DLTS Experiments with Schottky Diodes
(14.3.3 and 14.3.4; 22.2.1)
(a) Vol. **89**, No. 2, p. 693—698 (1985).

B. Olejníková
*A Method of Determination of the Trap Depth of Deep Centres by Capacitance Measurements
(13.4 and 14.3.4; 22.1.2)
(a) Vol. **89**, No. 2, p. K197—K200 (1985).

J. Koshy
*Measurements of the Rectifying Barrier Height of Sputter Deposited Bi_2Te_3 Contacts on p-Silicon
(14.3.4; 22.1.2)
(a) Vol. **89**, No. 2, p. K219—K222 (1985).

V. I. Sokolov, T. P. Surkova, M. P. Kulakov, and A. V. Fadeev
New Experimental Evidence Concerning the Nature of L, M, and N Lines in ZnSe:Co
(13.4 and 13.5.1; 14.3.4; 20.1; 22.4.2)
(b) Vol. **130**, No. 2, p. 267—272 (1985).

B. Ya. Shapiro
*Induction of the Superconductivity by the Inner Contact Potential
(14.2 and 14.3.4; 13.3)
(b) Vol. **130**, No. 1, p. 359—368 (1985).

A. J. Muhammad, D. V. Morgan, and A. E. Guile
*Electronic Conduction in $Nb/Nb_2O_5/In$ Structures and the Effect of Space-Charge Overlap
(14.3.2 and 14.3.4; 22.6)
(a) Vol. **90**, No. 1, p. 371—381 (1985).

Y. Hoshino and T. Nishida
Transient Photocurrent Wave Form Deformation in Amorphous Silicon by Successive Light Pulse Irradiations
(2 and 16; 14.3.4; 22.1.2)
(a) Vol. **90**, No. 1, p. K101—K102 (1985).

N'guyen Duc Tien
*Characteristics with Negative Differential Resistance of a Device Using $Si–SiO_2$ Structure
(14.3.2 and 14.3.4; 22.1.2)
(a) Vol. **90**, No. 1, p. K103—K105 (1985).

O. ZIEP, M. MOCKER, and R. KEIPER
*New Aspects in the Theory of Impact Ionization
(13 and 14.3.4; 22.2.1)
(b) Vol. **131**, No. 2, p. 775–784 (1985).

K. MAASS, K. IRMSCHER, and H. KLOSE
On the Determination of Deep Level Concentration Profiles by DLTS Measurements
(13.4; 11; 14.3.4; 22.1.2)
(a) Vol. **91**, No. 2, p. 667–675 (1985).

V. M. EFIMOV, E. E. MEERSON, and A. A. EVTUKH
*Study of Tunnel Currents of Electrons and Holes in Thermal SiO_2 with Charge Accumulation in the Dielectric
(14.3.4; 13.1; 22.6)
(a) Vol. **91**, No. 2, p. 693–703 (1985).

I. L. BAGINSKII and E. G. KOSTSOV
*Transient Currents in Dielectrics at Non-Linear Law of Single Injection
(14.3.4 and 14.4; 22)
(a) Vol. **91**, No. 2, p. 705–714 (1985).

A. N. ANAGNOSTOPOULOS, C. MANOLIKAS, and A. KAROUTIS
*Electrical Effects Associated with the Ordering Process in $CdInGaS_4$ Crystals. II. Electron Traps Determined with Conductivity Measurements
(13.4 and 14.3.4; 22.8)
(a) Vol. **92**, No. 1, p. 231–235 (1985).

H. VON BERLEPSCH
*Field-Dependent Injection in HDPE
(14.3.4; 13.3; 22.9)
(a) Vol. **92**, No. 2, p. 285–292 (1985).

I. P. MIKHAILOVSKII and A. E. EPOV
*A Model of Field (Tunnel) Generation of Positive Charge in Thermal Silicon Dioxide of Si–SiO_2 Structures
(14.3.4; 22.1.2)
(a) Vol. **92**, No. 2, p. 615–622 (1985).

14.4 Dielectrics

Original Papers and Short Notes

E. A. Secco
*Electrical Conductivity Measurements to Test for Rotating Sulfate Ions in Fast Ion Conductors
(10.2 and 14.4; 22.8.1)
(a) Vol. **88**, No. 1, p. K75–K77 (1985).

L. Latie, G. Villeneuve, Ch. Cros, and P. Hagenmuller
Influence of Cationic Vacancies on the Mobility of Li^+ Ions in Some Cation-Deficient Materials. An NMR Study
(10.2; 9; 14.4; 19; 22.5.4)
(b) Vol. **128**, No. 2, p. 475–482 (1985).

Kaiyi Zhang and K. A. Chao
*Nonlinear Thomas-Fermi Dielectric Function in Semiconductors
(14.4; 22)
(b) Vol. **128**, No. 2, p. 723–729 (1985).

V. L. Aksenov and M. Bobeth
Structural Glass in Solid Solutions with Competing Interactions
(1.2; 14.4; 22.8.1)
(b) Vol. **128**, No. 2, p. K105–K109 (1985).

I. L. Baginskii, and E. G. Kostsov
Theory of Transient Currents in Dielectrics at a Limited Level of Single Injection. II. The Case of Weak Injection
(14.3.4; 14.4; 22)
(a) Vol. **88**, No. 2, p. 637–646 (1985).

A. Subrahmanyam, K. R. Murali, B. S. V. Gopalam, and J. Sobhanadri
*Dielectric Properties of Zinc Phosphide
(14.4; 14.3; 22.3)
(a) Vol. **88**, No. 2, p. 681–686 (1985).

V. N. Chebotin and V. A. Mezrin
Ordering of Defects, Thermodynamic and Transport Properties of Solid Oxide Electrolytes with Fluorite Structure
(10.2; 14.4; 22.6.1)
(a) Vol. **89**, No. 1, p. 199–211 (1985).

S. C. Sabharwal and B. Ghosh
*Electrical Conductivity of $NH_4H_2PO_4$ Single Crystal
(11 and 14.4; 22.8.1)
(a) Vol. **89**, No. 1, p. K83–K87 (1985).

A. Fontana, G. Mariotto, and F. Rocca
*Low-Frequency Light Scattering in Superionic Glasses $(AgI)_x(Ag_201B_2O_3)_{1-x}$
(2 and 14.4; 20.1; 22.8)
(b) Vol. **129**, No. 2, p. 489–496 (1985).

L. Wendler
Electron–Phonon Interaction in Dielectric Bilayer Systems. Effect of the Electronic Polarizability
(6 and 13; 13.5.3; 14.4; 22.2.1; 22.2.2)
(b) Vol. **129**, No. 2, p. 513–530 (1985).

D. BINGHAM and C. H. BURTON
Fast-Ion Conduction in the Presence of a Lattice Phase Transition
(1.2; 6; 14.4)
(b) Vol. **130**, No. 2, p. 483–487 (1985).

V. V. KRASNOGOLOVETS and P. M. TOMCHUK
*The Coulomb Correlations and the Electric Field Local Heterogeneities on the Proton Conductivity of Hydrogen Bonded Chains
(14.4; 6; 22)
(b) Vol. **130**, No. 2, p. 807–817 (1985).

A. TORRES, J. JIMENEZ, B. VEGA, and J. A. DE SAJA
*On the Influence of Heating Rate in the Dipolar Relaxation of $Ce_2(SO_4)_3 \cdot 9\,H_2O$ Single Crystals
(14.4; 22.8.2)
(a) Vol. **90**, No. 2, p. 749–754 (1985).

S. M. BABULANAM
*AC and DC Superimposed AC Conduction in Sodium Fluoride Thin Films
(14.4; 22.5.2)
(a) Vol. **90**, No. 2, p. 755–762 (1985).

F. BECHSTEDT and R. ENDERLEIN
Dielectric Screening, Polar Phonons, and Longitudinal Electronic Excitations of Quantum Well Double Heterostructures. Application to Light Scattering from Charge Density Fluctuations
(6 and 13.3; 14.4; 20; 22.2.4)
(b) Vol. **131**, No. 1, p. 53–66 (1985).

S. IHARA and K. SUZUKI
*Molecular Dynamical Study of β-$AgI_{1-x}Cl_x$
(10.2 and 14.4; 1.2; 22.5.4)
(b) Vol. **131**, No. 1, p. 97–103 (1985).

J. M. KELLER and S. C. DATT
*Thermally Stimulated Depolarization Currents in Polypropylene Foils
(14.4; 22.9)
(a) Vol. **91**, No. 1, p. 205–210 (1985).

T. YAMAMOTO, S. KIKKAWA, and M. KOIZUMI
*The Effect of Dispersed PLZT Particles on the Electrical Conductivity of Lithium Iodide
(14.4; 22.5)
(a) Vol. **91**, No. 1, p. K67–K70 (1985).

N. DUTT, G. G. AGRAWAL, and J. SHANKER
Volume Dependence of Dielectric Properties of Some Fluorite-Structure Crystals
(4 and 6; 14.4; 22.5.3; 22.8.1)
(b) Vol. **131**, No. 2, p. 505–510 (1985).

V. V. KRASNOGOLOVETS and P. M. TOMCHUK
*The Anharmonicity Influence on the Proton Conductivity of an Hydrogen-Bonded Chain
(14.4; 6; 22.8.1)
(b) Vol. **131**, No. 2, p. K177–K179 (1985).

14.4.1 Ferroelectrics

Original Papers and Short Notes

N. KRISTOFFEL
*On the Possibility of Anomalous Bulk Photovoltaic Effect and of Induced Absorption in the Nominal Transparency Region of a Noncentrosymmetric Crystal in a Magnetic Field
(14.4.1 and 16)
(b) Vol. **127**, No. 1, p. 413—418 (1985).

G. CH. CHRISTAKUDIS, S. K. PLACHKOVA, E. S. AVILOV, L. E. SHELIMOVA, and N. KH. ABRIKOSOV
Transport Phenomena of Ferroelectric Solid Solutions
$(GeTe)_{1-x}((Ag_2Te)_{1-y}(Sb_2Te_3)_y)_x$ with $y = 0.60$, $0 \leqq x \leqq 1$ at 300 K
(14.3; 8; 14.4.1; 15; 22.7.1)
(b) Vol. **127**, No. 2, p. 419—426 (1985).

B. KOSTUREK
*Depolarization Field Effect on Domain Wall Motion in TGS Crystals
(14.4.1; 22.9)
(a) Vol. **87**, No. 1, p. K93—K95 (1985).

G. V. KOZLOV, E. B. KRYUKOVA, B. WYNCKE, F. BREHAT, and M. E. SHERIF
Low-Frequency Lattice Dynamics of $NaNO_2$ Crystals
(6 and 20.1; 14.4.1; 22.8.1)
(b) Vol. **127**, No. 2, p. 465—472 (1985).

R. BLINC, P. CEVC, and N. I. SILKIN
*EPR Study of the Pseudo-One-Dimensional Ferroelectric Phase Transition in Tl^{2+} Doped CsH_2PO_4
(14.4.1 and 19; 22.8.1)
(b) Vol. **127**, No. 2, p. K147—K149 (1985).

H. CERVA and W. GRAEFF
Contrast Investigations of Surface Acoustic Waves by Stroboscopic Topography. II. Wavefield Deviation Contrast
(7; 14.4.1; 22.8.1)
(a) Vol. **87**, No. 2, p. 507—516 (1985).

R. STYRKOWIEC
*The Interaction between Moving Domain Walls in Rochelle Salt Crystals
(14.4.1; 22.8.1)
(a) Vol. **87**, No. 2, p. K135—K138 (1985).
Erratum
(a) Vol. **89**, No. 1, p. K117 (1985).

F. BREHAT and B. WYNCKE
Soft Mode Spectroscopy in Potassium Dihydrogen Phosphate
(6 and 20.1; 14.4.1; 22.8.1)
(b) Vol. **128**, No. 1, p. 83—92 (1985).

M. K. Kerimov
*Electron Paramagnetic Resonance of Stabilized Electrons in PTFE Coronoelectrets
(14.4.1 and 19; 22.9)
(a) Vol. **90**, No. 1, p. K71–K74 (1985).

J. M. Wesselinowa
*Critical Behaviour of the Ising Model with a Transverse Field
(14.4.1; 18.2)
(b) Vol. **130**, No. 2, p. 689–697 (1985).

I. H. Ismailzade and O. A. Samedov
*Influence of Permanent Electric Fields on the Successive Phase Transitions in the Solid Solutions of the System $(Pb_{1-x}Ba_x)ZrO_3$.
(1.2 and 14.4.1; 22.8.1)
(a) Vol. **90**, No. 2, p. 445–448 (1985).

S. Mahmoud, E. M. Kassem, H. Müller, and E. Hegenbarth
*Thermal Conductivity of $LiKSO_4$ at Low Temperatures
(8 and 14.4.1; 6; 22.8.1)
(a) Vol. **90**, No. 2, p. K135–K138 (1985).

E. Anachkova and I. Savatinova
Isotopic Shift and Mode Interaction in $K(D_xH_{1-x})_2PO_4$ Crystals
(6 and 20.1; 14.4.1; 22.8.1)
(b) Vol. **131**, No. 2, p. K101–K105 (1985).

M. Happ and K.-P. Möllmann
**I–U* Characteristics of Strongly Indium-Doped Lead Tin Telluride at Low Temperatures
(14.4.1 and 16; 22.8)
(a) Vol. **91**, No. 2, p. K159–K164 (1985).

A. H. Piekara
*A Theoretical Model of Ferroelectrics Exhibiting One, Two, or Three Curie Points
(14.4.1; 22.8.1)
(b) Vol. **132**, No. 1, p. 277–282 (1985).

V. V. Klubovich, N. S. Buinov, and S. R. Syrtsov
*The Influence of Impurities on the Phase Transition in Ferroelectrics of Restricted Dimensions
(6 and 14.4.1)
(b) Vol. **132**, No. 1, p. K19–K22 (1985).

15. Thermoelectric and Thermomagnetic Properties

S. A. MAZEN, A. A. GHANI, and A. H. ASHOUR
*Charge Transport in Cu–Cd Ferrospinels
(14.3 and 15; 22.8.2)
(a) Vol. **88**, No. 1, p. 343–346 (1985).

A. BATH, J. HUGEL, and R. KLEIM
*Thermoelectric Power of Liquid Germanium and Tin
(2 and 15; 14.1; 22.1; 22.1.1)
(b) Vol. **128**, No. 2, p. 761–767 (1985).

I. A. SMIRNOV, A. V. GOLUBKOV, E. V. GONCHAROVA, V. M. SERGEEVA, V. V. POPOV, M. V. ROMANOVA, B. M. BUTTAEV, and V. S. OSKHOTSKII
Electrical and Galvanomagnetic Properties of $Tm_{1-x}La_xS$
(14.3; 15; 22; 22.8)
(a) Vol. **88**, No. 2, p. K201–K203 (1985).

J. HORÁK, P. LOŠŤÁK, and M. MATYÁŠ
Point Defects in Sn-Doped Sb_2Te_3 Single Crystals
(13.4 and 14.3; 15; 20.1; 22.7)
(b) Vol. **129**, No. 1, p. 381–386 (1985).

I. MERTIG, E. MROSAN, R. ZELLER, P. H. DEDERICHS, and P. ZIESCHE
*Calculation of the Thermoelectric Power of Ga, Ni, Rh, and Pd Impurities in Copper
(13.4 and 15; 14.1; 21.1)
(b) Vol. **129**, No. 1, p. 407–414 (1985).

H. ISHIKAWA, M. AOYAMA, O. UEMURA, and T. SATOW
Electronic Properties of the Liquid Tl–$GeTl_2Te_4$ and Tl–$SnTl_2Te_4$ Systems
(2 and 14.3; 15; 18; 22.7.1)
(b) Vol. **129**, No. 1, p. 427–434 (1985).

V. NARASIMHAN, H. V. KEER, and D. K. CHAKRABARTY
Structural and Electrical Properties of the $LaCo_{1-x}Mn_xO_3$ and $LaCo_{1-x}Fe_xO_3$ Systems
(1.2. and 14.3; 15; 22.8.2)
(a) Vol. **89**, No. 1, p. 65–71 (1985).

L. G. KHVOSTANTSEV, A. I. ORLOV, N. KH. ABRIKOSOV, and L. D. IVANOVA
*Kinetic Properties and Phase Transitions in Sb_2Te_3 under Hydrostatic Pressure up to 9 GPa
(1.2 and 15; 14.3; 22.7)
(a) Vol. **89**, No. 1, p. 301–309 (1985).

B. POUMELLEC, J. F. MARUCCO, and F. LAGNEL
Electron Transport in TiO_{2-x}at Intermediate Temperatures $300\ K < T < 1500\ K$
(13.4 and 14.3; 15; 22.6)
(a) Vol. **89**, No. 1, p. 375–382 (1985).

M. P. BOIKO, N. A. REDKO, P. P. BODIUL, and N. A. RODIONOV
*Phonon Drag of L-Holes in Ultraquantum Magnetic Fields in $Bi_{0.88}Sb_{0.12}$ Alloy
(6 and 15; 22.7.1)
(b) Vol. **130**, No. 1, p. K85–K88 (1985).

A. Casanovas, A. Cantarero, A. Segura, and A. Chevy
*Thermoelectric Power in n-InSe
(15; 22.3)
(a) Vol. **92**, No. 2, p. K155--K159 (1985).

16. Photoconductivity. Photovoltaic Effects

Review Articles

Original Papers and Short Notes

Y. Fournier and Nguyen Van Dong
Properties of Sputtered Amorphous Silicon Films Prepared under a Reduced Pumping Speed. The Effect of Hydrogen
(14.3.1; 2; 16; 20.1; 22.1.2)
(a) Vol. **88**, No. 1, p. 309—314 (1985).

S. Fung and R. J. Nicholas
*Double Source Photoconductivity of InP:Fe
(16; 22.2.2)
(a) Vol. **88**, No. 1, p. 347—353 (1985).

G. S. Korotchenkov, A. A. Gutkin, and A. M. Marinova
The Characteristics of Thin-Film Gallium Phosphide MOS Structures with a Tunnel Transparent Native Anodic Oxide
(14.3.3; 16; 22.2.2)
(a) Vol. **88**, No. 1, p. 355—363 (1985).

O. Gilar and I. Petr
*Combination of a Silicon Photodiode and a Scintillator as a Dose Rate Detector
(14.3.3 and 16; 22.1.2)
(a) Vol. **88**, No. 1, p. 375—380 (1985).

S. Brehme and R. Pickenhain
*A Simple Method for the Determination of Both Photoionization and Photoneutralization Cross Sections of Deep Levels by Optical DLTS
(13.4 and 16; 22.2.4)
(a) Vol. **88**, No. 1, p. K63—K66 (1985).
Erratum
(a) Vol. **90**, No. 1, p. K119 (1985).

S. Shigetomi, T. Ikari, Y. Koga, and S. Shigetomi
Drift Mobility of Electrons Parallel to the *c*-Axis in InSe
(14.3; 13.4; 16; 22.3)
(a) Vol. **88**, No. 1, p. K71—K73 (1985).

P. I. Stoianova and R. E. Andreichin
*Coordinate Sensitive Photoelement on the Basis of the Photoelectret State in Vitreous Semiconductors
(14.3.3 and 16; 2; 22)
(a) Vol. **88**, No. 1, p. K91—K99 (1985).

U. Braatz, F. Fink, and P. Klinger
Optoelectric Switches on the Basis of Mercuric Iodide (Red Phase, γ-HgI_2)
(14.3.4; 16; 20.2; 22.5)
(a) Vol. **88**, No. 1, p. K95—K97 (1985).

R. Gulati, S. B. Kaushik, P. L. Trivedi, H. S. Sharma, I. Chandra, and B. L. Sharma
Optical Effect in Normally-Off GaAs MESFET
(14.3.3; 16; 22.2.1)
(a) Vol. **88**, No. 1, p. K99—K103 (1985).

S. KUŹMIŃSI and A. T. SZAYNOK
Shallow Energy States in CdTe
(13.3 and 14.3.2; 16; 22.4.3)
(a) Vol. **89**, No. 2, p. 623–628 (1985).

W. BREMSER, K. H. HERRMANN, and F. RIETENBACH
*Minority Carrier Lifetimes in $PbS_{0.1}Se_{0.9}$ Determined from Diffusion Length by a Scanning Laser Microprobe
(14.3 and 16; 22.8)
(a) Vol. **89**, No. 2, p. 653–656 (1985).

R. OBERSCHMID
Conductivity Instabilities and Polarization Effects of $Bi_{12}(Ge, Si)\ O_{20}$ Single-Crystal Samples
(14.3.4; 16; 22.8)
(a) Vol. **89**, No. 2, p. 657–671 (1985).

K. MOCHIZUKI, H. ONODERA, and K. IGAKI
*Photoconductivity of CdS_xSe_{1-x} with Controlled Deviation from Stoichiometry
(16; 13.4; 20.3; 22.4.4)
(a) Vol. **89**, No. 2, p. 673–678 (1985).

D. A. ARONOV, P. I. KNIGIN, and B. R. MAMATKULOV
The Enhancement of Exclusion Effect in Compensated Semiconductors under the Action of Impurity Illumination Generating Minority Current Carriers
(14.3.3 and 14.3.4; 16; 22)
(a) Vol. **89**, No. 2, p. 683–691 (1985).

WANG HUAFU, SHI GUOTONG, and WU ZHONGKANG
*Photovoltaic Effect in $LiNbO_3$:Mg
(16; 14.4.1; 22.8.1)
(a) Vol. **89**, No. 2, p. K211–K213 (1985).

S. A. AZIMOV and T. M. RAZYKOV
Energy Band Model of Ultrathin Solar Cells with Thickness up to 5 μm
(13.1 and 14.3.3; 16; 22; 22.4.4)
(a) Vol. **89**, No. 2, p. K223–K226 (1985).

A. M. ANDRIESH, V. V. BIVOL, V. I. VERLAN, B. V. EFREMUSHKIN, M. S. IOVU, and E. G. KHANCHEVSKAYA
*Influence of Electrode Material on the Electrical and Photoelectrical Properties of p-InP/As_2Se_3 Heterostructures
(14.3.3 and 16; 2; 14.3.1; 22; 22.2.2)
(a) Vol. **89**, No. 2, p. K231–K234 (1985).

N. A. AVDEEV, YU. E. GARDIN, V. A. GURTOV, and S. N. KUZNETSOV
Study of Charge Accumulation Kinetics in MOS Structures with Intrinsic Anodic Oxide on InSb by Measuring Transient Photocurrents
(13.3 and 14.3.3; 16; 22.2.3)
(a) Vol. **90**, No. 1, p. 393–400 (1985).

S. A. MIRSAGATOV and K. SABIROV
*Investigation of Illuminance–Voltage Characteristics for Thin-Film p-CdTe–n-CdTe–n-CdS Structures
(14.3.3 and 16; 14.3.1; 22.4.1; 22.4.3)
(a) Vol. **90**, No. 2, p. K219–K223 (1985).

A. I. LEBEDEV and KH. A. ABDULLIN
*Impurity Photoconductivity and Electrical Properties of $Pb_{1-x-y}Ge_xSn_yTe$ Doped with Indium
(14.3 and 16; 13.4; 22)
(a) Vol. **91**, No. 1, p. 225–234 (1985).

S. SCHRADER and M. KRYSZEWSKI
Multiple Trapping Approach to the TSC Drift Experiment
(13.4 and 14.3.4; 16; 22.9)
(a) Vol. **91**, No. 1, p. 243–255 (1985).

K.-P. MÖLLMANN and M. HAPP
*Impurity Photoconductivity of Indium Doped Lead Tin Telluride
(16; 22)
(a) Vol. **91**, No. 1, p. K71–K74 (1985).

W. ULRICI
Photoinduced Optical Absorption of Cr^{4+} in Semi-Insulating GaAs:Cr
(13.4; 16; 19; 20.1; 22.2.1)
(b) Vol. **131**, No. 2, p. 707–717 (1985).

K. JÓŹWIKOWSKI, Z. ORMAN, and A. ROGALSKI
*On the Performance or Non-Cooled (In, As)Sb Photoelectromagnetic Detectors for 10.6 μm Radiation
(16; 22.2.4)
(a) Vol. **91**, No. 2, p. 745–751 (1985).

K. H. HERMANN and K.-P. MÖLLMANN
*Long-Wavelength Intrinsic Photoconductivity in PbSnTe:In
(16; 22.8)
(a) Vol. **91**, No. 2, p. K147–K150 (1985).

M. HAPP and K.-P. MÖLLMANN
*$I - U$ Characteristics of Strongly Indium-Doped Lead Tin Telluride at Low Temperatures
(14.4.1 and 16; 22.8)
(a) Vol. **91**, No. 2, p. K159–K164 (1985).

W. KNAP, R. STĘPNIEWSKI, and E. FANTNER
Optically Induced Nernst-Ettinghausen Effect in the Far Infrared and Strong Magnetic Fields in HgTe and InSb
(13.1; 16; 22.2.3; 22.4.3)
(b) Vol. **132**, No. 1, p. 133–140 (1985).

YE MING, E. KRÄTZIG, and R. ORLOWSKI
Photorefractive Effects in $LiNbO_3$:Cr Induced by Two-Step Excitation
(20.1; 13.4; 16; 22.8.1)
(a) Vol. **92**, No. 1, p. 221–229 (1985).

17. Emission of Electrons and Ions

Original Papers and Short Notes

H.-J. Fitting, K. König, and D. Hecht
*Exoelectron Emission Due to Alkali Ions on Insulating Layers
(17; 1.5; 21.2; 22.1.2; 22.6)
(a) Vol. **87**, No. 1, p. K77–K79 (1985).

F. Bell
On the Measurement of Solid State Compton Profiles by Inelastic Fast-Ion–Atom Collisions under Channeling Conditions. A Case Study
(11; 13.1; 17)
(b) Vol. **127**, No. 2, p. 599–603 (1985).

V. Drchal and J. Kudrnovský
The Theory of Auger Spectra from Alloys. The Effect of Electron Correlations in Partially Filled Narrow Bands
(13; 17; 21.1)
(b) Vol. **127**, No. 2, p. 611–619 (1985).

H. Düsterhöft, R. Daugs, and R. Pippig
*A New Method for the Interpretation of Energy Distributions Measured on Secondary Ions
(17; 1.6; 11; 21.1; 21.1.1; 22.1.2)
(a) Vol. **87**, No. 2, p. 499–506 (1985).

V. S. Kortov, T. S. Bessonova, M. S. Akselrod, and I. I. Milman
*Hole-Induced Exoelectron Emission and Luminescence of Corundum Doped with Mg
(17 and 20.3; 10.2; 13.4; 22.6)
(a) Vol. **87**, No. 2, p. 629–639 (1985).

H.-J. Fitting and J. Reinhardt
*Monte-Carlo Simulation of keV-Electron Scattering in Solid Targets
(17; 11; 22.1.2)
(a) Vol. **88**, No. 1, p. 245–259 (1985).

A. B. Gerasimov, M. M. Mer'kin, and A. A. Tsertsvadze
*Calculation of the EEE Parameters at Annealing of the F-Centers in NaCl Crystals
(17; 22.5.2)
(b) Vol. **128**, 2, p. 693–700 (1985).

H. Pagnia, J. Schnellbächer, and N. Sotnik
*Correlation between Current and Induced Emission Processes in MIM Diodes
(17 and 20.3; 14.3.3)
(a) Vol. **88**, No. 2, p. K209–K212 (1985).

Yu. M. Shulga, G. L. Gutsev, and V. I. Rubtsov
A Comparison of the C KVV Auger Spectra of NbC and $NbN_{0.7}C_{0.3}$
(13.1; 17; 22)
(b) Vol. **129**, No. 2, p. 683–686 (1985).

I. N. Shabanova
Chemical Bond Peculiarities of Elements in Amorphous Alloy Surface Layers on the Basis of Fe
(1.6 and 2; 17; 21.1; 21.1.1)
(a) Vol. **91**, No. 1, p. K13—K17 (1985).

W. Frentrup, M. Griepentrog, and U. Müller-Jahreis
*Negative Secondary Ion Emission Influenced by Alkali Atoms
(17; 1.6; 22.1.1; 22.1.2; 22.2.1; 22.2.2)
(a) Vol. **91**, No. 2, p. 447—452 (1985).

T. Chassé, U. Berg, and O. Brümmer
*Photoelectron Diffraction and Band Structure Effects in ARXPS from the Valence Bands of GeS
(13.1 and 17; 22)
(b) Vol. **132**, No. 1, p. 141—144 (1985).

17.1 Field Emission Microscope Investigations

Short Note

18. Magnetic Properties

Review Articles

H. W. Streitwolf
Physical Properties of Polyacetylene
(13.5 and 13.5.3; 6; 18; 19; 20.1; 22.9)
(b) Vol. **127**, No. 1, p. 11–54 (1985).

E. P. Wohlfarth
*Magnetic Materials for Recording
(18; 18.3; 21.1; 21.1.1; 22.6; 22.8.2)
(a) Vol. **91**, No. 2, p. 339–350 (1985).

Original Papers and Short Notes

G. Hein
Analysis of Shubnikov-De Haas Oscillations in HgSe
(13.1; 18; 22.4.2)
(b) Vol. **127**, No. 1, p. 227–236 (1985).

E. A. Goremychkin, E. Mühle, B. Lippold, O. D. Chistyakov, and E. M. Savitskii
Crystal Field Study of $HoNi_5$ by Neutron Scattering
(13.4; 8; 18; 21.4)
(b) Vol. **127**, No. 1, p. 371–375 (1985).
Erratum
(b) Vol. **128**, No. 2, p. 807 (1985).

Z. Onyszkiewicz
*Theory of Heat Capacity of the Mn_3GaC Metamagnet
(8 and 18; 22.8)
(b) Vol. **127**, No. 1, p. K37–K40 (1985).

G. D. Khattak, A. Twardowski, and R. R. Gałązka
*Magnetization of $Cd_{1-x}Mn_xTe$
(18; 22.4.4)
(a) Vol. **87**, No. 1, p. K57–K60 (1985).

G. F. Kventsel and J. Katriel
*Two-State Mean-Field Theory of Reentrant Phase Transitions
(2 and 18)
(b) Vol. **128**, No. 1, p. 169–175 (1985).

K. Bortel
*Magnetic Properties of the Transition Metals in a Superconductor
(14.2 and 18; 21.1)
(b) Vol. **128**, No. 1, p. 177–185 (1985).

V. Z. Polinger, L. F. Chibotaru, and I. B. Bersuker
*The Effect of Spin–Vibrational Interaction in Magnetic Properties of Exchange Tetraclusters
(13.4 and 18; 22.8)
(b) Vol. **129**, No. 2, p. 615–624 (1985).

U. Neitzel and K. Bärner
*Optical and Magnetic Investigations of Some NiAs-Type Transition Metal Compounds
(18 and 20.1; 13.1; 21.1)
(b) Vol. **129**, No. 2, p. 707–715 (1985).

M. Mondal and K. P. Ghatak
Influence of Magnetic Quantization on the Effective Electron Mass in Ternary Semiconductors
(13.1; 18; 22.4.4)
(b) Vol. **129**, No. 2, p. 745–753 (1985).

C. Yu. Shunyaev, M. T. Varshavskii et. A. N. Men
*Influence de l'ordre à courte distance sur le moment magnétique des solutions solides quasibinaires à la structure du grenat avec la substitution dans un sous-réseau
(Influence of Short-Range Order on the Magnetic Moment of Quasibinary Granat-Type Solid Solutions with Sublattice Substitution)
(1.4 and 18; 22.8.2)
(b) Vol. **129**, No. 2, p. K131–K134 (1985).

M. Piotrowski and A. Murasik
*Exchange and Dipolar Interactions in TbF_3
(18; 1.2; 22.5)
(a) Vol. **89**, No. 2, p. 571–580 (1985).

H. Matsuda, O. Uemura, and T. Satow
The X-Ray Diffraction Study of Molton Tl_2X and $AgTl_3X_2$ (X: Se and Te)
(2; 18; 22.3; 22.8)
(a) Vol. **90**, No. 1, p. 77–82 (1985).

W. Nolting and U. Dubil
*$T = 0$ Magnetic Polaron in f-Systems with Antiferromagnetic s–f Exchange
(13 and 18)
(b) Vol. **130**, No. 2, p. 561–573 (1985).

B. R. Sood
*Magnetoconductance in Metallic Granular Gallium Films
(14.1 and 18; 2; 21)
(b) Vol. **130**, No. 2, p. 801–806 (1985).

F. W. O. da Silva and A. S. T. Pires
*Short-Range Order in the Quasi-One-Dimensional Classical Heisenberg Model
(18)
(b) Vol. **130**, No. 2, p. K117–K120 (1985).

V. N. Vivcharuk, Yu. V. Gudyma, V. N. Ermakov, and V. V. Nitsovich
Exciton Mechanism of Exchange Interaction in Semiconductors with Magnetic Impurities
(13.4 and 13.5.1; 18; 22.4.3)
(b) Vol. **130**, No. 2, p. K131–K134 (1985).

A. Gilabert, F. de Schutter, and Y. Bruynseraede
*Magnetic Properties of Superconducting Oxygen-Doped Vanadium Foils
(14.2 and 18)
(a) Vol. **90**, No. 2, p. 651–656 (1985).

M. Krause, C. Michalk, M. Benedix, and J. Suwalski
Low-Temperature Mössbauer Measurement of a Polymeric Oxalate (2,2′-Bipyridine) Iron Compound
(5; 18; 22.9)
(a) Vol. **90**, No. 2, p. 657–660 (1985).

N. Suárez, J. L. Sánchez, and S. García
*Cd_2–W: a New Material for Permanent Magnets?
(5 and 18; 22.8.2)
(a) Vol. **90**, No. 2, p. K171–K174 (1985).

M. Damnjanović and M. Vujičić
*Reality Condition on Order Parameter in Magnetic Phase Transitions
(18)
(b) Vol. **131**, No. 1, p. 113–115 (1985).

J. T. Lin and Z. Y. Li
*Microscopic Model for the Relaxation and Excitation of Localized Spin Waves
(13.5.4 and 18; 13.4)
(b) Vol. **131**, No. 1, p. 141–149 (1985).

D. Neogy and T. Purohit
The Behavior of Active Centers in a Laser Host. A. Crystal Field Investigation on Nd^{3+} and Er^{3+} in LaF_3 Single Crystals
(13.4; 8; 18; 19; 22.5)
(b) Vol. **131**, No. 1, p. 329–338 (1985).

Sanjeev Kumar, P. K. Ahluwalia, K. C. Sharma, and R. S. Tripathi
*Metallic Systems Containing Crystal Field Split Impurities. Free Energy in the Presence of Magnetic Field – a Contribution from the Loop Diagrams
(8 and 18; 21)
(b) Vol. **131**, No. 1, p. K49–K53 (1985).

H. Stramska
*Magneto-Optical Effects on the Generation of the Third Harmonic in Semiconductors
(18 and 20; 13; 22)
(b) Vol. **131**, No. 1, p. K77–K79 (1985).

Ch. Kuhrt, Th. Schittny, and K. Bärner
*Magnetic *B–T* Phase Diagramm of Anion Substituted MnAs. Magnetocaloric Experiments
(18; 8; 21.1)
(a) Vol. **91**, No. 1, p. 105–113 (1985).

Huei Li Huang and Ming Yen Lin
*Remanent Magnetization in Randomly Distributed Cubic Particles
(18; 21.1; 21.1.1)
(a) Vol. **91**, No. 1, p. 123–127 (1985).

F. Leccabue, J. L. Sánchez, L. Pareti, F. Bolzoni, and R. Panizzieri
*Curie Temperatures of Some Substituted Nd_2Fe_{14} B Tetragonal Alloys
(18; 21.1.1)
(a) Vol. **91**, No. 1, p. K63–K65 (1985).

Th. Nattermann
*The Random Field Ising Model in 2 + ε Dimensions. Critical Behaviour from Interface Fluctuations
(18; 1.6)
(b) Vol. **131**, No. 2, p. 563–571 (1985).

B. V. Karpenko, A. V. Kuznetsov, L. D. Falkovskaya, and G. L. Budrina
*On the Atomic Limit of the s–f Model
(13 and 18)
(b) Vol. **131**, No. 2, p. 585–593 (1985).

B. Mengel, H.-U. Müller, F. Ludwig, V. L. Kostantinov, and R. Herrmann
*Magneto-Optical Investigations in the Submillimeter Wavelength Range on p-InSb Bicrystals
(18 and 20.1; 13.4; 22.2.3)
(b) Vol. **131**, No. 2, p. K145–K147 (1985).

F. Ludwig, H.-U. Müller, B. Mengel, and R. Herrmann
*Magneto-Optical Investigations on n-InSb in the Far Infrared
(18 and 20.1; 13.4; 22.2.3)
(b) Vol. **131**, No. 2, p. K149–K152 (1985).

R. W. Chantrell and E. P. Wohlfarth
*Rate Dependence of the Field-Cooled Magnetisation of a Fine Particle System
(18; 22.8.1)
(a) Vol. **91**, No. 2, p. 619–626 (1985).

K. Bärner and E. Gmelin
*Coupling of a Structural and a High-Spin–Low-Spin Transiton in $MnAs_{1-x}P_x$ Compounds
(18; 8; 12.1; 21.1)
(b) Vol. **132**, No. 2, p. 431–436 (1985).

E. Kolley and W. Kolley
*Dynamic and Static Aspects in the Theory of Itinerant Spin Glasses
(18)
(b) Vol. **132**, No. 2, p. 437–448 (1985).

18.1 Paramagnetic Properties

Original Papers and Short Notes

J. Schreiber and D. Schulz
*Self-Consistent Dynamical Theory for Heisenberg Spin Glasses
(2 and 18.1; 8; 18.2; 18.4)
(b) Vol. **127**, No. 1, p. 139—147 (1985).

P. R. Silva
*The Site-Diluted Ising Model in the Simple Cubic Lattice
(18.1 and 18.2)
(b) Vol. **127**, No. 1, p. K41—K44 (1985).

H. Kido, T. Hoshiwaka, M. Shimada, and M. Koizumi
*Crystal Structure and Magnetic and Electrical Properties of Pseudo-Ternary $HoCu_{1-x}Zn_xSi$ and $NdCu_{0.6}Zn_{0.4}Si$ Compounds
(14.1 and 18.1; 1.3; 18.2; 22.8)
(a) Vol. **87**, No. 1, p. 273—277 (1985).

H. Kido, T. Hoshikawa, M. Shimada, and M. Koizumi
*Preparation and Magnetic Properties of $YMn_2(Si_{1-x}Ge_x)_2$ Compounds
(18.1 and 18.2; 1.3; 21.1; 21.4)
(a) Vol. **87**, No. 1, p. K61—K64 (1985).

K. Iwahana
Theory for the Magnetic Circular Polarization of the F-Center Luminesence
(13.4 and 20.3; 18.1; 22.5.2)
(b) Vol. **128**, No. 1, p. 297—311 (1985).

K. Sierański and J. Szatkowski
*Free-Carrier Absorption in Semimagnetic Semiconductors
(18.1 and 20.1; 13.1; 22.4.4)
(b) Vol. **128**, No. 1, p. K33—K35 (1985).

H. Kido, T. Hoshikawa, M. Shimada, and M. Koizumi
Preparation and Magnetic Properties of YMnSi
(18.2 and 18.4; 18.1; 21.1; 21.4)
(a) Vol. **88**, No. 1, p. K39—K43 (1985).

A. Birczyński and Z. T. Lalowicz
^{1}H-NMR Spectra of Molecular Groups in a Paramagnetic Matrix
(19; 18.1; 18.4; 22)
(b) Vol. **128**, No. 2, p. 545—552 (1985).

K. Kiymaç
*Giant Moment, Superparamagnetic, Spin-Glass, and Quasiferromagnetic Properties of Dilute Pt–Fe Alloys
(18.1; 18.2; 21.1.1)
(b) Vol. **128**, No. 2, p. 553—562 (1985).

S. Fakioğlu
Short-Range Spin Interaction and Critical Exponents
(18.2; 18.1)
(b) Vol. **128**, No. 2, p. K123—K124 (1985).

S. Pokrzywnicki and E. Solik
The Magnetic Resonance of Gd^{3+} in Gd_2Se_3 near the Phase Transition and up to 300 K
(19; 18.1; 18.4; 22)
(a) Vol. **88**, No. 2, p. 619—623 (1985).

O. Žižić, M. Stojić, and B. Stošić
Hall Effect and Electrical Resistivity in the Paramagnetic Region of $Hg_{1-x}Mn_xSe$ Crystals
(14.3; 18.1; 22.4.4)
(a) Vol. **88**, No. 2, p. K191—K196 (1985).

R. Sikdar and A. K. Pal
EPR Studies of Dimensionality in Copper Calcium Acetate Hexahydrate
(19; 18.1; 22.9)
(a) Vol. **89**, No. 1, p. 243—248 (1985).

A. Akroune, J. Claverie, A. Tazairt, G. Villeneuve and A. Casalot
Structural, Magnetic, and Electrical Properties of $V_{1-x}M_xO_{2-2x}F_{2x}$ (M = Mg, Ni) Oxyfluorides
(1.2 and 14; 8; 18.1; 22.8)
(a) Vol. **89**, No. 1, p. 271—282 (1985).

H. Kido, T. Hoshikawa, M. Shimada, and M. Koizumi
*Crystal Structure and Magnetic and Electrical Properties of Pseudo-Ternary $DyCu_{1-x}Zn_xSi$ Compounds
(14.1 and 18.1; 1.1; 18.2; 21)
(a) Vol. **90**, No. 1, p. 307—311 (1985).

H. Fehske
*On a Simple Functional Moment Approach to Itinerant Magnetism. Application to Ni
(13 and 18.1; 18.2; 21.1)
(b) Vol. **130**, No. 2, p. K121—K125 (1985).

Mbungu-Tsumbu, A. Van den Bosch, D. Segers, M. Dorikens, and L. Dorikens-Vanpraet
Positron Annihilation and Magnetic Susceptibility Measurements on Annealed Fast-Neutron Irradiated Crystalline Quartz
(11; 18.1; 22.6)
(a) Vol. **90**, No. 2, p. 595—606 (1985).

R. Z. Sadykhov, L. M. Valiev, and A. D. Namazov
*Magnetic Susceptibility and the Hall Effect in the $FeCr_{2-x}In_xS_4$ ($0 \leqq x \leqq 1$) System
(14.3 and 18.1; 22.8)
(a) Vol. **90**, No. 2, p. K189—K191 (1985).

D. S. Haroutunian, A. G. Karagyozyan, and G. N. Karajian
Anomalous Paramagnetic State in AgMn Spin Glass and the Kondo Effect
(1.3 and 14.1; 18.1; 21)
(b) Vol. **131**, No. 1, p. K87–K90 (1985).

A. Lipowski and Z. Onyszkiewicz
*The Theory of Hydrostatic Pressure Dependence of the Tricritical Temperature of the $FeCl_2$ Metamagnet
(18.1; 22.5)
(b) Vol. **131**, No. 2, p. K117–K122 (1985).

18.2 Ferromagnetic Properties

Review Articles

K. H. Fischer
*Spin Glasses (II)
(18.2 and 18.4; 2; 5; 7; 8; 13; 14; 21; 21.1; 21.6; 22.6; 22.8
(b) Vol. **130**, No. 1, p. 13–71 (1985).

Original Papers and Short Notes

J. Schreiber and D. Schulz
Self-Consistent Dynamical Theory for Heisenberg Spin Glasses
(2 and 18.1; 8; 18.2; 18.4)
(b) Vol. **127**, No. 1, p. 139–147 (1985).

J. A. Hołyst, A. Zagórski, and A. Sukiennicki
*On the Possibility of Deterministic Chaos in a Driven Damped Heisenberg Chain
(18.2)
(b) Vol. **127**, No. 1, p. K29–K32 (1985).

P. R. Silva
*The Site-Diluted Ising Model in the Simple Cubic Lattice
(18.1 and 18.2)
(b) Vol. **127**, No. 1, p. K41–K44 (1985).

H. Kido, T. Hoshikawa, M. Shimada, and M. Koizumi
Crystal Structure and Magnetic and Electrical Properties of Pseudo-Ternary $HoCu_{1-x}Zn_xSi$ and $NdCu_{0.6}Zn_{0.4}Si$ Compounds
(14.1 and 18.1; 1.3; 18.2; 22.8)
(a) Vol. **87**, No. 1, p. 273–277 (1985).

J. Ihlemann, H. J. Krokoszinski, E. Gmelin, and K. Bärner
*The Detection of Magnetic Order–Order Transitions in $MnAs_{1-x}P_x$ Crystals
(18.2 and 18.4; 8; 22.8)
(a) Vol. **87**, No. 1, p. 279–282 (1985).

T. Tarnóczi
*Structural Relaxation and Curie Point of a Metallic Glass
(2 and 18.2; 6; 21.1.1)
(a) Vol. **87**, No. 1, p. 283–291 (1985).

H. Kido, T. Hoshikawa, M. Shimada, and M. Koizumi
*Preparation and Magnetic Properties of $YMn_2(Si_{1-x}Ge_x)_2$ Compounds
(18.1 and 18.2; 1.3; 21.1; 21.4)
(a) Vol. **87**, No. 1, p. K61–K64 (1985).

Z. Frait, D. Fraitová, and N. Zárubová
*Observation of FMR Surface Spin Wave Mode in Bulk Amorphous Ferromagnets
(18.2 and 19; 2; 21.1; 21.1.1)
(b) Vol. **128**, No. 1, p. 219–222 (1985).

I. A. Vakarchuk, Yu. K. Rudavskii, and G. V. Ponedilok
*Free Energy of the Amorphous Ferromagnets with Heisenberg Exchange Interaction and Liquid-Like Disorder
(2 and 18.2)
(b) Vol. **128**, No. 1, p. 231–242 (1985).

A. M. van der Kraan, D. B. de Mooij, and K. H. J. Buschow
*Magnetic Properties and ^{57}Fe Mössbauer Effect in $V_{1-x}Fe_x$ Alloys
(18.2; 5; 21.1; 21.1.1)
(a) Vol. **88**, No. 1, p. 231–237 (1985).

H. Kido, T. Hoshikawa, M. Shimada, and M. Koizumi
*Preparation and Magnetic Properties of YMnSi
(18.2 and 18.4; 18.1; 21.1; 21.4)
(a) Vol. **88**, No. 1, p. K39–K43 (1985).

M. Yu. Nikolaev, N. V. Ryzhanova, A. V. Vedyaev, and S. M. Zubritskii
*Ferromagnetism and Elementary Excitations in the Hubbard Model
(13.5 and 18.2)
(b) Vol. **128**, No. 2, p. 513–523 (1985).

M. A. Yurishchev ·
*Critical Temperature of a Quasi-One-Dimensional Ising Model. The Bethe-Peierls Approximation
(18.2)
(b) Vol. **128**, No. 2, p. 537–543 (1985).

K. Kiymač
Giant Moment, Supermagnetic, Spin-Glass, and Quasiferromagnetic Properties of Dilute Pt–Fe Alloys
(18.1; 18.2; 21.1.1)
(b) Vol. **128**, No. 2, p. 553–562 (1985).

S. Fakioğlu
*Short-Range Spin Interaction and Critical Exponents
(18.2; 18.1)
(b) Vol. **128**, No. 2, p. K123–K124 (1985).

A. V. Dobromyslov, A. E. Ermakov, N. I. Taluts, and M. A. Uimin
Electron Microscopy Investigation of Phase Transformations in Mn–Al–C Alloy
(1.2; 1.4; 18.2; 21.1)
(a) Vol. **88**, No. 2, p. 443–454 (1985).

H. J. BLYTHE and F. WALZ
*The Magnetic Relaxation Spectrum of Plastically Deformed Non-Oriented Iron–Silicon Steel in the Temperature Range 200 to 800 K
(18.2; 10.1; 21.1.1)
(a) Vol. **89**, No. 1, p. 213–224 (1985).

G. WIATROWSKI
*Soliton Excitations in Terms of Pseudo-Riemannian Manifolds
(13.5 and 18.2)
(b) Vol. **129**, No. 2, p. 607–614 (1985).

G. DIENER and M. POSSELT
*A Self-Consistent Approach to the Ideal Magnetization in Heterogeneous Ferromagnetics
(18.2)
(b) Vol. **129**, No. 2, p. 631–640 (1985).

K. G. FISCHER and G. HEBER
*On Static Kink Solutions for an Anisotropic Heisenberg Ferromagnetic Chain
(18.2)
(b) Vol. **129**, No. 2, p. 649–656 (1985).

P. PAWLICKI
*Temperature Induced Metamagnetic Transitions in the Magnetoelastic Spin Model
(18.2 and 18.4)
(b) Vol. **129**, No. 2, p. K127–K130 (1985).

T. G. PETROVA and E. S. SYRKIN
*Effect of Uniaxial Long-Range Exchange Interaction on the Magnetic Properties of the Biferromagnetic Interface
(18.2)
(b) Vol. **129**, No. 2, p. K147–K151 (1985).

F. WALZ, H. J. BLYTHE, and M. FÖLDEAKI
*Magnetic After-Effects in Nitrogen-Charged α-Fe Following Low-Temperature Electron- and Neutron-Irradiation
(11 and 18.2; 10.1; 21.1.1)
(a) Vol. **89**, No. 2, p. 581–594 (1985).

K. PÁTEK and I. TOMÁŠ
*A Quantitative Characterization of Magnetic Defects by Domain Nucleation
(18.2; 22.8.2)
(a) Vol. **89**, No. 2, p. 595–599 (1985).

T. BALCERZAK, A. BOBÁK, J. MIELNICKI, and V. H. TRUONG
*A New Approach for the Diluted Ising Ferromagnet Description
(18.2; 8)
(b) Vol. **130**, No. 1, p. 183–190 (1985).

H. KRONMÜLLER
*The Nucleation Fields of Uniaxial Ferromagnetic Crystals
(18.2)
(b) Vol. **130**, No. 1, p. 197–203 (1985).

W. Fernengel, M. Rapp, and H. Kronmüller
*The Coercive Field of the Partially Crystallized Amorphous Alloy $Fe_{40}Ni_{40}P_{14}B_6$
(2 and 18.2; 21.1; 21.1.1)
(a) Vol. **90**, No. 2, p. 637–642 (1985).

H. J. Blythe, F. Walz, and E. Klugmann
*The Influence of Neutron Irradiation on the Magnetic Relaxation Spectrum of Commercial Non-Oriented Iron–Silicon Steel
(18.2; 10.1; 11; 21.1.1)
(a) Vol. **90**, No. 2, p. 643–650 (1985).

E. E. Shaisha and A. A. Bahgat
*Magnetic and Thermal Behaviour of the Amorphous Ferromagnet $Fe_{79}B_{16}Si_5$
(13.5.4 and 18.2; 5; 8; 21.1.1)
(a) Vol. **90**, No. 2, p. 661–668 (1985).

W. Zych and J. J. Milczarek
*Magnetic Properties of $Fe_{80-x}TM_xB_{20}$ and $Fe_{80-x}Mn_xSi_{12}B_8$ Amorphous Alloys
(2 and 18.2; 21.1.1)
(a) Vol. **90**, No. 2, p. K165–K170 (1985).

D. K. Chakrabarty, M. K. Jha, and S. Unnikrishnan
*Preparation and Properties of Barium Substituted Gadolinium Cobaltite
(18.2; 1.1; 14.3; 22.8.1)
(a) Vol. **90**, No. 2, p. K175–K177 (1985).

G. A. Govor
*Peculiarities of the First-Order Phase Transition in MnAs
(18.2; 22)
(a) Vol. **90**, No. 2, p. K185–K188 (1985).

K. G. Fischer and G. Heber
*Magnons in an Anisotropic Ferromagnetic Heisenberg Chain with One Static π-Kink
(13.5.4 and 18.2)
(b) Vol. **131**, No. 1, p. 117–128 (1985).

K. Witański
*On the Magnetic Properties of $Gd_{1-x}RE_xAl_2$ Alloys
(18.2; 21.4)
(b) Vol. **131**, No. 1, p. 129–136 (1985).

G. S. Ameta and U. N. Upadhyaya
Electronic Properties of Ternary Superconductors
(14.2; 18.2; 21.4)
(b) Vol. **131**, No. 1, p. 369–386 (1985).

A. Mody and A. A. Rangwala
*Magnetization and Susceptibility of Magnetic Systems
(18.2)
(b) Vol. **131**, No. 1, p. K45–K48 (1985).

A. H. Wafik
*Effect of Annealing Temperature and Magnetizing Frequency on Barkhausen Jumps in Rolled Steel
(18.2; 21.1.1)
(a) Vol. **91**, No. 2, p. K133–K137 (1985).

P. R. Silva
*Nearly Exact Critical Exponents through Landau's Free Energy Expansion for Two- and Three-Dimensional Cooperative Systems
(18.2)
(b) Vol. **132**, No. 1, p. K27–K31 (1985).

T. Tsuboi and R. Laiho
*Blackening of $CsFeCl_3$ by External Magnetic Field
(18.2 and 20.1; 13.5; 22.5)
(b) Vol. **132**, No. 1, p. K33–K35 (1985).

P. Pawlicki
Calculation of the Magnon Spectrum for Mn_3GaC Metamagnets
(13.5.4; 18.2; 18.4; 22.8)
(b) Vol. **132**, No. 1, p. K37 K40 (1985).

G. J. Muench, S. Arajs, and E. Matijević
*The Morin Transition in Small α-Fe_2O_3 Particles
(18.2; 18.4; 22.6)
(a) Vol. **92**, No. 1, p. 187–192 (1985).

H. J. Blythe
*An Automated Technique for the Measurement of the AC Initial Susceptibility and Its Disaccommodation in Ferromagnetic Materials
(18.2; 21.1.1)
(a) Vol. **92**, No. 1, p. 193–203 (1985).

L. Pareti, H. Szymczak, and H. K. Lachowicz
*High Field Magnetization Processes in $Pr_2Fe_{14}B$
(18.2; 22.8.1)
(a) Vol. **92**, No. 1, p. K65–K68 (1985).

R. Z. Sadykhov, L. M. Valiev, and A. I. Akhmedov
*Magnetic Properties of the $Ni_{0.25}Cu_xFe_{0.75-x}Cr_2S_4$ ($0 \leqq x \leqq 0.25$) System
(18.2; 22.8.2)
(a) Vol. **92**, No. 1, p. K69–K72 (1985).

W. Beck and H. Kronmüller
*Magnetic Properties of Amorphous $Fe_{90.6}Zr_{9.4}$
(2 and 18.2; 21.1.1)
(b) Vol. **132**, No. 2, p. 449–458 (1985).

E. A. Popov, M. M. Kotlyarskii, and I. S. Edelman
Temperature and Field Dependence of Exciton-Magon Absorption of $NaMnCl_3$
(13.5.1 and 13.5.4; 18.2; 20.1; 22.5.4)
(b) Vol. **132**, No. 2, p. 465–470 (1985).

18.2.1 Ferromagnetic Films

Original Papers and Short Notes

D. KONTOS
*Spin Green Functions for Ferromagnetic Films with Asymmetric Boundary Conditions
(18.2.1; 13.5.4)
(b) Vol. **127**, No. 1, p. 149–156 (1985).

J. DUSZEWSKI
*Influence of Perpendicular Field on Magnetic Fluctuations in Thin Films
(18.2.1)
(b) Vol. **127**, No. 1, p. 157–162 (1985).

A. BORKOWSKA
*Magnetooptical Diffraction and Modulation in Thin Garnet Films
(18.2.1 and 20; 22.8.2)
(a) Vol. **88**, No. 2, p. 601–610 (1985).

R. ŚWIRKOWICZ
*The General Theory of the Dynamics Susceptibility in Thin Ferromagnetic Films
(18.2.1)
(b) Vol. **129**, No. 2, p. 641–648 (1985).

W. ANDRÄ and H. DANAN
*Remanent Magnetization and Wall Coercivity of Continuous Films with Perpendicular Anisotropy
(18.2.1)
(a) Vol. **89**, No. 2, p. K173–K176 (1985).

NGUYEN CHI THANH and R. KRISHNAN
*Anisotropy of Magnetostriction in Coevaporated $Co_{1-x}Zr_x$ Amorphous Thin Films
(18.2.1; 2; 21.1)
(a) Vol. **90**, No. 1, p. 313–323 (1985).

K. SCHRÖDER
*The Effect of a Cu Overlayer on the Magnetization of an Fe Film
(18.2.1; 21.1; 21.1.1)
(b) Vol. **130**, No. 2, p. 581–585 (1985).

V. I. MALYUTIN, V. G. KAZAKOV, and D. V. KORYCHEV
*On the Mechanism of Domain Wall Break-Through from Localized Obstacles in Thin Magnetic Films
(18.2.1)
(a) Vol. **90**, No. 2, p. K161–K163 (1985).

18.3 Ferrimagnetic Properties

Review Article

E. P. WOHLFARTH
Magnetic Materials for Recording
(18; 18.3; 21.1; 21.1.1; 22.6; 22.8.2)
(a) Vol. **91**, No. 2, p. 339–350 (1985).

Original Papers and Short Notes

E. M. SMOKOTIN, O. P. KVASHNINA, and A. M. KAPITONOV
*Acoustic Properties of Terbium Ferrite Garnet in the Region of the Compensation Temperature Point
(7 and 18.3; 22.8.2)
(a) Vol. **87**, No. 1, p. K53–K56 (1985).

K. IWAUCHI and N. KOIZUMI
*Magnetic and Dielectric Relaxations of Fe_3O_4 and Some Ferrites
(14.4 and 18.3; 22.6; 22.8.2)
(a) Vol. **87**, No. 2, p. 617—621 (1985).

V. K. VLASKO-VLASOV and A. F. KHAPIKOV
*Reorientation of Domain Walls in a Garnet Film under the Action of an AC Magnetic Field
(18.3; 22.8.2)
(a) Vol. **87**, No. 2, p. K157–K162 (1985).

P. C. RIEDI and L. ZAMMIT-MANGION
*Difference between Domain Wall Motion in Single Crystal and Polycrystalline YIG Observed by Pulsed NMR
(18.3 and 19; 22.8.2)
(a) Vol. **87**, No. 2, p. K163–K166 (1985).

V. I. MALTSEV
*Umbrella-Type Magnetic Structures in Diluted Ordered Lithium Ferrospinels
(18.3; 22.8.2)
(b) Vol. **128**, No. 1, p. 193–200 (1985).

S. R. MURTHY and T. S. RAO
*Elastic Behaviour of Mixed Li–Ni Ferrites
(12.2 and 18.3; 22.8.2)
(a) Vol. **88**, No. 1, p. 239–243 (1985).

Z. ŠIMŠA, Y. KAWAI, and V. A. M. BRABERS
*Anelasticity and Relaxation Effects in Single Crystal Mn Ferrites
(12.2 and 18.3; 7; 22.8.2)
(a) Vol. **88**, No. 2, p. 611–617 (1985).

I. GUERASIMENKO and A. GONŹALEZ-ARIAS
*Thermal Treatments and Magnetic Properties in High Zn Content Ni–Zn Ferrites
(18.3; 22.8.2)
(a) Vol. **88**, No. 2, p. K159–K163 (1985).

A. A. SIDOROV, V. P. SUETIN, V. G. POKAZANIEV, and B. V. KARPENKO
*Determination of Effective Indirect Exchange Interaction Integrals in Ferrimagnets
(18.3; 22.8.2)
(b) Vol. **129**, No. 2, p. 625–629 (1985).

I. O. TROYANCHUK, L. A. BASHKIROV, L. V. BALYKO, V. M. PAVLOV, A. N. PIROGOV, A. P. VOKHMYANIN, and O. A. ZALUKOVSKAYA
*Magnetic Ordering in the System $Ca(Cu_{1-x}Mn_x)_3Mn_4O_{12}$
(18.3; 22.8.2)
(a) Vol. **89**, No. 2, p. 601–608 (1985).

C. MICHALK and W. THIEL
Garnet Formation of Al-Substituted Yttrium Iron Garnet Ceramics Studied by Mössbauer Effect and Magnetization Measurements
(1.2 and 5; 18.3; 22.8.2)
(a) Vol. **90**, No. 1, p. 325–330 (1985).

E. Z. KATSNELSON and A. G. KAROZA
*Direct Observation of Ferrite Photomagnetization
(18.3; 22.8.2)
(a) Vol. **90**, No. 1, p. K49–K53 (1985).

P. C. M. GUBBENS, A. M. VAN DER KRAAN, and K. H. J. BUSCHOW
*^{161}Dy Mössbauer Effect in $Dy_2Fe_{14}B$
(5 and 18.3; 22.8)
(b) Vol. **130**, No. 2, p. 575–580 (1985).

S. R. MURTHY and T. S. RAO
*Magnetostriction of Ni–Zn and Co–Zn Ferrites
(18.3; 22.8.2)
(a) Vol. **90**, No. 2, p. 631–635 (1985).

E. Z. KATSNELSON and A. G. KAROZA
*Peculiarities of Photomagnetization and Magnetoelectric Interactions in Ferrites
(18.3; 22.8.2)
(a) Vol. **90**, No. 2, p. K179–K184 (1985).

S. LIGENZA
*Dynamical Nature of Superexchange Interaction in $Y_3Fe_5O_{12}$ Studied by Neutron Spectroscopy
(13 and 18.3; 22.8.2)
(b) Vol. **131**, No. 1, p. 105–112 (1985).

F. M. MOSTAFA, H. PTASIEWICZ-BAK, and S. LIGENZA
*Neutron Diffraction Study of the Spinel $CoAlFeO_4$
(18.3; 22.8.2)
(a) Vol. **91**, No. 1, p. 99–103 (1985).

18.4 Antiferromagnetic Properties

Review Article

K. H. FISCHER
*Spin Glasses (II)
(18.2 and 18.4; 2; 5; 7; 8; 13; 14; 21; 21.1; 21.6; 22.6; 22.8)
(b) Vol. **130**, No. 1, p. 13—71 (1985).

Original Papers and Short Notes

J. SCHREIBER and D. SCHULZ
Self-Consistent Dynamical Theory for Heisenberg Spin Glasses
(2 and 18.1; 8; 18.2; 18.4)
(b) Vol. **127**, No. 1, p. 139—147 (1985).

P. H. GREENWOOD and P. DE V. DU PLESSIS
*Critical Behaviour of Antiferromagnetic Fe_2As
(18.4; 21.1.1)
(b) Vol. **127**, No. 1, p. K19—K23 (1985).

A. S. T. PIRES and M. E. GOUVEA
*Non-Linear Dynamics of a Classical One-Dimensional Antiferromagnet with Two Anisotropies
(13.5.4 and 18.4)
(b) Vol. **127**, No. 1, p. K25—K28 (1985).

M. OGATA, H. KOZAKA, and N. URYÛ
*Magnetic Interactions in Spinel Compound $CoGa_2O_4$
(18.4; 22.8.2)
(b) Vol. **127**, No. 1, p. K33—K36 (1985).

J. IHLEMANN, H. J. KROKOSZINSKI, E. GMELIN, and K. BÄRNER
*The Detection of Magnetic Order–Order Transitions in $MnAs_{1-x}P_x$ Crystals
(18.2 and 18.4; 8; 22.8)
(a) Vol. **87**, No. 1, p. 279—282 (1985).

P. COLOMBET and W. S. GLAUNSINGER
Anomalous High-Temperature EPR Linewidths in the Spinel Solid Solution $Cu_{2x}Cr_{2x}Sn_{2-2x}S_4$
(19; 18.2; 18.4; 22.8.2)
(b) Vol. **128**, No. 1, p. 141—149 (1985).

G. ÜFFINGER and A. KASTEN
*Specific Heat and Magnetocaloric Effects in $TbPO_4$. I. Experiments
(1.2 and 18.4; 8; 22.8.1)
(b) Vol. **128**, No. 1, p. 201—207 (1985).

H. KIDO, T. HOSHIKAWA, M. SHIMADA, and M. KOIZUMI
*Preparation and Magnetic Properties of YMnSi
(18.2 and 18.4; 18.1; 21.1; 21.4)
(a) Vol. **88**, No. 1, p. K39—K43 (1985).

N. B. IVANOV
*Virtual Dipole-Exchange Surface Magnons on Semi-Infinite Uniaxial Antiferromagnets
(13.5.4 and 18.4)
(b) Vol. **129**, No. 2, p. K141 – K145 (1985).

P. J. BECKER, H. G. KAHLE, and E. KELLER
*The Phase Diagramm of $TbPO_4$ Measured by Linear Optical Birefringence
(1.2 and 18.4; 20; 22.8.1)
(b) Vol. **130**, No. 1, p. 191 – 196 (1985).

D. HAJDUKOVIĆ and I. M. SAVIĆ
*On the Properties of Two-Dimensional Ising Antiferromagnets with General Spin S at Zero Temperature
(18.4)
(b) Vol. **130**, No. 1, p. K33 – K36 (1985).

B. CHENEVIER, M. BACMANN, D. FRUCHART, J. P. SENATEUR et R. FRUCHART
*Propriétés magnétiques de MnRhAs. Structures magnétiques
(Magnetic Properties of MnRhAs. Magnetic Structures)
(18.2 and 18.4; 21)
(a) Vol. **90**, No. 1, p. 331 – 341 (1985).

YU. M. SEIDOV and A. M. SULEIMANOV
Theory of Absorption and Scattering of Light by Singlet–Triplet Magnetic Systems
(13.5.4 and 20.1; 18.2; 18.4)
(b) Vol. **131**, No. 2, p. 545 – 554 (1985).

M. AVEROUS, B. A. LOMBOS, C. FAU, E. ILBNOUELGHAZI, J. C. TEDENAC, G. BRUN, and M. A. BARTKOWSKI
*$Pb_{1-x}Gd_xTe$ a New Semimagnetic Semiconductor with Rare Earths
(14.3 and 18.4; 19; 22.8)
(b) Vol. **131**, No. 2, p. 759 – 768 (1985).

M. NOGUES, M. HAMEDOUN, J. L. DORMANN, and G. VILLERS
*Magnetic Domains in Antiferromagnetic $ZnCr_2X_4$ ($X = S, Se$) as Seen by Magnetization Measurements
(18.4; 22.8.2)
(a) Vol. **91**, No. 2, p. 597 – 602 (1985).

G. A. GOVOR
*Crystal Structure and Magnetic Properties of $MnAs_{0.95}P_{0.05}$
(18.2 and 18.4; 21)
(a) Vol. **91**, No. 2, p. K129 – K132 (1985).

T. NATTERMANN
*Some Remarks on the Random Field Ising Model on a Lattice
(18.4)
(b) Vol. **132**, No. 1, p. 125 – 131 (1985).

P. PAWLICKI
Calculation of the Magnon Spectrum for Mn_3GaC Metamagnets
(13.5.4; 18.2; 18.4; 22.8)
(b) Vol. **132**, No. 1, p. K37 – K40 (1985).

19. Magnetic Resonance

Review Articles

H. W. Streitwolf
Physical Properties of Polyacetylene
(13.5 and 13.5.3; 6; 18; 19; 20.1; 22.9)
(b) Vol. **127**, No. 1, p. 11–54 (1985).

M. Godlewski
*On the Application of the Photo-EPR Technique to the Studies of Photoionization, DAP Recombination, and Non-Radiative Recombination Processes
(13.4 and 19; 16; 22.2; 22.4.1; 22.5)
(a) Vol. **90**, No. 1, p. 11–52 (1985).

Original Papers and Short Notes

P. Reineker, J. Köhler, and U. Schmid
Triplet Excitons in Non-Resonant Molecular Pairs. Saturation Effects in Electron Spin Resonance
(13.5.1; 19; 22.9)
(b) Vol. **127**, No. 1, p. 237–244 (1985).

Prem Chand and M. Umar
*Electron Paramagnetic Resonance Study of Copper(II) Ions in Single Crystal of Strontium Formate Dihydrate
(13.4 and 19; 22.9)
(b) Vol. **127**, No. 1, p. 279–286 (1985).

G. Gerbeth, E. Mrosan, R. Zeller, and P. H. Dederichs
Calculation of the Change in Knight Shift Due to Zn, Ga, and Ge Impurities in Copper
(13.4; 19; 21.1)
(b) Vol. **127**, No. 1, p. 365–369 (1985).

M. N. Aliev and L. R. Tagirov
*Theory of Temperature Dependent EPR Linewidth in Semimagnetic Semiconductors
(19; 22.4.4)
(b) Vol. **127**, No. 1, p. K61–K64 (1985).

B. Berulava and R. Mirianashvili
*Radio-Frequency Discrete Saturation Measurement of the Signs of Ligand Hyperfine Interaction Tensor Components for Ce^{3+} in CaF_2
(19; 22.5.3)
(b) Vol. **127**, No. 1, p. K69–K72 (1985).

P. J. Alonso and R. Alcalá
*EPR Study of Mn^+ Ions in SrF_2
(11 and 19; 22.5.3)
(b) Vol. **127**, No. 1, p. K77–K79 (1985).

O. Z. ALEKPEROV, V. G. GOLUBEV, V. I. IVANOV-OMSKII, and A. SH. MEKHTIEV
*The Effect of Electron–Electron Interaction on the Cyclotron Resonance Linewidth in n-GaAs
(14.3.4 and 19; 22.2.1)
(b) Vol. **127**, No. 2, p. K171–K174 (1985).

G. A. PETRAKOVSKII, G. S. PATRIN, and N. V. VOLKOV
*Change of Magnetic Resonance Parameters in $FeBO_3$:NiO Induced by Radiation ($\lambda = 1.06\ \mu m$)
(10.2 and 19; 22.8.1)
(a) Vol. **87**, No. 2, p. K153–K156 (1985).

P. C. RIEDI and L. ZAMMIT-MANGION
*Difference between Domain Wall Motion in Single Crystal and Polycrystalline YIG Observed by Pulsed NMR
(18.3 and 19; 22.8.2)
(a) Vol. **87**, No. 2, p. K163–K166 (1985).

P. COLOMBET and W. S. GLAUNSINGER
*Anomalous High-Temperature EPR Linewidths in the Spinel Solid Solution $Cu_{2x}Cr_{2x}Sn_{2-2x}S_4$
(19; 18.2; 18.4; 22.8.2)
(b) Vol. **128**, No. 1, p. 141–149 (1985).

Z. FRAIT, D. FRAITOVÁ, and N. ZÁRUBOVÁ
*Observation of FMR Surface Spin Wave Mode in Bulk Amorphous Ferromagnets
(18.2 and 19; 2; 21.1; 21.1.1)
(b) Vol. **128**, No. 1, p. 219–222 (1985).

S. D. LETYUCHENKO, E. I. SLINKO, K. D. TOVSTYUK, A. G. KHANDOZHKO, and N. K. TOVSTYUK
*Jahn-Teller Effect in Li-Doped SnTe
(13.4 and 19; 10.2; 22)
(b) Vol. **128**, No. 1, p. K87–K92 (1985).

A. A. ALYBAKOV, O. M. ARBOTOEV, V. A. GUBANOVA, and K. KUDABAEV
Cr^{5+} Paramagnetic Centers in X-Irradiated KH_2PO_4:Cr^{3+} Single Crystals
(13.4; 11; 19; 20.1; 22.8.1)
(b) Vol. **128**, No. 1, p. K93–K96 (1985).
Erratum
(b) Vol. **130**, No. 1, p. K89 (1985).

L. LATIE, G. VILLENEUVE, CH. CROS, and P. HAGENMÜLLER
Influence of Cationic Vacancies on the Mobility of Li^+ Ions in Some Cation-Deficient Materials. An NMR Study
(10.2; 9; 14.4; 19; 22.5.4)
(b) Vol. **128**, No. 2, p. 475–482 (1985).

A. BIRCZYŃSKI and Z. T. LALOWICZ
*^{1}H-NMR Spectra of Molecular Groups in a Paramagnetic Matrix
(19; 18.1; 18.4; 22)
(b) Vol. **128**, No. 2, p. 545–552 (1985).

R. C. DUVARNEY, J. R. NIKLAS, and J.-M. SPAETH
*Electron Nuclear Double Resonance of the Trapped Hole Center in α-Al_2O_3:Mg^{2+}
(13.4 and 19; 22.6)
(b) Vol. **128**, No. 2, p. 673–681 (1985).

P. J. Alonso and R. Alcala
*EPR Study of a New Trigonal Mn^{2+} Defect in SrF_2
(19; 13.4; 22.5.3)
(b) Vol. **128**, No. 2, p. K153–K155 (1985).

S. Pokrzywnicki and E. Solik
*The Magnetic Resonance of Gd^{3+} in Gd_2Se_3 near the Phase Transition and up to 300 K
(19; 18.1; 18.4; 22)
(a) Vol. **88**, No. 2, p. 619–623 (1985).

A. Edgar, J. M. Spaeth, and G. J. Troup
*EPR and Optical Fluorescence Spectra of Cr^{3+} Ions in Alexandrite
(19 and 20.3; 22.8)
(a) Vol. **88**, No. 2, p. K175–K178 (1985).

D. Backs
*Determination of the Spin-Hamiltonian Parameter of Mn-Doped Wurtzite (ZnS Platelets)
(19; 22.4.1)
(a) Vol. **88**, No. 2, p. K183–K186 (1985).

V. P. Seth, V. K. Jain, and R. S. Bansal
*Electron Paramagnetic Resonance of Mn^{2+} in Double Nitrate Crystals
(13.4 and 19; 22.8.1)
(b) Vol. **129**, No. 1, p. 375–380 (1985).

A. M. Hassib, A. A. S. Musmus, and M. A. A. Issa
*Conduction Electron Spin Resonance of Blue Sodalite
(19; 22.8)
(a) Vol. **89**, No. 1, p. 147–153 (1985).

R. Sikdar and A. K. Pal
*EPR Studies of Dimensionality in Copper Calcium Acetate Hexahydrate
(19; 18.1; 22.9)
(a) Vol. **89**, No. 1, p. 243–248 (1985).

S. A. Y. Al-Ismail, K. Arshak, and C. A. Hogarth
Electron Spin Resonance Measurements and Electrical Characteristics, before and after Electroforming, of Thin Films of SiO/Nb_2O_5 and Nb_2O_5
(14.3.1 and 14.3.4; 2; 19; 22.6; 22.6.1)
(a) Vol. **89**, No. 1, p. 363–374 (1985).

D. Backs
*Axial $\Delta M = \pm 3$ and Cubic $\Delta m = \pm 1$ Forbidden Transitions in the EPR Spectrum of Mn^{2+} in ZnS
(19; 22.4.1)
(a) Vol. **89**, No. 1, p. K75–K78 (1985).

K. E. Tarkpea and A. Ots
Thermal Stability of Fe^{+} Centres and Thermoluminescence of ZnS Single Crystals
(13.4 and 20.3; 19; 22.4.1)
(b) Vol. **129**, No. 2, p. 799–803 (1985).

N. SIELAFF, H. NELKOWSKI, H. PRADELLA, J. SAHM, and K. ZINK
*EPR Measurements on Fe–Cu Complexes in ZnSe
(19; 13.4; 22.4.2)
(b) Vol. **129**, No. 2, p. K177–K179 (1985).

G.-H. CHEN, A. MORIMOTO, M. KUMEDA, and T. SHIMIZU
*ESR Studies on Doped Hydrogenated Amorphous Silicon–Carbon
(2 and 19; 13.4; 22.1.2)
(a) Vol. **89**, No. 2, p. K181–K183 (1985).

A. L. ERZINKYAN, V. V. MURAVEVA, V. P. PARFENOVA, and V. V. TUROVTSEV
*Magnetic Hyperfine Interaction of ^{92m}Nb in $ZrFe_2$
(19; 21.1.1)
(a) Vol. **89**, No. 2, p. K201–K205 (1985).

V. M. POLYANOVSKII
Harmonics of Cyclotron and Cyclotron–Phonon Resonances in Semiconductors with Superlattice
(6 and 13; 19; 20.1; 22)
(b) Vol. **130**, No. 1, p. 95–101 (1985).

S. V. NISTOR, I. HEYNDERICKX, E. GOOVAERTS, A. BOUWEN, and D. SCHOEMAKER
*ESR and Optical Absorption Study of the $Tl^0(1)$ Center in Na Cl. A Stable Laser-Active Type Defect
(102 and 19; 20.1; 22.5.2)
(b) Vol. **130**, No. 1, p. 175–182 (1985).

A. DE KEERSGIETER, F. CALLENS, M. BAROEN, P. MATTHYS, and E. BOESMAN
*ESR of Isolated Mn^{2+} Centers in Alkali Chlorides
(13.4 and 19; 22.5.2)
(b) Vol. **130**, No. 1, p. 297–301 (1985).

H. SÖTHE, P. STUDZINSKI, and J.-M. SPAETH
*Pairs of F-Centers and F^- Impurities in KCl Studied by ENDOR
(13.4 and 19; 10.2; 22.5.2)
(b) Vol. **130**, No. 1, p. 339–346 (1985).

YU. V. GORELKINSKII, N. N. NEVINNYI, and A. A. KIM
*New EPR Spectrum in Helium-Implanted Silicon
(19; 11; 22.1.2)
(b) Vol. **130**, No. 1, p. K51–K53 (1985).

M. K. KERIMOV
*Electron Paramagnetic Resonance of Stabilized Electrons in PTFE Coronoelectrets
(14.4.1 and 19; 22.9)
(a) Vol. **90**, No. 1, p. K71–K74 (1985).

M. KORKMAZ and B. AKTAŞ
*EPR of Mn^{2+} in Strontium Tartrate Trihydrate Single Crystals
(13.4 and 19; 22.8.1)
(b) Vol. **130**, No. 2, p. 743–748 (1985).

T. SH. ABESADZE and M. PUNKINEN
Discrete Saturation of the Optical Absorption Line of Pr^{3+} in CaF_2 under the Condition of Pumping
(13.4 and 20.1; 19; 22.5.3)
(b) Vol. **131**, No. 2, p. K153—K155 (1985).

R. BLINC, P. CEVC, J. SELIGER, and M. KOREN
*EPR of Tl^{2+} Doped Incommensurate K_2SeO_4
(13.5.4 and 19; 22.8.1)
(b) Vol. **131**, No. 2, p. K167—K172 (1985).
Erratum
(b) Vol. **132**, No. 2, p. K143 (1985).

M. M. EL-ZAIDIA, A. A. AMMAR, and R. A. EL-MALLWANY
Infra-Red Spetra, Electron Spin Resonance Spectra, and Density of $(TeO_2)_{100-x}(WO_3)_x$ and $(TeO_2)_{100-x}(ZnCl_2)_x$ Glasses
(2 and 20.1; 19; 22.6)
(a) Vol. **91**, No. 2, p. 637—642 (1985).

J. CORBERO, D. J. PUSIOL, A. E. WOLFENSON, and A. H. BRUNETTI
*Possible Incommensurate Structure in 4,4'-Dichlorobiphenyl Sulphone as Studied by the ^{35}Cl Nuclei NQR
(1.2 and 19; 22.9)
(a) Vol. **91**, No. 2, p. K97—K99 (1985).

L. VAN ROBBROECK, E. GOOVAERTS, and D. SCHOEMAKER
*Electron-Spin-Resonance Study of Co^{2+} and Ni^{+} Centers in AgCl (Cu, Co, Ni)
(13.4 and 19; 22.5.1)
(b) Vol. **132**, No. 1, p. 179—187 (1985).

V. P. SETH, A. YADAV, and R. S. BANSAL
*Solution of the Spin-Hamiltonian with Orthorhombic HF and g-Tensors ($S = 1/2$). I. Theory
(19; 22.8)
(b) Vol. **132**, No. 1, p. 255—259 (1985).

E. L. FRANKEVICH, D. I. KADYROV, I. A. SOKOLIK, A. I. PRISTUPA, V. M. KOBRYANSKII. and N. Y. ZURABYAN
On the Conductivity Mechanism of Weakly Doped Polyacetylene
(14.3.1; 19; 22.9)
(b) Vol. **132**, No. 1, p. 283—294 (1985).

P. C. SCHMIDT, W. BADEN, N. WEIDEN, and A. WEISS
*The Intermetallic System $NaHg_{1-x}Tl_x$. X-Ray Investigations and Measurements of the Knight Shift of Na, Hg, and Tl
(19; 21)
(a) Vol. **92**, No. 1, p. 205—212 (1985).

A. V. DVURECHENSKII and V. V. SUPRUNCHIK
*Low Symmetry Centre in Silicon
(11 and 19; 22.1.2)
(a) Vol. **92**, No. 1, p. K53—K56 (1985).

D. K. Fowler, D. C. Creagh, R. W. N. Kinnear, and G. V. H. Wilson
*Two-Pulse Stimulated Echo in Ferromagnetic Metals
(19; 21.1; 21.1.1)
(a) Vol. **92**, No. 2, p. 545–553 (1985).

20. Optical Properties

Original Papers and Short Notes

F. EGGERT and W. SCHOLZ
*A Rapid Deconvolution Method Based on the Bayesian Theorem Applied to Energy Dispersive X-Ray Emission Analysis
(1 and 20)
(a) Vol. **88**, No. 2, p. K123—K125 (1985).

E. L. ALBUQUERQUE, N. S. ALMEIDA, and M. C. OLIVEROS
*Light Scattering by Surface Ripples in a Spatially Dispersive Medium
(13.5.1 and 20; 7)
(b) Vol. **129**, No. 1, p. 177—186 (1985).

I. V. STASYUK and S. S. KOTSUR
*The Ionic Gyration and Electrogyration in Dielectric Crystals
(14.4 and 20; 22)
(b) Vol. **129**, No. 2, p. 577—586 (1985).

C. SUGIURA and S. MURAMATSU
*Sulfur K X-Ray Absorption-Edge Structures and Electronic States of Transition-Metal Sulfides MS and MS_2 (M = Fe, Co, Ni)
(13.1 and 20; 22)
(b) Vol. **129**, No. 2, p. K157—K161 (1985).

V. HIETSCHOLD and G. SEIFERT
*Soft X-Ray Spectra of Amorphous Hydrogenated Silicon
(13 and 20; 2; 22.1.2)
(b) Vol. **129**, No. 2, p. K163—K166 (1985).

K. CSACH, J. MIŠKUF, and V. KAREL
*Angular Distribution of Emitted X-Rays from an Ion-Implanted Layer
(11 and 20; 21.1.1)
(a) Vol. **89**, No. 2, p. K153—K155 (1985).

I. V. STASYUK and S. S. KOTSUR
*On the Theory of Electrogyration in Dielectric Crystals. Sheelite-Type Crystals
(14.4 and 20; 22.8.1)
(b) Vol. **130**, No. 1, p. 103—113 (1985).

P. J. BECKER, H. G. KAHLE, and E. KELLER
The Phase Diagram of $TbPO_4$ Measured by Linear Optical Birefringence
(1.2 and 18.4; 20; 22.8.1)
(b) Vol. **130**, No. 1, p. 191—196 (1985).

M. S. KUSHWAHA
*Helicon Wave Propagation in Semiconductor Magnetoplasma
(13 and 20; 22.2.3)
(b) Vol. **130**, No. 1, p. K37—K41 (1985).

R. POPRAWSKI, J. SHALDIN, and S. MATYJASIK
*Pyroelectric and Optical Properties of α-$LiIO_3$ Crystals
(14.4 and 20; 22.8.1)
(a) Vol. **90**, No. 1, p. 167—171 (1985).

L. A. Bugaev and R. V. Vedrinskii
Hartree-Fock Approximation in the Scattering Wave Method. K-Absorption of Ionic Crystals in the Single-Scattering Formalism (XANES and EXAFS)
(13.1; 20; 22.5.2)
(b) Vol. **132**, No. 2, p. 459–464 (1985).

M. S. Islam
*Chemical Shifts and EXAFS of the L_{III} Absorption Discontinuity of Dysprosium in Some of Its Compounds
(20; 21.4)
(b) Vol. **132**, No. 2, p. 471–476 (1985).

K. Hemachandran and A. R. Chetal
*X-Ray Spectroscopic Studies of Azurite Mineral
(13.1 and 20; 22.8.1)
(b) Vol. **132**, No. 2, p. 503–507 (1985).

A. Gras-Marti, V. E. Anderson, R. H. Ritchie, and J. C. Ashley
*Influence of Damping on the Energy Loss Straggling of Low Velocity Protons in an Electron Gas
(13 and 20)
(b) Vol. **132**, No. 2, p. 509–513 (1985).

Yu. V. Sukhetskii, A. V. Soldatov, A. N. Gusatinskii, K. Žičkus, and A. Audzijonis
Energie Band Structure of $SbS_xSe_{1-x}I$
(13.1; 20; 22.8)
(b) Vol. **132**, No. 2, p. K103–K106 (1985).

C. Sugiura and S. Muramatsu
*K-Edge X-Ray Absorption Spectra from Potassium Compounds
(20; 22)
(b) Vol. **132**, No. 2, p. K111–K115 (1985).

S. Götz
*Measurement of Anomalous Dispersion Effects around an Absorption Edge in the Soft X-Ray Region
(1.1 and 20; 22.9)
(a) Vol. **92**, No. 2, p. 407–412 (1985).

20.1 Spectra, Optical Constants (X-Ray Spectra See 20)

Review Articles

H. W. Streitwolf
Physical Properties of Polyacetylene
(13.5 and 13.5.3; 6; 18; 19; 20.1; 22.9)
(b) Vol. **127**, No. 1, p. 11–54 (1985).

A. A. Kaminskii
Achievements in the Field of Physics and Spectroscopy of Activated Laser Crystals
(20.2; 20.1; 20.3; 22.8)
(a) Vol. **87**, No. 1, p. 11–57 (1985).

S. Sugai
Lattice Vibrations in the Charge-Density-Wave States of Layered Transition Metal Dichalcogenides
(6 and 13; 1.2; 20.1; 22)
(b) Vol. **129**, No. 1, p. 13–39 (1985).

T. S. Moss
*Relations between the Refractive Index and Energy Gap of Semiconductors
(13.1 and 20.1; 22; 22.2.1; 22.2.3)
(b) Vol. **131**, No. 2, p. 415–427 (1985).

Original Papers and Short Notes

P. Kleinert
Self-Consistent Field Approximation for the Vibrational Spectra of Quaternary Alloys
(6; 20.1; 22.2.4)
(b) Vol. **127**, No. 1, p. 109–119 (1985).

C. Trallero Giner and O. Sotolongo Costa
*One-Phonon Resonance Raman Scattering
(6 and 20.1; 13.5.1; 22.4.1)
(b) Vol. **127**, No. 1, p. 121–130 (1985).

B. G. Dick
*Electric Field-Induced First Order Raman Scattering in Alkali Halides
(6 and 20.1; 22.5.2)
(b) Vol. **127**, No. 1, p. 131–138 (1985).

K. Kambas, A. Anagnostopoulos, S. Ves, B. Ploss, and J. Spyridelis
*Optical Absorption Edge Investigation of $CdIn_2S_4$ and β-In_2S_3 Compounds
(20.1; 22.3; 22.8)
(b) Vol. **127**, No. 1, p. 201–208 (1985).

YU. A. EKMANIS and I. S. RADCHENKO
*Light Extinction by Colloidal Thallium in Ionic Crystals
(20.1; 22.5.2)
(b) Vol. **127**, No. 1, p. 287–298 (1985).

G. I. PILIPENKO, D. V. OPARIN, O. I. TYUTYUNNIK, and F. F. GAVRILOV
Photostimulated Transformations of Complex Colour Centres in Lithium Hydride Single Crystals
(6 and 10.2; 20.1; 20.3; 22)
(b) Vol. **127**, No. 1, p. 299–306 (1985).

A. G. MIRONOV, V. A. MOROZOVA, and V. S. VAVILOV
*Some Peculiarities of the Electrooptical Spectra of n-GaAs(Cr) at Liquid Helium Temperatures (Wannier Levels?)
(13.4 and 20.1; 14.3.4; 22.2.1)
(b) Vol. **127**, No. 1, p. 359–363 (1985).

H. NEUMANN, W. KISSINGER, FALAH S. HASOON, B. R. PAMPLIN, H. SOBOTTA, and V. RIEDE
*Infrared Reflectivity Spectra of $CuSi_2P_3$
(6 and 20.1; 22.8)
(b) Vol. **127**, No. 1, p. K9–K14 (1985).

K. M. PARASKEVOPOULOS
*Electron Effective Mass Dependence on Carrier Concentration in $TlBiTe_2$ Monocrystals
(13.1 and 20.1; 14.3; 22.8)
(b) Vol. **127**, No. 1, p. K45–K50 (1985).

K. R. ALLAKHVERDIEV, T. G. MAMEDOV, V. E. SHTEINSHREIBER, and I. K. EFENDIEVA
*Fundamental Absorption Edge of TlSe and $TlInTe_2$ Crystals
(13.1 and 20.1; 13.5.1; 22; 22.8)
(b) Vol. **127**, No. 1, p. K55–K60 (1985).

N. A. MIRONOVA and V. N. SKVORTSOVA
*Fine Structure of the NiO: Co Infrared Absorption Spectrum
(13.5.3 and 20.1; 22.6)
(b) Vol. **127**, No. 1, p. K81–K82 (1985).

R. P. GOYAL, RAVIENDRA D., and B. R. K. GUPTA
Electroless Deposition of In_2O_3 and In_2O_3:Sn (ITO)
(1.5 and 14.3.1; 20.1; 22.6)
(a) Vol. **87**, No. 1, p. 79–83 (1985).

S. K. J. AN-ANI and C. A. HOGARTH
*Optical Absorption Edge in Amorphous $In_{30}Se_{70}$ Films
(2 and 20.1; 22.3)
(a) Vol. **87**, No. 1, p. K65–K67 (1985).

K. R. ALLAKHVERDIEV, U. A. ALESHENKO, N. A. BAKHYSHOV, L. K. VODOPYANOV, F. M. GASHIMZADE, R. M. SARDARLY, and V. YA. SHTEINSHRAIBER
*Vibrational Spectrum of TlS_xSe_{1-x} Solid Solutions
(6 and 20.1; 22.3.1)
(b) Vol. **127**, No. 2, p. 459–464 (1985).

K. M. MUKIMOV, SH. M. SHARIPOV, and L. A. ERNAZAROVA
*The Magneto-Optical Spectra and Behaviour of the Joint Density of States of the Intermetallic Compounds RFe_2(R = Gd, Tb, Dy, Ho, Er) and YF_2
(13.1 and 20.1; 21.1.1; 21.4)
(b) Vol. **127**, No. 2, p. K129–K131 (1985).

G. D. BAGRATISHVILI, R. B. DZHANELIDZE, T. V. ETERASHVILI, D. A. JISHIASHVILI, and E. R. KUTELIA
*Properties and Structure of Silicon Oxynitride Films Obtained in a Hydrazine Plasma
(1.5 and 20.1; 22.8)
(a) Vol. **87**, No. 2, p. 435–440 (1985).

S. MOCHIZUKI
*Radiative Optical Modes of Vibration in CeO_2 Thin Films
(6 and 20.1; 13.5; 22.6)
(b) Vol. **128**, No. 1, p. 23–30 (1985).

V. L. BONCH-BRUEVICH and T. Z. KAKHLISHVILI
*On the Nonlinear Light and Sound Absorption in Materials with Weakly Bound Atoms
(7 and 20.1; 22)
(b) Vol. **128**, No. 1, p. 39–48 (1985).

A. MEMON and D. B. TANNER
*Far-Infrared Dielectric Function of Zincblende ZnS
(6 and 20.1; 22.4.1)
(b) Vol. **128**, No. 1, p. 49–52 (1985).

B. OREL, M. KLANJŠEK, V. MOISEENKO, M. VOLNYANSKII, N. LE CALVÉ, B. PASQUIER, and A. NOVAK
Isotopic Impurity Modes and Polariton Dispersion in Li_2GeO_3 Crystals
(6 and 13.5; 20.1; 22.8.1)
(b) Vol. **128**, No. 1, p. 53–67 (1985).

F. BREHAT and B. WYNCKE
*Soft Mode Spectroscopy in Potassium Dihydrogen Phosphate
(6 and 20.1; 14.4.1; 22.8.1)
(b) Vol. **128**, No. 1, p. 83–92 (1985).

R. VOSS, R. KOTZOTT, and L. MERTEN
*Dispersion of Magnon-Polaritons in Uniaxial Ferrites
(13.5.4 and 20.1; 22.8.2)
(b) Vol. **128**, No. 1, p. 159–167 (1985).

R. SUDHARSANAN and B. P. CLAYMAN
Frequency-Dependent Conductivity Studies of the Ordered and Disordered Phases of Ag_2HgI_4 and Cu_2HgI_4
(10.2 and 14.4; 20.1; 22.5.4)
(b) Vol. **128**, No. 1, p. 329–336 (1985).

K. SIERAŃSKI and J. SZATKOWSKI
*Free-Carrier Absorption in Semimagnetic Semiconductors
(18.1 and 20.1; 13.1; 22.4.4)
(b) Vol. **128**, No. 1, p. K33–K35 (1985).

J. A. Duffy
*The Absorption Edge of Lead(II) in Oxide Systems
(20.1; 2; 22.6.1)
(a) Vol. **88**, No. 1, p. K55—K57 (1985).

Raviendra D̃., Sundeep, and J. K. Sharma
In_2O_3 and In_2O_3:Sn (ITO) Films by Post-Oxidation of Metal Films
(14.3.1; 3; 20.1; 22.6)
(a) Vol. **88**, No. 1, p. K83—K86 (1985).

V. M. Burlakov, A. P. Litvinchuk, V. N. Pyrkov, G. G. Tarasov, and N. I. Vitrikhovskii
*Optical Properties of the Quaternary II–VI Mixed Crystals in the Far Infrared Region
(6 and 20.1; 22.4.4)
(b) Vol. **128**, No. 2, p. 389—400 (1985).

A. G. Abdullaev, K. R. Allakhverdiev, T. D. Ibragimov, and R. M. Sardarly
*Two-Phonon Processes in TlSe and $TlInSe_2$
(6 and 20.1; 22.3; 22.3.1)
(b) Vol. **128**, No. 2, p. 401—409 (1985).

L. Wendler
*Phonon-Polaritons in Bilayer Systems. s-Polarized Guided Wave Phonon-Polaritons
(6 and 20.1; 13.5; 22.2.1; 22.2.2)
(b) Vol. **128**, No. 2, p. 425—437 (1985).

A. M. Basu, A. Choudhury, S. Chaudhuri, P. Mukhopadhyay, and B. B. Baliga
The Influence of Impurities on the Range of 28 MeV α-Particles in NaCl Single Crystals
(10.2 and 11; 20.1; 22.5.2)
(b) Vol. **128**, No. 2, p. 483—488 (1985).

N. I. Grigorchuk
*Energy Spectrum and Light Absorption Coefficient for Bound Electron–Exciton States in One- and Two-Dimensional Molecular Crystals
(13.5.1 and 20.1; 22.9)
(b) Vol. **128**, No. 2, p. 599—610 (1985).

I. Rückmann, K. Jarašiūnas, and E. Gaubas
*Transient Light-Induced Gratings Due to Excitons in CdSe
(13.5.1 and 20.1; 22.4.2)
(b) Vol. **128**, No. 2, p. 627—641 (1985).

Yu. K. Pozhela, E. V. Starikov, and P. N. Shiktorov
*Far Infrared Absorption by Hot Holes in p-Ge under $\boldsymbol{E} \perp \boldsymbol{B}$ Fields
(14.3.4 and 20.1; 22.1.1)
(b) Vol. **128**, No. 2, p. 653—661 (1985).

B. Atoussi, C. Pedrini, B. Moine, and C. Madej
Spectroscopic Investigation and Polarized Luminescence of Silver Pair Centers in Sodium Chloride
(13.4 and 20.3; 20.1; 22.5.2)
(b) Vol. **128**, No. 2, p. 683—692 (1985).

R. C. Bhuniya, P. Sridaran, and V. V. Ratnam
Emission Processes in Luminescence and Thermally Stimulated Luminescence of Ag- and Ca-Doped KCl
(13.4; 20.1; 20.3; 22.5.2)
(b) Vol. **129**, No. 1, p. 331–338 (1985).

W. Ulrici and P. Kleinert
*Optical Absorption of Cr^{4+} in GaAs:Cr
(13.4 and 20.1; 6; 22.2.1)
(b) Vol. **129**, No. 1, p. 339–347 (1985).

P. Triggs and F. Lévy
*Optical and Electrical Properties of Ruthenium-Doped TiO_2
(13.4 and 20.1; 14.3; 22.6)
(b) Vol. **129**, No. 1, p. 363–374 (1985).

J. Horák, P. Lošťák, and M. Matyáš
Point Defects in Sn-Doped Sb_2Te_3 Single Crystals
(13.4 and 14.3; 15; 20.1; 22.7)
(b) Vol. **129**, No. 1, p. 381–386 (1985).

T. R. Gawron and J. Mycielski
Erratum to: Shallow Impurity States in Strong Magnetic Field
(13.4; 20.1; 22)
(b) Vol. **129**, No. 1, p. 448 (1985).

A. F. Goncharov, I. N. Makarenko, and N. M. Gasanly
*Effect of Pressure on the Optical Absorption Edge in InS Monocrystals
(1.2 and 20.1; 13.1; 22.3)
(b) Vol. **129**, No. 1, p. K79–K81 (1985).

K. Atobe, N. Nishimoto, and M. Nakagawa
*Irradiation-Induced Aggregate Centres in Single Crystal Al_2O_3
(10.2 and 20.1; 11; 22.6)
(a) Vol. **89**, No. 1, p. 155–162 (1985).

A. K. Abass, F. Y. M. Al-Eithan, and R. H. Misho
*Indirect Electronic Transitions in Single Crystals of Triglycine Sulfate
(20.1; 6; 13.1; 22.9)
(a) Vol. **89**, No. 1, p. 225–229 (1985).

D. Lapraz, F. Gaume, and M. Barland
On the Thermoluminescent Mechanism of a Calcium Fluorapatite Single Crystal Doped with Mn^{2+}
(13.4 and 20.3; 14.3; 20.1; 22.8)
(a) Vol. **89**, No. 1, p. 249–253 (1985).

U. K. Reddy
Electrical and Optical Properties of Non-Crystalline As–Se–Cd Thin Films
(2 and 14.3.1; 13.4; 20.1; 22.8)
(a) Vol. **89**, No. 1, p. 255–261 (1985).

R. KAWASHIMA, K. KATSUKI, and K. SUZUKI
Temperature Dependence of Optical Absorption near the Structural Phase Transition of Potassium Nitrate
(1.2 and 13.1; 14.4.1; 20.1; 22.8.1)
(b) Vol. **129**, No. 2, p. 697–705 (1985).

U. NEITZEL and K. BÄRNER
*Optical and Magnetic Investigations of Some NiAs-Type Transition Metal Compounds
(18 and 20.1; 13.1; 21.1)
(b) Vol. **129**, No. 2, p. 707–715 (1985).

R. STEARNS, R. BRAUNSTEIN, and L. MULDAWER
*Optical Properties of β'-AuZn by Wavelength-Modulated Derivative Spectroscopy
(13.1 and 20.1; 21.1; 21.6)
(b) Vol. **129**, No. 2, p. 735–743 (1985).

G. JACOBS
*Transient I_c Centres in Alkali Halides
(6 and 20.1; 13.4; 22.5.2)
(b) Vol. **129**, No. 2, p. 755–762 (1985).

M. SELG
Theory of Vibrational Relaxation of R_2^* Centres in Rare Gas-Solids. Application to Ne_2^*
(13.5.1 and 20.3; 6; 20.1; 23)
(b) Vol. **129**, No. 2, p. 775–784 (1985).

R. K. CHAUDHARY, A. V. KRISHNA RAO, and K. V. RAO
*Optical Anisotropy and Dichroism of A-Centres in NaCl and KCl Single Crystals Irradiated with X-Rays under DC Fields
(10.2 and 20.1; 22.5.2)
(b) Vol. **129**, No. 2, p. 785–788 (1985).

R. CAPELLETTI, M. MANFREDI, and M. SOLZI
*Energy Transfer Mechanisms in the NaCl:Pb^{++}, Mn^{++} System along Aggregation Processes
(13.4 and 20.1; 10.2; 22.5.2)
(b) Vol. **129**, No. 2, p. 789–798 (1985).

M. G. KAPLUNOV, E. B. YAGUBSKII, L. P. ROSENBERG, and YU. G. BORODKO
*Optical Properties of the Two Crystal Modifications of the Organic Conductor $(BEDT\text{-}TTF)_2I_3$
(20.1; 6; 13; 22.9)
(a) Vol. **89**, No. 2, p. 509–516 (1985).

B. T. DESHMUKH, B. H. GHIYA, and R. B. PODE
On the Dose Dependence of the Formation of Z_1 Centres in KCl:Ca
(10.2 and 20.3; 11; 20.1; 22.5.2)
(a) Vol. **89**, No. 2, p. 517–520 (1985).

C. ASCHERON, C. BAUER, H. SOBOTTA, and V. RIEDE
Investigations of Hydrogen Implanted GaP Single Crystals by Means of Particle Induced γ-Spectroscopy, Infrared Spectroscopy, and Rutherford Backscattering Channelling Technique
(11; 20.1; 22.2.2)
(a) Vol. **89**, No. 2, p. 549–557 (1985).

V. I. Sokolov, T. P. Surkova, M. P. Kulakov, and A. V. Fadeev
New Experimental Evidence Concerning the Nature of L, M, and N Lines in ZnSe:Co
(13.4 and 13.5.1; 14.3.4; 20.1; 22.4.2)
(b) Vol. **130**, No. 1, p. 267–272 (1985).

G. I. Pilipenko, V. G. Plekhanov, V. I. Tyutyunnik, O. I. Tyutyunnik, and F. F. Gavrilov
*Direct Observation of Bound Excitons in the Spectra of Resonance Secondary Emission of LiH:Mg Crystals
(13.5.1 and 20.1; 20.3; 22.5)
(b) Vol. **130**, No. 1, p. 273–280 (1985).

I. V. Brovchenko, N. I. Ostapenko, M. T. Shpak, L. L. Valkūnas, and G. B. Vektaris
Resonance Interaction between Impurities and Crystal Molecules. Application to Benzselenophene in Naphthalene
(13.4; 20.1; 20.3; 22.9)
(b) Vol. **130**, No. 1, p. 281–291 (1985).

V. N. Salomatov, N. N. Kristoffel, and I. A. Parfianovich
*The $F_H(H^-)$ Center Energy Parameters and the Nature of F Center Unrelaxed Ground State in KCl
(13.4 and 20.1; 10.2; 22.5.2)
(b) Vol. **130**, No. 1, p. 293–296 (1985).

H. Zimmermann and R. Boyn
*Site Selective Spectroscopy of ZnS:Tm^{3+} Crystals
(13.4 and 20.1; 20.2; 22.4.1)
(b) Vol. **130**, No. 1, p. 315–324 (1985).

S. W. Biernacki
The Splitting of Impurity d-States in Zincblende Crystals
(13.4; 20.1; 22.4.2)
(b) Vol. **130**, No. 1, p. K59–K63 (1985).

J. Berdowski and M. Hajduk
*The Influence of Isotope Effects on the Interaction of Light and Surface Acoustic Waves in DADA Crystals
(7 and 20.1; 14.4.1; 22.8.1)
(a) Vol. **90**, No. 1, p. 157–165 (1985).

P. D. Cikmach, J. J. Kleperis, A. R. Lusis, and G. M. Ramāns
*Influence of Ordering of Structure on the Optical Properties of WO_3 Thin Film
(1.5 and 20.1; 22.6)
(a) Vol. **90**, No. 1, p. K1–K4 (1985).

K. Yasuda, A. Yoshida, M. Takeda, H. Masuda, and I. Akasaki
Boron Nitride Films Prepared by RF Sputtering
(2 and 3; 6; 20.1; 22)
(a) Vol. **90**, No. 1, p. K7–K9 (1985).

V. A. Bibik, I. V. Blonskii, M. S. Brodin, and N. A. Davydova
Structural Phase Transition in the Layer Semiconductor PbI_2 Induced by Laser Irradiation
(1.2 and 13.5.1; 20.1; 22.5)
(a) Vol. **90**, No. 1, p. K11–K14 (1985).

C. S. SUNANDANA and K. S. RAO
ESR and Other Properties of Copper-Substituted Tellurium Vanadate Glasses
(2; 14; 19; 20.1; 22.6.1)
(a) Vol. **90**, No. 2, p. 681–690 (1985).

P. GAWORZEWSKI, E. HILD, and K. SCHMALZ
Oxygen Donor Formation and Oxygen Precipitation in Czochralski-Silicon Due to Heat Treatment at 600 to 800 °C
(10.2; 14.3; 20.1; 22.1.2)
(a) Vol. **90**, No. 2, p. K151–K156 (1985).

A. J. WILLIS and M. A. MARAIS
*Deduction of Semiconductor Parameters from the Frequency Dependence of Free Carrier Reflectivity
(20.1; 22.2.1)
(a) Vol. **90**, No. 2, p. K193–K197 (1985).

V. GHIORDĂNESCU and L. DOLLINGHER
*Some Observations on the Absorption of Copper in Photochromic Glass
(20.1; 13.4; 22.8)
(a) Vol. **90**, No. 2, p. K199–K204 (1985).

C. N. AFONSO, C. ORTIZ, and G. J. CLARK
Defect Formation in LiF by Low Energy Ion Implantation
(10.2 and 11; 20.1; 22.5.2)
(b) Vol. **131**, No. 1, p. 87–96 (1985).

A. L. TRONCONI and O. A. C. NUNES
*Spin-Flip Intraband Absorption of Intense Laser Field in the Europium Compounds
(13.5.4 and 20.1; 22.6)
(b) Vol. **131**, No. 1, p. 137–140 (1985).

S. F. TEREKHOVA, N. A. ONISHCHENKO, O. N. TALENSKII, and S. A. PENDUR
*The Refractive Index and the Exciton Transition Oscillator Strength of CdSe Single Crystals
(13.5.1 and 20.1; 22.4.2)
(b) Vol. **131**, No. 1, p. 207–212 (1985).

A. V. SHERMAN
*Exciton Absorption Spectrum
(13.5.1 and 20.1; 6)
(b) Vol. **131**, No. 1, p. 225–233 (1985).

E. G. TSITSISHVILI
*Piezo-Optical Activity of Cubic Crystals
(12.2 and 20.1; 13.1; 13.5.1; 22.5)
(b) Vol. **131**, No. 1, p. 279–290 (1985).

R. C. BHUNIYA and S. E. RAO
*Colour Centres and Luminescence in KCl:Zn Crystals
(20.1 and 20.3; 10.2; 13.4; 22.5.2)
(b) Vol. **131**, No. 1, p. 339–348 (1985).

G. Contreras, L. Tapfer, A. K. Sood, and M. Cardona
Physical Properties of Ion-Implanted Laser Annealed n-Type Germanium
(6 and 13.4; 14.3; 20.1; 22.1.1)
(b) Vol. **131**, No. 2, p. 475–487 (1985).

T. S. Shi, L. M. Xie, G. R. Bai, and M. W. Qi
The Nature of Two Intense Si–H IR Stretching Bands in FZ-Si:H
(6; 10.2; 20.1; 22.1.2)
(b) Vol. **131**, No. 2, p. 511–517 (1985).

Yu. M. Seidov and A. M. Suleimanov
*Theory of Absorption and Scattering of Light by Singlet–Triplet Magnetic Systems
(13.5.4 and 20.1; 18.2; 18.4)
(b) Vol. **131**, No. 2, p. 545–554 (1985).

M. S. Brodin, I. V. Blonskii, A. O. Gushcha, and V. V. Tishenko
Secondary Emission Spectra and Energy Relaxation of Polaritons in Layer Polar Semiconductors
(6 and 13.5.1; 20.1; 20.3; 22.5)
(b) Vol. **131**, No. 2, p. 615–620 (1985).

F. V. Kusmartsev
*The Narrow Bands of Exciton Absorption in Polar Solids and Alloys. Urbach's Rule
(13.5.1 and 20.1; 22.5; 22.5.4)
(b) Vol. **131**, No. 2, p. 621–629 (1985).

R. Stearns, R. Braunstein, and L. Muldawer
*Wavelength Modulated Spectra of α-$Ag_{0.7}Zn_{0.3}$ near the Optical Absorption Edge
(13.1 and 20.1; 21.1; 21.6)
(b) Vol. **131**, No. 2, p. 659–665 (1985).

W. Ulrici
Photoinduced Optical Absorption of Cr^{4+} in Semi-Insulating GaAs:Cr
(13.4; 16; 19; 20.1; 22.2.1)
(b) Vol. **131**, No. 2, p. 707–717 (1985).

W. Ulrici, K. Friedland, L. Eaves, and D. P. Halliday
Optical and Electrical Properties of Vanadium-Doped GaAs
(13.4; 14.3; 20.1; 22.2.1)
(b) Vol. **131**, No. 2, p. 719–728 (1985).

A. Schenk, R. Enderlein, and D. Suisky
*Electroabsorption for Deep Level to Band Transitions Accompanied by Multiphonon Processes
(13.4 and 20.1; 6; 14.3.4; 22)
(b) Vol. **131**, No. 2, p. 729–739 (1985).

T. Schröder, O. Dössel, H. Nahme, and N. Schwentner
*Transient Absorption of Xe and Kr Guest Atoms in Ar Crystals
(13.4 and 20.1; 23)
(b) Vol. **131**, No. 2, p. 741–746 (1985).

H. B. Gon, S. K. Gupta, and K. V. Rao
*Optical Absorption and Thermoluminescence of AC Field-Treated LiF Single Crystals later Irradiated with X-Rays
(20.1 and 20.3; 10.2; 22.5.2)
(a) Vol. **91**, No. 2, p. K143—K146 (1985).

F. Bolduan, W. Hönle, H. D. Hochheimer, and W. Henkel
Temperature and High Pressure Raman Study of M_3P_7 (M = Li, Na, K, Rb, Cs) Compounds
(1.2; 20.1; 21.2)
(b) Vol. **132**, No. 1, p. 41—50 (1985).

D. Duval and R. A. Condrate, Sr.
Temperature Dependence of the Raman Spectra of Calcium Oxalate Monohydrate
(1.2; 6; 20.1; 22.9)
(b) Vol. **132**, No. 1, p. 83—92 (1985).

G. Irmer, A. Heinrich, and J. Monecke
*First Order Raman Scattering in $CdSnP_2$
(6 and 20.1; 22.8)
(b) Vol. **132**, No. 1, p. 93—98 (1985).

V. May
*Light Transmission at the ls-Exciton Resonance of CdS in the Presence of a Density Profile of Excitons
(13.5.1 and 20.1; 22.4.1)
(b) Vol. **132**, No. 1, p. 145—151 (1985).

T. Tsuboi and R. Laiho
*Blackening of $CsFeCl_3$ by External Magnetic Field
(18.2 and 20.1; 13.5; 22.5)
(b) Vol. **132**, No. 1, p. K33—K35 (1985).

I. G. Oganova and I. M. Aliev
*On the Exponential Shape of the Absorption Edge in Heteroepitaxial n-InSb Layers
(20.1; 14.3.1; 22.2.3)
(b) Vol. **132**, No. 1, p. K41—K44 (1985).

A. Brau, J. P. Farges, and P. Dupuis
*Induced Effect of Charge Localisation in Complex TCNQ Salts. Evidence from Powder Infrared Absorption Spectra
(20.1; 14.3; 22.9)
(b) Vol. **132**, No. 1, p. K67—K71 (1985).

Ye Ming, E. Krätzig, and R. Orlowski
*Photorefractive Effects in $LiNbO_3$:Cr Induced by Two-Step Excitation
(20.1; 13.4; 16; 22.8.1)
(a) Vol. **92**, No. 1, p. 221—229 (1985).

M. Quintero and J. C. Wolley
Crystallography and Optical Energy Gap Values for $Cd_{2x}(AgIn)_yMn_{2z}Te_2$ Alloys
(1.1; 13.1; 20.1; 22.8)
(a) Vol. **92**, No. 2, p. 449–456 (1985).

K. Y. Lay, H. Neff, and K. J. Bachmann
*Photoreflectance and Photoluminescence Studies of $Cd_{1-x}Mn_xTe$: Broadening Effects
(20.1 and 20.3; 22.4.3)
(a) Vol. **92**, No. 2, p. 567–571 (1985).

L. Tarricone, E. Grilli, M. Guzzi, and L. Zanotti
Optical Absorption and Recombination Processes in $CdGaInS_4$ Layered Crystals
(13.4 and 20.3; 16; 20.1; 22.8)
(a) Vol. **92**, No. 2, p. 595–602 (1985).

G. M. Mileva and L. I. Pavlov
Third-Order Optical Polarizability of ZnS at Low Temperature
(13.4; 20.1; 22.4.1)
(a) Vol. **92**, No. 2, p. 603–607 (1985).

20.2 Lasers

Review Articles

A. A. Kaminskii
*Achievements in the Field of Physics and Spectroscopy of Activated Laser Crystals
(20.2; 20.1; 20.3; 22.8)
(a) Vol. **87**, No. 1, p. 11–57 (1985).

Yu. E. Perlin and A. A. Kaminskii
Nonradiative Transitions of the Trivalent Lanthanides in Insulating Laser Crystals
(13.4 and 20.3; 6; 20.2; 22.8.1)
(b) Vol. **132**, No. 1, p. 11–40 (1985).

Original Papers and Short Notes

U. Braatz, F. Fink, and P. Klinger
Optoelectronic Switches on the Basis of Mercuric Iodide (Red Phase, α-HgI_2)
(14.3.4; 16; 20.2; 22.5)
(a) Vol. **88**, No. 1, p. K95–K97 (1985).

S. P. Chernov, L. I. Devyatkova, O. N. Ivanova, A. A. Kaminskii, V. V. Mikhailin, S. N. Rudnev, and T. V. Uvarova
*$5d^14f^{N-1}$–$4f^N$ Absorption and Luminescence of Ce^{3+}, Pr^{3+}, and Nd^{3+} Ions in BaY_2F_8 Single Crystals
(20.2 and 20.3; 13.4; 22.8)
(a) Vol. **88**, No. 2, p. K169–K173 (1985).

H. Zimmermann and R. Boyn
Site Selective Spectroscopy of $ZnS:Tm^{3+}$ Crystals
(13.4 and 20.1; 20.2; 22.4.1)
(b) Vol. **130**, No. 1, p. 315–324 (1985).

B. R. Sood
*Rate Equations for Phasers
(6 and 20.2)
(a) Vol. **90**, No. 1, p. 143–150 (1985).

R.-L. Ma and P. Y. Yu
*A New Technique for Measuring Temporal Profiles of Picosecond and Femtosecond Laser Pulses Using Self-Induced Deflection in a Semiconductor Wedge
(20.2; 13.5.2; 22.1.2)
(a) Vol. **90**, No. 1, p. 343–348 (1985).

A. A. Kaminskii, K. Kurbanov, S. E. Sarkisov, M. M. Sattarova, T. V. Uvarova, and P. P. Fedorov
*Stimulated Emission of Nd^{3+} Ions in Non-Stoichiometric $Cd_{1-x}Ce_xFe_{2+x}$ and $Cd_{1-x}Nd_xFe_{2+x}$ Fluorides with Fluorite Structure
(20.2 and 20.3; 13.4; 22.8)
(a) Vol. **90**, No. 1, p. K55–K60 (1985).

L. K. Aminov, A. A. Kaminskii, and M. I. Chertanov
Theory of Luminescence Anisotropy of $YAlO_3:Nd^{3+}$ Single Crystals
(13.4; 20.2; 20.3; 22.8.1)
(b) Vol. **130**, No. 2, p. 757–766 (1985).

G. Beister, P. Krispin, J. Maege, H.-J. Bachert, and St. Stoeff
*Emission from a Semi-Insulating Layer in AlGaAs-GaAs Double Heterostructures
(14.3.3 and 20.2; 13.4; 22.2.4)
(a) Vol. **91**, No. 1, p. K93–K96 (1985).

R. H. Enns and S. S. Rangnekar
Solitary Wave Solutions of the Driven and Damped Three-Wave Problem
(13.5; 6; 20.2)
(b) Vol. **132**, No. 2, p. K121–K126 (1985).

A. P. Shkadarevich, M. V. Nikanovich, V. M. Reiterov, D. S. Umreiko, and A. R. Yarmolkevich
*Luminescence and Stimulated Emission of M_A Colour Centers in $MgF_2:Li$ Crystal
(20.2 and 20.3; 13.4; 22.5.3)
(a) Vol. **92**, No. 2, p. K135–K137 (1985).

20.3 Luminescence (X-Ray Spectra See 20)

Review Articles

Original Papers and Short Notes

W. RUDNICK, R. HAENSEL, H. NAHME, and N. SCHWENTNER
*Luminescence Efficiency and Grain Sizes of Ar Crystals
(1.4 and 20.3; 23)
(a) Vol. 87, No. 1, p. 319–322 (1985).

B. RHEINLÄNDER, G. OELGART, H. HAEFNER, and R. PICKENHAIN
*Degradation of Green Light Emitting Diodes LPE-GaP:N. II. Injection Luminescence Spectroscopy and Kinetics
(14.3.4 and 20.3; 22.2.2)
(a) Vol. 87, No. 1, p. 373–381 (1985).

N. A. DROZDOV, N. I. ZYATKOVA, A. A. PATRIN, and V. D. TKACHEV
*Photoluminescence of Silicon Annealed at Low Temperatures (420 to 550 °C)
(10.2 and 20.3; 13.4; 13.5.1; 22.1.2)
(a) Vol. 87, No. 1, p. K45–K48 (1985).

P. V. VODOLAZSKII, B. R. KIYAK, M. G. MATSKO, V. B. NOSOV, G. T. PETROVSKII, and A. V. SHATILOV
*Reversible and Irreversible Changes in Exciton Spectra of ZnSe Luminescence
(13.5.1 and 20.3; 22.4.2)
(a) Vol. 87, No. 1, p. K69–K71 (1985).

P. PIPINYS, A. RIMEIKA, and A. KIVERIS
*On a Trap Ionization Mechanism by Electric Field in CdF_2
(13.4 and 20.3; 14.3.4; 22.5.3)
(a) Vol. 87, No. 1, p. K89–K91 (1985).

M. J. J. LAMMERS and G. BLASSE
*Luminescence Properties of Rare-Earth-Activated Gadolinium Fluoride (GdF_3) and Oxyfluorides (GdOF)
(13.4 and 20.3; 22.8)
(b) Vol. 127, No. 2, p. 663–671 (1985).

A. REQUARDT and G. LEHMANN
A Hole Center Causing Rose Coloration in $AlPO_4$ and $GaPO_4$
(13.4; 11; 19; 20.1; 20.3; 22.8.1)
(b) Vol. 127, No. 2, p. 695–701 (1985).

V. S. KORTOV, T. S. BESSONOVA, M. S. AKSELROD, and I. I. MILMAN
*Hole-Induced Exoelectron Emission and Luminescence of Corundum Doped with Mg
(17 and 20.3; 10.2; 13.4; 22.6)
(a) Vol. 87, No. 2, p. 629–639 (1985).

R. STEGMANN, G. OELGART, and F. THUSELT
*The Influence of Coulomb Screening and of NN_i-Pairs on the Recombination of Bound Excitons in $GaAs_{1-x}P_x$:N Mixed Crystals
(13.5.1 and 20.3; 22.2.4)
(a) Vol. 87, No. 2, p. 641–650 (1985).

N. B. Lukyanchikova, N. P. Garbar, and M. I. Lisyanskii
*Study of Degradation of Diode Structures by the Use of Probes
(14.3.3 and 20.3; 22.2.2)
(a) Vol. **87**, No. 2, p. 693–698 (1985).

D. Lapraz and P. Iacconi
*Photo-Induced Thermoluminescence of X-Irradiated α-Al_2O_3
(11 and 20.3; 22.6)
(a) Vol. **87**, No. 2, p. K167–K169 (1985).

G. Massé
*Time Resolved Spectra in $CuGaS_2$
(13.4 and 20.3; 22.8)
(a) Vol. **87**, No. 2, p. K171–K173 (1985).

G. Blasse and G. J. Dirksen
*The Influence of the Phase Transformation on the Luminescence of $NbPO_5$
(1.2 and 20.3; 22.8)
(a) Vol. **87**, No. 2, p. K181–K184 (1985).

K. Iwahana
*Theory for the Magnetic Circular Polarization of the F-Center Luminescence
(13.4 and 20.3; 18.1; 22.5.2)
(b) Vol. **128**, No. 1, p. 297–311 (1985).

B. G. Tagiev, G. M. Niftiev, and F. Sh. Aidaev
*Photoluminescence of GaSe:Tm Single Crystals
(20.3; 13.4; 22.3)
(b) Vol. **128**, No. 1, p. K65–K68 (1985).

E. Mugeński and R. Cywiński
*Low-Temperature Photoluminescence of Eu^{2+} Aggregate Centres in NaCl Matrix
(20.3; 22.5.2)
(b) Vol. **128**, No. 1, p. K75–K79 (1985).

B. Ray and P. Sephton
*Reflection and Fluorescence Emission Spectra in the $Ag_{1-x}Cu_xI$ System
(20.1 and 20.3; 13.4; 22.5.4)
(a) Vol. **88**, No. 1, p. 261–265 (1985).

W. Hohenau
*On the Kinetics of X-Ray Induced Luminescence in Natural Quartz
(13.4 and 20.3; 22.6)
(a) Vol. **88**, No. 1, p. 267–276 (1985).

J. W. Tomm K. Herrmann, A. E. Yunovich, and S. I. Solotov
*Photoluminescence Spectra of $Pb_{1-x}Sn_xTe$ Epitaxial Layers and Heterostructures up to Room Temperature
(20.3; 14.3.1; 22.8)
(a) Vol. **88**, No. 1, p. 277–281 (1985).

A. I. Fiksel, Yu. I. Aristov, and V. N. Parmon
*Interpretation of the Curves of Thermally Stimulated Luminescence in the Case of Tunneling Recombination
(20.3; 22.6; 22.9)
(a) Vol. **88**, No. 1, p. 283–289 (1985).

P. Silfsten, P. Ketolainen, and M. Lehkonen
*On the Thermal'Resistance and Its Possible Role in Thermoluminescence Measurements
(8 and 20.3; 22.5.2; 22.5.4)
(a) Vol. **88**, No. 1, p. K67–K70 (1985).

R. Mićić and B. Drašković
*Some Photoluminescence Properties of Mn^{2+} in Magnesium Bromide
(10.2 and 20.3; 22.5.3)
(b) Vol. **128**, No. 2, p. 489–494 (1985).

E. K. Arushanov, L. L. Kulyuk, L. N. Lukyanova, A. N. Nateprov, S. I. Radautsan, and A. A. Shtanov
*Stimulated Emission Spectra and Valence Band Structure in the $Cd_3(As_xP_{1-x})_2$ System
(13.1 and 20.3; 22.3.1)
(b) Vol. **128**, No. 2, p. 583–589 (1985).

B. Atoussi, C. Pedrini, B. Moine, and C. Madej
*Spectroscopic Investigation and Polarized Luminescence of Silver Pair Centers in Sodium Chloride
(13.4 and 20.3; 20.1; 22.5.2)
(b) Vol. **128**, No. 2, p. 683–692 (1985).

C. Benecke, W. Busse, H.-E. Gumlich, and U. W. Pohl
*Systems of Zero-Phonon Lines Due to Mn^{2+} Pairs in ZnS
(13.4 and 20.3; 22.4.1)
(b) Vol. **128**, No. 2, p. 701–707 (1985).

R. Capelletti, R. Cywiński, M. Manfredi, and M. Solzi
Radiative Energy Transfer Process in the KCl:Pb^{++}, Eu^{++} System
(13.4; 20.3; 22.5.2)
(b) Vol. **128**, No. 2, p. 717–722 (1985).

K. Iwahana, P. Knoll, and H. Kuzmany
Effect of Irradiation with Intense Laser Light on Trans-Polyacetylene
(11; 20.3; 22.9)
(b) Vol. **128**, No. 2, p. 735–741 (1985).

Th. Kleinefeld and W. von der Osten
*Erratum to: Localized and Free Biexcitons in Silver Bromide
(13.5.1 and 20.3; 22.5.1)
(b) Vol. **128**, No. 2, p. 808 (1985).

H. Kobayashi, S. Tanaka, V. Shanker, M. Shiiki, T. Kunou, J. Mita, and H. Sasakura
*Multicolor Electroluminescent ZnS Thin Films Doped with Rare Earth Fluorides
(20.3; 14.3.1; 22.4.1)
(a) Vol. **88**, No. 2, p. 713–720 (1985).

A. B. AHMED and R. K. GARTIA
*Application of the Fractional Glow Technique in the Analysis of a Complex Thermoluminescence Pattern
(13.4 and 20.3; 22.8.1)
(a) Vol. **89**, No. 1, p. 231–235 (1985).

J. A. GARCÍA, A. REMÓN, and J. PIQUERAS
*Annealing-Induced Changes in the Photoluminescence of Deformed CaO Single Crystals
(20.3; 22.6)
(a) Vol. **89**, No. 1, p. 237–241 (1985).

D. LAPRAZ, F. GAUME, and M. BARLAND
*On the Thermoluminescent Mechanism of a Calcium Fluorapatite Single Crystal Doped with Mn^{2+}
(13.4 and 20.3; 14.3; 20.1; 22.8)
(a) Vol. **89**, No. 1, p. 249–253 (1985).

YU. V. ZHILYAEV, V. V. KRIVOLAPCHUK, A. V. RODIONOV, V. V. ROSSIN, T. V. ROSSINA, and YU. N. SVESHNIKOV
*The Investigation of a Transition Layer in Epitaxial GaAs by the Low Temperature Photoluminescence Technique
(20.3; 13.4; 22.2.1)
(a) Vol. **89**, No. 1, p. K61–K64 (1985).

S. SHIGETOMI and T. MATSUMORI
*Photoluminescence Spectra of Si-Implanted GaAs
(20.3; 10.2; 13.4; 22.2.1)
(a) Vol. **89**, No. 1, p. K79–K82 (1985).

R. CYWIŃSKI, E. MUGEŃSKI, W. NOWY-WIECHUŁA, and J. WIECHUŁA
*Aggregation of Mn^{2+}-Cation Vacancy Dipoles in KCl Matrix
(10.2 and 20.3; 22.5.2)
(b) Vol. **129**, No. 2, p. 601–605 (1985).

A. A. O'CONNEL-BRONIN
*Resonance LO Phonon-Assisted Luminescence and Light and Heavy Excitons in NaI Crystals
(6 and 20.3; 13.5.1; 22.5.2)
(b) Vol. **129**, No. 2, p. 675–681 (1985).

CH. NÖLDEKE, W. METZGER, R. P. HUEBENER, and H. SCHNEIDER
*Exciton Impact Ionization and Electric Conductivity in Germanium at Low Temperature
(13.5.1 and 20.3; 14.3; 22.1.1)
(b) Vol. **129**, No. 2, p. 687–695 (1985).

M. SELG
*Theory of Vibrational Relaxation of R_2^* Centres in Rare Gas-Solids. Application to Ne_2^*
(13.5.1 and 20.3; 6; 20.1; 23)
(b) Vol. **129**, No. 2, p. 775–784 (1985).

E. Radzhabov
*Photodissociation of $O^{--}V_a^+$ Centers in Lithium Fluoride
(13.4 and 20.3; 10.2; 22.5.2)
(b) Vol. **130**, No. 1, p. K55—K58 (1985).

V. V. Vasilev and I. P. Mikhailovskii
*Effect of High-Temperature Annealing on PL Spectra of Silicon Nitride
(20.3; 22)
(a) Vol. **90**, No. 1, p. 355—358 (1985).

J. Llopis, C. Ballesteros, and J. Piqueras
*Cathodoluminescence from Deformed SrO
(20.3; 22.6)
(a) Vol. **90**, No. 1, p. 359—364 (1985).

A. A. Kaminskii, K. Kurbanov, S. E. Sarkisov, M. M. Sattarova, T. V. Uvarova, and P. P. Fedorov
*Stimulated Emission of Nd^{3+} Ions in Non-Stoichiometric $Cd_{1-x}Ce_xFe_{2+x}$ and $Cd_{1-x}Nd_xF_{2+x}$ Fluorides with Fluorite Structure
(20.2 and 20.3; 13.4; 22.8)
(a) Vol. **90**, No. 1, p. K55—K60 (1985).

A. V. Egranov and A. I. Nepomnyachikh
*Emission Spectra of NaCl:Mg Crystals
(13.4 and 20.3; 10.2; 22.5.2)
(a) Vol. **90**, No. 1, p. K65—K69 (1985).

I. J. Booth and C. F. Schwerdtfeger
*Optically Detected Cyclotron Resonance in AgBr
(19 and 20.3; 13.4; 22.5.1)
(b) Vol. **130**, No. 2, p. 749—756 (1985).

L. K. Aminov, A. A. Kaminskii, and M. I. Chertanov
Theory of Luminescence Anisotropy of $YAlO_3:Nd^{3+}$ Single Crystals
(13.4; 20.2; 20.3; 22.8.1)
(b) Vol. **130**, No. 2, p. 757—766 (1985).

N. Takeuchi, K. Ohtaki, and S. Nakamura
*Kinetics of Thermoluminescence in X-Irradiated KCl:Ag Crystals
(20.3; 11; 22.5.2)
(b) Vol. **130**, No. 2, p. K145—K148 (1985).

A. Opanowicz
*Simple Model for Thermoluminescence in Conductive Semiconductors
(13.4 and 20.3; 22.4.2)
(a) Vol. **90**, No. 2, p. 675—680 (1985).

E. Grilli, M. Guzzi, E. Camerlenghi, and F. Pio
*On the Radiative Recombination in $ZnIn_2S_4$
(13.4 and 20.3; 22.8)
(a) Vol. **90**, No. 2, p. 691—701 (1985).

HJ. BERNHARDT
*Investigations of the Orange Luminescence of $PbMoO_4$ Crystals
(20.1 and 20.3; 13.4; 22.8.1)
(a) Vol. **91**, No. 2, p. 643–648 (1985).

S. KEMMLER-SACK, A. EHMANN, and D. WOLF
*Sensitized IR-Emission of Ho^{3+}, Er^{3+}, and Tm^{3+} in 12R-$Ba_3LaBW_2O_{12}$ (B = Er, Tm, Yb) and 12R-$Sr_3LaYbW_2O_{12}$
(13.4 and 20.3; 22.8)
(a) Vol. **91**, No. 2, p. 661–665 (1985).

H. B. GON, S. K. GUPTA, and K. V. RAO
*Optical Absorption and Thermoluminescence of AC Field-Treated LiF Single Crystals later Irradiated with X-Rays
(20.1 and 20.3; 10.2; 22.5.2)
(a) Vol. **91**, No. 2, p. K143–K146 (1985).

F. F. CHARFI and M. ZOUAGHI
Hot-Exciton Orientation in Indirect-Gap $GaAs_{1-x}P_x$ Alloys
(6 and 13.5.1; 20.3; 22.2.4)
(b) Vol. **132**, No. 1, p. 109–115 (1985).

M. J. SPRINGIS and J. A. VALBIS
*Red Luminescence of Color Centres in Sapphire
(13.4 and 20.3; 10.2; 22.6)
(b) Vol. **132**, No. 1, p. K61–K65 (1985).

E. GRILLI, M. GUZZI, E. CAMERLENGHI, and F. PIO
*Photoluminescence of $Zn_xCd_{1-x}In_2S_4$ Layered Alloys ($0.7 < x < 1$)
(13.4 and 20.3; 22.8)
(a) Vol. **92**, No. 1, p. 213–220 (1985).

V. K. JAIN and M. S. JAHAN
*Low-Temperature Thermoluminescent Characteristics of CaF_2:Mn
(13.4 and 20.3; 22.5.3)
(a) Vol. **92**, No. 1, p. 237–248 (1985).

J. P. LEYRIS and J. P. AICARDI
Photoconductivity Measurements on $AgGaS_2$ Crystals
(16; 20.3; 22.8)
(a) Vol. **92**, No. 1, p. 279–284 (1985).

V. A. BESPALOV, P. K. KASHKAROV, V. F. KISELEV, V. A. MATVEEV, and G. S. PLOTNIKOV
The Study of Slow Relaxation of Charged Centres in Insulator–Semiconductor Structures Using Fluorescent Molecule Probes
(14.3.2; 13.3; 20.3; 22.1.1; 22.6)
(a) Vol. **92**, No. 1, p. 315–326 (1985).

I. B. Bakhtiyarov, S. M. Nakhmetov, P. G. Rustamov, I. M. Niftiev, and O. B. Tagiev
Physical Properties of $Ga_2S_3 \cdot Eu_2O_3$-Based Glasses
(2; 14.3; 20.3; 22.3; 22.6)
(a) Vol. **92**, No. 1, p. K31—K35 (1985).

M. Sarkar
*Secondary Fluorescence Effect for Low Z Elements in Electron-Probe Microanalysis
(20.3; 21.1.1).
(a) Vol. **92**, No. 1, p. K73—K76 (1985).

H.-J. Müller
Influence of Local Environment on Bound Excitons and Bound Excitonic Molecules in III–V Alloys. II. Numerical Results for $Ga_xIn_{1-x}P$:N and the Role of Disorder Topology for the Optical Spectrum
(13.4 and 13.5.1; 20.3; 22.2.4)
(b) Vol. **132**, No. 2, p. 607—617 (1985).

J. P. Jouart, C. Bissieux, G. Mary, and M. Egée
*Study of the Sensitization Effect of Eu^{2+} Ions on the Trivalent Rare-Earth Fluorescence in CdF_2 Crystals
(13.4 and 20.3; 22.5.3)
(b) Vol. **132**, No. 2, p. 619—625 (1985).

A. B. Henriques and J. P. von der Weid
*Decay Kinetics of Triplet Molecular Excitons in Alkali Cyanides
(13.5.1 and 20.3; 22.9)
(b) Vol. **132**, No. 2, p. 627—633 (1985).

R. Cywiński, E. Mugeński, W. Nowy-Wiechuła, and J. Wiechuła
*Aggregation and Precipitation of Mn^{2+}–Cation Vacancy Dipoles in NaCl Matrix
(10.2 and 20.3; 22.5.2)
(b) Vol. **132**, No. 2, p. K91—K94 (1985).

K. Y. Lay, H. Neff, and K. J. Bachmann
*Photoreflectance and Photoluminescence Studies of $Cd_{1-x}Mn_xTe$: Broadening Effects
(20.1 and 20.3; 22.4.3)
(a) Vol. **92**, No. 2, p. 567—571 (1985).

J. K. Srivastava and S. J. Supe
Thermoluminescence of $CaSO_4$:Dy. A Semiempirical Approach for a Qualitative Analysis of Nonlinear Thermoluminescence Effects
(13.4; 20.3; 22.8.1)
(a) Vol. **92**, No. 2, p. 573—588 (1985).

L. Tarricone, E. Grilli, M. Guzzi, and L. Zanotti
*Optical Absorption and Recombination Processes in $CdGaInS_4$ Layered Crystals
(13.4 and 20.3; 16; 20.1; 22.8)
(a) Vol. **92**, No. 2, p. 595—602 (1985).

P. Trung Dung and M. Láznička
*Outdiffusion of Mn into GaAs Films Grown on Semi-Insulating GaAs Substrates by MBE
(9 and 20.3; 22.2.1)
(a) Vol. **92**, No. 2, p. K113–K116 (1985).

A. P. Shkadarevich, M. V. Nikanovich, V. M. Reiterov, D. S. Umreiko, and A. R. Yarmolkevich
*Luminescence and Stimulated Emission of M_A Colour Centers in MgF_2: Li Crystal
(20.2 and 20.3; 13.4; 22.5.3)
(a) Vol. **92**, No. 2, p. K135–K137 (1985).

K. D. Glinchuk, V. I. Guroshev, A. V. Prokhorovich, and N. S. Zayats
*Photoluminescence in Electron-Irradiated Unannealed and Annealed GaAs: Zn — Irradiation-Induced 1.26 and 1.39 eV Emission Bands
(11 and 20.3; 13.4; 22.2.1)
(a) Vol. **92**, No. 2, p. K145–K149 (1985).

21. Metals. Alloys

Review Articles

K. H. Fischer
Spin Glasses (II)
(18.2 and 18.4; 2; 5; 7; 8; 13; 14; 21; 21.1; 21.6; 22.6; 22.8)
(b) Vol. **130**, No. 1, p. 13–71 (1985).

V. Christoph and G. Röpke
Theory of Inverse Linear Response Coefficients
(13 and 14; 15; 21; 22)
(b) Vol. **131**, No. 1, p. 11–42 (1985).

S. V. Lubenets, V. I. Startsev, and L. S. Fomenko
Dynamics of Twinning in Metals and Alloys
(10.1; 21)
(a) Vol. **92**, No. 1, p. 11–55 (1985).

Original Papers and Short Notes

M. Tomak, Hanuman Singh, B. K. Sharma, and S. Manninen
Electron Momentum Distribution in Niobium and the Renormalised Free Atom Model
(13.1; 21)
(b) Vol. **127**, No. 1, p. 221–225 (1985).

D. Sen and S. K. Sen
Transport in Simple Metals in the Framework of a Unified Pseudopotential Study
(8 and 14.1; 15; 21; 21.2)
(b) Vol. **127**, No. 1, p. 377–383 (1985).

G. Schubert and V. Christoph
Calculation of Thermopower and Thermal Conductivity Using Correlation Functions
(14.1 and 15; 21)
(b) Vol. **127**, No. 1, p. K89–K93 (1985).

B. A. Men and M. L. Katsnelson
The Ground State Superstructures in Ordered Binary S.C., B.C.C., and F.C.C. Alloys with Pair Interactions up to Third Neighbours
(1.3; 21)
(a) Vol. **87**, No. 1, p. 93–108 (1985).

W. Wasserbäch
Electron–Phonon Scattering and Low-Temperature Thermal Conductivity of Niobium and Tantalum Single Crystals
(6 and 8; 21)
(b) Vol. **127**, No. 2, p. 481–492 (1985).

D. Schryvers, G. van Tendeloo, and S. Amelinckx
On the Ordering Mechanism of Ni_3Mo. A High Resolution Electron Microscopy Study
(1.3 and 1.4; 21; 21.1)
(a) Vol. **87**, No. 2, p. 401–415 (1985).

G. Neubert and S. Rogaschewski
Measurements of the Backscattering of 20 to 60 keV Electrons from Double Layers at Various Angles of Incidence
(11; 21; 21.3; 21.6)
(a) Vol. **87**, No. 2, p. 451–457 (1985).

M. Takeda, H. Yoshida, and H. Hashimoto
Local Tetragonality and Atomic Structure in Nb_3Sn Superconductor Studied by High Resolution Electron Microscopy
(1.4 and 14.2; 21)
(a) Vol. **87**, No. 2, p. 473–482 (1985).

A. V. Dobromyslov, G. V. Dolgikh, and I. A. Pereturina
Effect of Normal Stresses on Yield Stress of Molybdenum Single Crystals at Slip on the Plane $(\bar{2}11)$
(10.1; 12.1; 21)
(a) Vol. **87**, No. 2, p. 559–563 (1985).

L. P. Mokhracheva, Yu. G. Listopadov, and P. V. Geld
Frequency Ratio Change of Localized Hydrogen Atom Vibrations in Nb and Ta on Isotope Substitution
(6; 21)
(b) Vol. **128**, No. 1, p. K9–K14 (1985).

M. R. Nagy, M. S. Sakr, M. M. Mostafa, and R. Kamel
Effect of Structure Transformation on Creep and Creep Recovery of Al–14 wt% Zn
(10.1; 1.2; 21; 21.1)
(a) Vol. **88**, No. 1, p. 61–67 (1985).

R. Kuśka
The Application of Landau's Theory for Studying First-Order Phase Transitions by the Internal Friction Method
(1.2; 10.1; 12.1; 21)
(a) Vol. **88**, No. 1, p. 151–156 (1985).

D. O. Northwood and I. O. Smith
Steady-State Creep and Strain Transients for Stress Change Tests in an Al–0.4 wt% Li Solid Solution Alloy
(10.1 and 12.1; 21)
(a) Vol. **88**, No. 1, p. 181–191 (1985).

A Nohara
Dislocation-Structure Changes during the Dip Tests and the Dynamical Internal Stress in Al–Mg Alloys at High Temperatures
(10.1; 12.1; 21; 21.3)
(a) Vol. **88**, No. 1, p. 213–222 (1985).

M. ARITA, H.-U. NISSEN, and W. SCHAUER
Structure Images of the Hexagonal Nb_5Ge_3 ($D8_8$) Structure
(1.4; 14.2; 21)
(a) Vol. **88**, No. 1, p. K1–K4 (1985).

L. N. ALEKSANDROV and B. K. BOGOMOLOV
Mass-Spectrometry of Tungsten–Silicon System
(1.5 and 17.1; 21)
(a) Vol. **88**, No. 1, p. K15–K18 (1985).

R. TOGNATO
On the Free Energies of Hexagonal Dislocation Loops at High Temperatures
(10.1; 21)
(a) Vol. **88**, No. 1, p. K23–K25 (1985).

A. RAJPUT
Phonon Dispersion in Platinum and Palladium
(6; 8; 12.1; 21)
(b) Vol. **128**, No. 2, p. 411–417 (1985).

W. WASSERBÄCH
Phonon–Dislocation Scattering and Low-Temperature Lattice Thermal Conductivity of Niobium and Tantalum Single Crystals Deformed at Low Temperatures
(8; 6; 10.1; 21)
(b) Vol. **128**, No. 2, p. 453–466 (1985).

W. EBELING and W. RICHERT
Thermodynamic Properties of Liquid Hydrogen Metal
(2; 8; 21; 23)
(b) Vol. **128**, No. 2, p. 467–474 (1985).

T. HARA, H. TAKAHASHI, and S.-C. CHEN
Arsenic Implantation in CVD Tungsten Silicide
(11; 21)
(a) Vol. **88**, No. 2, p. K131–K136 (1985).

S. CHAKRABORTY, A. MANNA, and A. K. GHOSH
Atomic Scattering Factors of Solid Elements
(13; 21; 21.1; 21.1.1; 21.3; 22.1.1; 22.1.2)
(b) Vol. **129**, No. 1, p. 211–220 (1985).

E. CANESSA
Note on the Temperature Dependence of the Liquid Structure Factors
(2; 21; 21.1)
(b) Vol. **129**, No. 1, p. K1–K4 (1985).

V. FREI
Born-von Kármán Force Constants of Eight Coordination Spheres of Atoms in Face-Centred Cubic Elements Studied by Means of Phonon Frequency Sums
(6; 21; 23)
(b) Vol. **129**, No. 1, p. K17–K20 (1985).

N. J. Castellani, D. B. Leroy, and W. Lambrecht
Core-Hole Screening in Small Metallic Particles as a Function of Core-Hole Position
(13; 21; 21.2)
(b) Vol. **129**, No. 1, p. K69—K73 (1985).

E. A. Koptelov and A. A. Semenov
Instability Conditions for a Spatially Homogeneous Void Distribution in Irradiated Metals
(10.1; 1.4; 11; 21)
(a) Vol. **89**, No. 1, p. 117—122 (1985).

L. T. Miyada-Naborikawa and R. de Batist
Low-Temperature Internal Friction Peaks in Pure Zirconium Deformed at 300 K
(10.1 and 12.1; 21)
(a) Vol. **89**, No. 1, p. 191—198 (1985).

B. Głowacki and M. Horobiowski
Influence of Temperature of Diffusion Growth and Morphology of Nb_3Sn Superconducting Layer on the Value of the Pinning Force
(1.4 and 14.2; 21)
(a) Vol. **89**, No. 1, p. K13—K16 (1985).

L. T. Miyada-Naborikawa, R. de Batist, and P. Delavignette
Transmission Electron Microscopy Observations of Dislocations and Twins in Polycrystalline Zirconium
(1.4; 10.1; 21)
(a) Vol. **89**, No. 2, p. 521—531 (1985).

V. F. Degtyareva, G. V. Chipenko, I. T. Belash, O. I. Barkalov, and E. G. Ponyatovskii
F.C.C. Solid Solutions in Al–Ge and Al–Si Alloys under High Pressure
(1.3; 14.2; 21; 22.1.1; 22.1.2)
(a) Vol. **89**, No. 2, p. K127—K131 (1985).

B. Saile
The Temperature Dependence of the Effective Threshold Energy for Atom Displacement in Tantalum
(11; 21)
(a) Vol. **89**, No. 2, p. K143—K145 (1985).

S.-L. Drechsler and A. P. Zhernov
Electron–Electron Interaction and Isotope Effects in Superconducting Metal–Hydrogen (Deuterium) Systems Like $PdH_x(D_x)$
(14.2; 21)
(b) Vol. **130**, No. 1, p. 381—386 (1985).

T. Soma, S. Nehashi, and H.-Matsuo Kagaya
Specific Heat and Thermal Expansion Coefficient of Al
(8; 6; 21)
(b) Vol. **130**, No. 1, p. K11—K14 (1985).

A. Pawełek
Erratum to: A Simple Model of Small Yield Point Due to Strain Rate Change Based on the Dynamics of the Source Generated Dislocation Group
(10.1 and 12.1; 21)
(a) Vol. **90**, No. 1, p. K118 (1985).

A. A. Rempel and A. I. Gusev
Relation between Short-Range and Long-Range Order in Solid Solutions with Basal B.C.C. and F.C.C. Structures
(1.3; 21)
(b) Vol. **130**, No. 2, p. 413–420 (1985).

L. Kalvoda
Computing Method of Elastic Moduli and Young's Modulus for Texturized Materials with Hexagonal Crystal Symmetry
(12.1; 21)
(b) Vol. **130**, No. 1, p. 451–456 (1985).

M. Omri and A. Vincent
Attenuation Stress Hysteresis Loops and Low-Frequency Internal Friction Associated with Thermomechanical Breakaway of Dislocations
(7 and 10.1; 12.1; 21)
(b) Vol. **130**, No. 2, p. 547–560 (1985).

R. E. Jewell, N. C. Halder, and W. E. Swartz, Jr.
X-Ray Photoelectron and Auger Electron Spectra of Cadmium Thin Films
(13.1 and 17; 21)
(b) Vol. **130**, No. 2, p. 699–709 (1985).

J. Petrů
Local-Density-Functional Approach to the Electronic Structure of Rhodium under High Pressure
(13.1; 21)
(b) Vol. **130**, No. 2, p. 711–718 (1985).

G. Schubert and W. Löser
Electrical Resistivity and Thermoelectric Power of Spinodal Decomposing Crystals
(14.1 and 15; 1.3; 21)
(b) Vol. **130**, No. 2, p. 775–780 (1985).

B. R. Sood
Magnetoconductance in Metallic Granular Gallium Films
(14.1 and 18; 2; 21)
(b) Vol. **130**, No. 2, p. 801–806 (1985).

E. A. Pamyatnykh and A. V. Poltavets
Effect of Thermal Charge Density Fluctuations on the Specific Heat of Normal Metals
(8; 21)
(b) Vol. **130**, No. 2, p. K101–K104 (1985).

W. Brauer and M. Rösler
On the Relation between Stopping Power and Proton-Induced Electron Emission for Intermediate Energies: Aluminum
(11 and 13.5.2; 17; 21)
(b) Vol. **131**, No. 1, p. 177–183 (1985).

Sanjeev Kumar, P. K. Ahluwalia, K. C. Sharma, and R. S. Tripathi
Metallic Systems Containing Crystal Field Split Impurities. Free Energy in the Presence of Magnetic Field – a Contribution from the Loop Diagrams
(8 and 18; 21)
(b) Vol. **131**, No. 1, p. K49–K53 (1985).

W. Henrion, G. Beddies, and H. Lange
Optical Characterization of Amorphous and Crystalline Molybdenum Silicide Films
(20.1; 2; 13.1; 21)
(b) Vol. **131**, No. 1, p. K55–K58 (1985).

S. Chandra and H. C. Verma
An Empirical Model for Electrical Field Gradients in Group IIIb and IVb Metals
(4 and 13; 21)
(b) Vol. **131**, No. 1, p. K73–K76 (1985).

D. S. Haroutunian, A. G. Karagyozyan, and G. N. Karajian
Anomalous Paramagnetic State in AgMn Spin Glass and the Kondo Effect
(1.3 and 14.1; 18.1; 21)
(b) Vol. **131**, No. 1, p. K87–K90 (1985).

F. C. Lovey, G. van Tendeloo, J. van Landuyt, L. Delaey, and S. Amelinckx
Erratum to: On the Nature of Various Stacking Defects in 18R Martensite in Cu–Al Alloys. A Study by High Resolution Electron Microscopy
(1.3; 10.1; 21; 21.1)
(a) Vol. **91**, No. 1, p. 301 (1985).

V. M. Gundyrev, M. A. Uimin, A. E. Ermakov, and O. B. Andreeva
$\varepsilon' \rightarrow \varepsilon$ Transformation in MnAl–C Alloys
(1.2 and 18.2; 21; 21.1)
(a) Vol. **91**, No. 1, p. K55–K58 (1985).

K. Urban, N. Moser, and H. Kronmüller
Phase Transitions between the Quasicrystalline, Crystalline, and Amorphous Phases in Al–14 at% Mn
(1.2; 2; 21; 21.1)
(a) Vol. **91**, No. 2, p. 411–422 (1985).

R. Schülbe and U. Schmidt
On the Influence of the Pre-Ageing Conditions on the Reversion Behaviour of an Al–1.87 at% Cu Alloy
(1.3; 12.1; 21; 21.1)
(a) Vol. **91**, No. 2, p. 433–438 (1985).

M. I. Darby, G. R. Evans, M. N. Read, and D. J. Shaughnessy
Interaction of Hydrogen Impurities with Edge Dislocations in α-Zirconium
(10.1 and 13.4; 21)
(a) Vol. **92**, No. 1, p. 149–158 (1985).

P. C. Schmidt, W. Baden, N. Weiden, and A. Weiss
The Intermetallic System $NaHg_{1-x}Tl_x$. X-Ray Investigations and Measurements of the Knight Shift of Na, Hg, and Tl
(19; 21)
(a) Vol. **92**, No. 1, p. 205–212 (1985).

R. A. Dunlap
The Formation of Icosahedral Al–Transition Metal Alloys
(1.2 and 1.3; 21; 21.1)
(a) Vol. **92**, No. 1, p. K11–K13 (1985).

T. N. Yang, G. P. Yu, and L. J. Chen
The Characterization of Nitrogen-Ion-Irradiation Induced Dislocation Loops in Zirconium
(1.4 and 11; 21)
(a) Vol. **92**, No. 2, p. 399–406 (1985).

G. Graiss and G. Saad
Recovery of Internal Friction in Cold-Worked Aluminium Containing 0.35 wt% Fe
(10.1 and 12.1; 21)
(a) Vol. **92**, No. 2, p. 489–493 (1985).

G. Hübner and H. Löffler
On the Quench-Sensitivity of the Decomposition Rate of Al–Zn Alloys
(1.3 and 10.1; 1.2; 21; 21.1)
(a) Vol. **92**, No. 2, p. 517–524 (1985).

Z. Czyżewski
Backscattering Factor of the ZAF-Correction Procedure in Quantitative Electron-Probe Microanalysis
(11; 13; 21; 21.1; 21.6)
(a) Vol. **92**, No. 2, p. 563–566 (1985).

S. Kiss, R. Schaller, and W. Benoit
Heating-Rate Dependent Internal Friction Background in Aluminium
(10.1; 21)
(a) Vol. **92**, No. 2, p. K109–K112 (1985).

21.1 Transition Metals (Group Sc to Zn) and Their Alloys

Review Articles

Original Papers and Short Notes

M. Singh, M. Barkei, G. Inden, and S. Bhan
High-Temperature X-Ray Diffraction Study on $Co_{75}Sn_{25}$ Alloy
(1.3; 21.1)
(a) Vol. **87**, No. 1, p. 165–168 (1985).

D. Rios Jara, C. Esnouf, and G. Guénin
A Quantitative Study of Stable Dislocations in the β_1-Phase of Cu–Zn–Al Alloys
(10.1; 21.1)
(a) Vol. **87**, No. 1, p. 187–197 (1985).

J. F. Dinhut, P. Bouillaud and J. P. Rivière
Ion Irradiation of Thin Films of Fe (65 at%)–Co (35 at%) Alloy. Production of a Metallic Fe (50 at%)–Co (30 at%)–P (20 at%) Glass
(11 and 14.1; 1.3; 21.1; 21.1.1)
(a) Vol. **87**, No. 1, p. 199–205 (1985).

H. Kido, T. Hoshikawa, M. Shimada, and M. Koizumi
Preparation of Magnetic Properties of $YMn_2(Si_{1-x}Ge_x)_2$ Compounds
(18.1 and 18.2; 1.3; 21.1; 21.4)
(a) Vol. **87**, No. 1, p. K61–K64 (1985).

G. Seifert and H. Eschrig
LCAO-Xα Calculations of Transition Metal Clusters
(13; 21.1; 21.6)
(b) Vol. **127**, No. 2, p. 573–585 (1985).

V. Drchal and J. Kudrnovský
The Theory of Auger Spectra from Alloys. The Effect of Electron Correlations in Partially Filled Narrow Bands
(13; 17; 21.1)
(b) Vol. **127**, No. 2, p. 611–619 (1985).

E. Kührt, R. Wedell, D. Semrad, and P. Bauer
Theoretical Description of the Stopping Power of Light Ions in the Intermediate Energy Range
(11; 13.5.2; 21.1; 22.1.2; 23)
(b) Vol. **127**, No. 2, p. 633–639 (1985).

P. Muth
A Common Maximum of the Minority Spin-Band Resistivities of Some Nickel Alloys
(14.1; 13.1; 21.1)
(b) Vol. **127**, No. 2, p. K151–K153 (1985).

D. Schryvers, G. van Tendeloo, and S. Amelinckx
On the Ordering Mechanism of Ni_3Mo. A High Resolution Electron Microscopy Study
(1.3 and 1.4; 21; 21.1)
(a) Vol. **87**, No. 2, p. 401–415 (1985).

V. VALVODA, V. ŠÍMA, Z. SMETANA, and P. ČAPKOVÁ
Neutron Powder Diffraction Study of the Debye-Waller Factor in TiC and TiN Compounds and Comparison with X-Ray Results
(4; 6; 21.1)
(a) Vol. **88**, No. 1, p. 157–163 (1985).

K. MIURA
Dissociation of Superdislocations and Damping in the Ll_2 Ordered Structure
(10.1; 21.1; 21.6)
(a) Vol. **88**, No. 1, p. 207–212 (1985).

A. M. VAN DER KRAAN, D. B. DE MOOIJ, and K. H. J. BUSCHOW
Magnetic Properties and ^{57}Fe Mössbauer Effect in $V_{1-x}Fe_x$ Alloys
(18.2; 5; 21.1; 21.1.1)
(a) Vol. **88**, No. 1, p. 231–237 (1985).

A. APOSTOLOV, CH. CHRISTOPH, M. MIKHOV, and V. SKUMRIEV
Atomic Arrangement in Amorphous Er_4Co_3
(2; 21.1; 21.4)
(a) Vol. **88**, No. 1, p. K9–K13 (1985).

H. KIDO, T. HOSHIKAWA, M. SHIMADA, and M. KOIZUMI
Preparation and Magnetic Properties of YMnSi
(18.2 and 18.4; 18.1; 21.1; 21.4)
(a) Vol. **88**, No. 1, p. K39–K43 (1985).

Z. MICHNO, A. BARANOWSKI, and W. ŚWIĄTKOWSKI
Some Remarks on the Applicability of the Positron Annihilation Technique for Studying the Electronic Structure of Metallic Glasses
(2 and 13.1; 21.1; 21.1.1)
(a) Vol. **88**, No. 1, p. K45–K49 (1985).

L. DĄBROWSKI
Short- and Long-Range Ordering in Binary Alloys
(1.3; 21.1)
(b) Vol. **128**, No. 2, p. 371–378 (1985).

T. BHATTACHARYA and A. K. GHOSH
Pair-Correlation Functions for Liquid Noble Metals through Percus-Yevick Theory
(2; 4; 21.1; 21.6)
(b) Vol. **128**, No. 2, p. 419–423 (1985).

BACH THANH CONG
On the Landau Phase Transition Theory for the Magnetic Itinerant Electron System Weakly Interacting with Localized Spins
(13 and 18; 21.1; 21.4)
(b) Vol. **128**, No. 2, p. 503–511 (1985).

A. Vilassakdanont and K. N. Subramanian
Effect of Temperature on Some of the Mechanical Properties of Cu–10Ni–6Sn Spinodal Alloy
(10.1 and 12.1; 21.1)
(a) Vol. **88**, No. 2, p. 553–560 (1985).

J. Sosnowski, R. Kuentzler, and R. Clad
Magnetic Losses in V_2Zr and V_2Hf Laves Phase Compounds
(14.2 and 18; 21.1)
(a) Vol. **88**, No. 2, p. 575–585 (1985).

M. A. Uimin, A. E. Ermakov, O. B. Andreeva, E. I. Teitel, and V. M. Gundyrev
Phase Transformations and Magnetic Properties of MnAl–C Single Crystals
(1.2 and 18.2; 21.1)
(a) Vol. **88**, No. 2, p. 587–594 (1985).

G. Mavrodiev, S. Koneska, and A. Szytuła
Dilatometric and DDC Measurements of $NiMnGe_{1-x}Si_x$ and $NiMn_{1-t}Ti_tGe$ Systems
(1.1 and 1.2; 21.1)
(a) Vol. **88**, No. 2, p. K105–K109 (1985).

R. Chakravarthy, S. K. Paranjpe, M. R. L. N. Murthy, L. Madhav Rao, and N. S. Satya Murthy
Magnetic Structure of $UCu_{5-x}Ni_x$
(18.4; 21.1; 21.5)
(a) Vol. **88**, No. 2, p. K155–K158 (1985).

S. Chakraborty, A. Manna, and A. K. Ghosh
Atomic Scattering Factors of Solid Elements
(13; 21; 21.1; 21.1.1; 21.3; 22.1.1; 22.1.2)
(b) Vol. **129**, No. 1, p. 211–220 (1985).

B. Pal, J. Singh, S. D. Raj, and S. Prakash
Electric Field Gradient in Dilute B.C.C. Alloys. Application to Vanadium Alloys
(13.4; 21.1)
(b) Vol. **129**, No. 1, p. 301–312 (1985).

I. Mertig, E. Mrosan, R. Zeller, P. H. Dederichs, and P. Ziesche
Calculation of the Thermoelectric Power of Ga, Ni, Rh, and Pd Impurities in Copper
(13.4 and 15; 14.1; 21.1)
(b) Vol. **129**, No. 1, p. 407–414 (1985).

E. Canessa
Note on the Temperature Dependence of the Liquid Structure Factors
(2; 21; 21.1)
(b) Vol. **129**, No. 1, p. K1–K4 (1985).

A. B. Zolotovitskii
Relaxation of Magnons in a Ferromagnetic Metal with Two Types of Conduction Electrons
(13.5.4 and 18.2; 21.1)
(b) Vol. **129**, No. 1, p. K59–K63 (1985).

T. Soma, K. Tanaka, and H.-Matsuo Kagaya
Pressure Derivative of Elastic Stiffness Constants for α-Phase Cu-Based Alloys
(12.1; 21.1)
(b) Vol. **129**, No. 2, p. K123 – K125 (1985).

V. A. Zhorin, D. P. Shashkin, and N. S. Enikolopian
Dynamics of Solid Solution Formation in Cu–Ni Mixture under Plastic Flow at High Pressure
(1.3 and 3; 10.1; 21.1)
(a) Vol. **89**, No. 2, p. 437 – 442 (1985).

L. Delaey, G. Guénin, Y. Murakami, P. F. Gobin, G. van Tendeloo, and J. van Landuyt
Strain Field Observation and Lattice Stability at the Tip of Thermoelastic Cu–Zn–Al Martensite Plates
(1.3; 1.4; 12.1; 21.1)
(a) Vol. **89**, No. 2, p. 457 – 465 (1985).

P. Merle and J. Merlin
Study of the Validity of a Lattice of Spherical Precipitates for the Description of θ-Precipitate Dissolution in Al–Cu Alloys
(1.3; 10.1; 21.1)
(a) Vol. **89**, No. 2, p. 483 – 492 (1985).

M. S. Sakr, A. E. E. Abdel-Raaheim, and A. A. El-Dally
Effect of Structure Transformation on the Transient Creep Characteristics of Zn–40 wt% Al and Zn–0.5 wt% Al Alloys
(1.2 and 10.1; 21.1)
(a) Vol. **89**, No. 2, p. K157 – K162 (1985).

H. J. van Rijn and H. L. Alberts
The Appearance of a Resistivity Anomaly at the Incommensurate–Commensurate SDW Transition of Cr + 0.7 at% Ge
(1.2 and 14.1; 21.1)
(b) Vol. **130**, No. 1, p. K81 – K84 (1985).

W. Püschl
Reactions between Glide Dislocations and Forest Dislocations in Anistropic B.C.C. Metals
(10.1; 21; 21.1; 21.1.1; 21.2)
(a) Vol. **90**, No. 1, p. 181 – 189 (1985).

M. Kobiyama and S. Takamura
Internal Friction and Electrical Resistivity Measurements in Cu and Ag Alloys after Neutron Irradiation at Low Temperature. I. Experiments
(10.1 and 14.1; 11; 21.1; 21.6)
(a) Vol. **90**, No. 1, p. 253 – 267 (1985).

S. Takamura and M. Kobiyama
Internal Friction and Electrical Resistivity Measurements in Cu and Ag Alloys after Neutron Irradiation at Low Temperature. II. Discussion
(10.1 and 14.1; 11; 21.1; 21.6)
(a) Vol. **90**, No. 1, p. 269 – 276 (1985).

J. MAYER and K. URBAN
Observation of Ni_8Mo Ordered Phase in Ni–Mo Alloys
(1.2; 1.3; 11; 21.1)
(a) Vol. **90**, No. 2, p. 469—475 (1985).

G. WENDROCK and H. LÖFFLER
Two-Step Ageing Treatments as a Means to Estimate the Zinc Content of Guinier-Preston Zones at Elevated Temperatures in an Al–15 at% Zn Alloy
(1.3; 21; 21.1)
(a) Vol. **90**, No. 2, p. 483—487 (1985).

K. Z. BOTROS and S. S. SHEININ
On the Applicability of Conventional Contrast Theory to Weak Beam Images of Stacking Faults
(1.4 and 10.1; 21.1)
(a) Vol. **90**, No. 2, p. 499—514 (1985).

A. PAWEŁEK, W. STRYJEWSKI, W. BOCHNIAK, and H. DYBIEC
Mobile Dislocation Density Variation during Strain Rate Change Evidenced by Acoustic Emission
(10.1; 7; 21.1)
(a) Vol. **90**, No. 2, p. 531—536 (1985).

W. FERNENGEL, M. RAPP, and H. KRONMÜLLER
The Coercive Field of the Partially Crystallized Amorphous Alloy $Fe_{40}Ni_{40}P_{14}B_6$
(2 and 18.2; 21.1; 21.1.1)
(a) Vol. **90**, No. 2, p. 637—642 (1985).

B. V. VASILEV, A. P. DOLGAREV, A. A. POTSELUIKO, and V. G. PYNKO
Transitional Epitaxial Layers in Nickel Films Deposited in Vacuum on Complex LiF/NaCl Substrates
(1.5 and 3; 21.1)
(a) Vol. **90**, No. 2, p. K129—K130 (1985).

I. BAKONYI, K.-S. HAN, and H. E. SCHONE
^{51}V NMR of Amorphous and Crystalline V–Zr Alloys
(19; 2; 21.1)
(b) Vol. **131**, No. 1, p. 249—254 (1985).

YU. A. BABANOV and V. R. SHVETSOV
EXAFS: Bond Length Determination for Multicomponent Systems
(1 and 20; 21.1)
(b) Vol. **131**, No. 1, p. K1—K4 (1985).

T. SOMA, M. TAKAHASHI, and M.-MATSUO KAGAYA
Compression Effect on the α–β Phase Transition of Cu-Based Binary Alloys
(1.2 and 12.1; 1.3; 21.1)
(b) Vol. **131**, No. 1, p. K11—K14 (1985).

A. R. ALI
On the Behaviour of Interaction of Copper Atoms with Lattice Defects in Quenched and Cold-Worked Ni–Cu Alloys
(10.1; 14.1; 21.1)
(a) Vol. **91**, No. 1, p. 51—56 (1985).

L. T. WILLE and J. VENNIK
Kinetics of Surface Segregation in Binary Alloys
(1.6; 21.1)
(b) Vol. **131**, No. 2, p. 443–449 (1985).

V. I. BOGDANOV and D. L. FUKS
Many-Body Contribution to the Energy of Metals with B.C.C. Lattice
(4; 1.3; 21.1; 21.2)
(b) Vol. **131**, No. 2, p. 495–503 (1985).

A. A. VASILYEV, V. V. SIROTINKIN, and A. I. MELKER
Small Vacancy Clusters in Nickel
(4 and 10.1; 21.1)
(b) Vol. **131**, No. 2, p. 537–544 (1985).

N. N. DELYAGIN, V. I. KRYLOV, G. T. MUJIRI, V. I. NESTEROV, and S. I. REIMAN
The Magnetic Hyperfine Field for ^{119}Sn Atoms in GdMnSi Intermetallic Compound. Zero Contribution from 3d Moments
(5 and 18.3; 21.1; 21.4)
(b) Vol. **131**, No. 2, p. 555–561 (1985).

K. BORTEL
*Electronic Contribution to the Ordering Energy of a Binary Alloy
(13; 1.3; 21.1)
(b) Vol. **131**, No. 2, p. 595–604 (1985).

R. STEARNS, R. BRAUNSTEIN, and L. MULDAWER
Wavelength Modulated Spectra of α-$Ag_{0.7}Zn_{0.3}$ near the Optical Absorption Edge
(13.1 and 20.1; 21.1; 21.6)
(b) Vol. **131**, No. 2, p. 659–665 (1985).

D. COLAITIS, D. VAN DYCK, and S. AMELINCKX
Electron Microscopic Study of the Domains and Interfaces in MV_4S_8 (M = Fe, Co, Ni)
(1.3 and 1.4; 21.1; 21.1.1)
(a) Vol. **91**, No. 2, p. 385–406 (1985).

J. ČERMÁK, M. ČERŇANSKÝ, P. DUHAJ, and J. HORVÁTH
Crystallization of the Metallic Glass $Fe_{40}Ni_{40}B_{20}$ during Isochronal Annealing in Argon and Hydrogen
(2 and 3; 21.1; 21.1.1)
(a) Vol. **91**, No. 2, p. 407–410 (1985).

K. URBAN, N. MOSER, and H. KRONMÜLLER
Phase Transitions between the Quasicrystalline, Crystalline, and Amorphous Phases in Al–14 at% Mn
(1.2; 2; 21; 21.1)
(a) Vol. **91**, No. 2, p. 411–422 (1985).

R. SCHÜLBE and U. SCHMIDT
On the Influence of the Pre-Ageing Conditions on the Reversion Behaviour of an Al–1.87 at% Cu Alloy
(1.3; 12.1; 21; 21.1)
(a) Vol. **91**, No. 2, p. 433–438 (1985).

W. J. Jiang, Z. Zhang, Dz. K. Hei, and K. H. Kuo
High Resolution Electron Microscopy of Translation Twins in NiZr
(1.3 and 1.4; 21.1)
(a) Vol. **92**, No. 1, p. K5–K6 (1985).

R. A. Dunlap
The Formation of Icosahedral Al–Transition Metal Alloys
(1.2 and 1.3; 21; 21.1)
(a) Vol. **92**, No. 1, p. K11–K13 (1985).

K. Bärner and E. Gmelin
Coupling of a Structural and a High-Spin–Low-Spin Transition in $MnAs_{1-x}P_x$ Compounds
(18; 8; 12.1; 21.1)
(b) Vol. **132**, No. 2, p. 431–436 (1985).

K. Yoshida
Displacement of the Fermi Level Due to Impurities in a Random Distribution
(13.4; 21.1)
(b) Vol. **132**, No. 2, p. 551–556 (1985).

I. Retat, Th. Steffens, and Ch. Schwink
Dislocation Arrangement in FeNi–Invar Single Crystals
(10.1; 18.2; 21.1; 21.1.1)
(a) Vol. **92**, No. 2, p. 507–510 (1985).

G. Hübner and H. Löffler
On the Quench-Sensitivity of the Decomposition Rate of Al–Zn Alloys
(1.3 and 10.1; 1.2; 21; 21.1)
(a) Vol. **92**, No. 2, p. 517–524 (1985).

D. Płusa, R. Pfranger, and B. Wysłocki
Dependence of Domain Width on Crystal Thickness in YCo_5 Single Crystals
(18.3; 21.1; 21.4)
(a) Vol. **92**, No. 2, p. 533–538 (1985).

D. K. Fowler, D. C. Creagh, R. W. N. Kinnear, and G. V. H. Wilson
Two-Pulse Stimulated Echo in Ferromagnetic Metals
(19; 21.1; 21.1.1)
(a) Vol. **92**, No. 2, p. 545–553 (1985).

Z. Czyżewski
Backscattering Factor of the ZAF-Correction Procedure in Quantitative Electron-Probe Microanalysis
(11; 13; 21; 21.1; 21.6)
(a) Vol. **92**, No. 2, p. 563–566 (1985).

R. Hergt and H. Pfeiffer
On Texture Formation in Metallic Films
(1.4 and 1.5; 21.1)
(a) Vol. **92**, No. 2, p. K89–K92 (1985).

21.1.1 Fe and Fe Alloys

Rievew Article

E. P. WOHLFARTH
Magnetic Materials for Recording
(18; 18.3; 21.1; 21.1.1; 22.6; 22.8.2)
(a) Vol. **91**, No. 2, p. 339–350 (1985).

Original Papers and Short Notes

P. H. GREENWOOD and P. DE V. DU PLESSIS
Critical Behaviour of Antiferromagnetic Fe_2As
(18.4; 21.1.1)
(b) Vol. **127**, No. 1, p. K19–K23 (1985).

J. E. FRĄCKOWIAK
Determination of the Long-Range Order Parameter in the Transition Metal Alloys Using Mössbauer Spectroscopy
(1.3 and 5; 21.1; 21.1.1)
(a) Vol. **87**, No. 1, p. 109–119 (1985).

J. A. CUSIDÓ, A. ISALGUÉ, and J. TEJADA
On the Amorphous to Crystalline Transformation of $Fe_{80}B_{20}$ by Means of Electrical and Thermal Conductivity, X-Ray Diffraction, and Mössbauer Measurements
(1.2 and 2; 5; 8; 14.1; 21.1.1)
(a) Vol. **87**, No. 1, p. 169–174 (1985).

J. F. DINHUT, P. BOUILLAUD, and J. P. RIVIÈRE
Ion Irradiation of Thin Films of Fe(65 at%)–Co(35 at%)Alloy. Production of a Metallic Fe(50 at%)–Co(30 at%)–P(20 at%)Glass
(11 and 14.1; 1.3; 21.1; 21.1.1)
(a) Vol. **87**, No. 1, p. 199–205 (1985).

A. PRASAD, R. M. SINGRU, and A. K. BISWAS
Study of the Roasting of Pyrite Minerals by Mössbauer Spectroscopy
(5; 1.3; 21.1.1)
(a) Vol. **87**, No. 1, p. 267–271 (1985).

T. TARNÓCZI
Structural Relaxation and Curie Point of a Metallic Glass
(2 and 18.2; 6; 21.1.1)
(a) Vol. **87**, No. 1, p. 283–291 (1985).

E. A. VASILEV and V. A. VIRCHENKO
Effect of Interstitial Atoms on the Quadrupole Interaction at ^{119}Sn Nuclei in Some B8 Type Solid Solutions
(1.3 and 5; 4; 21.1.1)
(a) Vol. **87**, No. 1, p. K19–K24 (1985).

S. G. Saxena and K. B. Garg
EXAFS Study of Some Iron Based Metallic Glasses
(2 and 4; 21.1.1)
(a) Vol. **87**, No. 1, p. K25–K28 (1985).

H. Kronmüller
The Rôle of Two-Level Systems in Amorphous Metallic Alloys. II. Micromagnetic Theory of Magnetic After-Effects in Amorphous Ferromagnets
(2 and 18.2; 21.1.1)
(b) Vol. **127**, No. 2, p. 531–541 (1985).

K. M. Mukimov, Sh. M. Sharipov, and L. A. Ernazarova
The Magneto-Optical Spectra and Behaviour of the Joint Density of States of the Intermetallic Compounds RFe_2 (R = Gd, Tb, Dy, Ho, Er) and YFe_2
(13.1 and 20.1; 21.1.1; 21.4)
(b) Vol. **127**, No. 2, p. K129–K131 (1985).

E. A. Little
Void Swelling of Ferritic Steel during 1 MeV Electron Irradiation
(1.4; 11; 21.1.1)
(a) Vol. **87**, No. 2, p. 441–449 (1985).

K. Urban
Determination of Long-Range Order Parameter in Alloys by Means of Electron Diffraction in the Electron Microscope
(1.3; 21.1; 21.1.1)
(a) Vol. **87**, No. 2, p. 459–471 (1985).

H. Düsterhöft, R. Daugs, and R. Pippig
A New Method for the Interpretation of Energy Distributions Measured on Secondary Ions
(17; 1.6; 11; 21.1; 21.1.1; 22.1.2)
(a) Vol. **87**, No. 2, p. 499–506 (1985).

W. Graeff and K. Wieteska
Observation of Domain Structure in a (111) Oriented Grain of Fe(Si) by X-Ray White Beam Topography
(1.4; 18.2; 21.1.1)
(a) Vol. **87**, No. 2, p. 517–525 (1985).

N. I. Noskova, N. F. Vildanova, Yu. I. Filippov, and A. P. Potapov
Strength, Plasticity, and Fracture of Ribbons of $Fe_5Co_{70}Si_{15}B_{10}$ Amorphous Alloy
(10.1 and 12.1; 2; 21.1; 21.1.1)
(a) Vol. **87**, No. 2, p. 549–557 (1985).

C. W. Lung and H. Gao
Analysis of K_{ic} and Its Temperature Dependence of Metals by a Simplified Dislocation Model
(10.1; 21.1.1)
(a) Vol. **87**, No. 2, p. 565–569 (1985).

M. Vázquez, W. Fernengel, and H. Kronmüller
Changes of the Magnetic Properties of the Nearly Non-Magnetostrictive Amorphous Alloy $Co_{58}Fe_5Ni_{10}Si_{11}B_{16}$ by Annealing under Tensile Stress
(10.1 and 18.2; 2; 21.1; 21.1.1)
(a) Vol. **87**, No. 2, p. 609–615 (1985).

T. Luciński and J. Baszyński
Thermopower in Amorphous $Fe_{1-x}Si_x$ Films
(15; 2; 21.1.1)
(a) Vol. **87**, No. 2, p. K191–K193 (1985).

Z. Frait, D. Fraitová, and N. Zárubová
Observation of FMR Surface Spin Wave Mode in Bulk Amorphous Ferromagnets
(18.2 and 19; 2; 21.1; 21.1.1)
(b) Vol. **128**, No. 1, p. 219–222 (1985).

P. Rochegude and J. Foct
The Transformation $\varepsilon \rightarrow \alpha''$ in Iron–Nitrogen Solid Solutions Studied by Mössbauer Spectrometry
(1.2 and 5; 21.1.1)
(a) Vol. **88**, No. 1, p. 137–142 (1985).

A. M. van der Kraan, D. B. de Mooij, and K. H. J. Buschow
Magnetic Properties and ^{57}Fe Mössbauer Effect in $V_{1-x}Fe_x$ Alloys
(18.2; 5; 21.1; 21.1.1)
(a) Vol. **88**, No. 1, p. 231–237 (1985).

Z. Michno, A. Baranowski, and W. Świątkowski
Some Remarks on the Applicability of the Positron Annihilation Technique for Studying the Electronic Structure of Metallic Glasses
(2 and 13.1; 21.1; 21.1.1)
(a) Vol. **88**, No. 1, p. K45–K49 (1985).

K. Kiymaç
Giant Moment, Superparamagnetic, Spin-Glass, and Quasiferromagnetic Properties of Dilute Pt–Fe Alloys
(18.1; 18.2; 21.1.1)
(b) Vol. **128**, No. 2, p. 553–562 (1985).

A. H. Phillips, N. K. Gobran, and R. Kamel
Structure Analysis of the Metallic Glass $Fe_{40}Ni_{40}B_{20}$ Using the Tangent Lattice Model
(1.2; 14.1; 21.1; 21.1.1)
(a) Vol. **88**, No. 2, p. 463–467 (1985).

P. Allia, E. Bonetti, E. Evangelista, L. Turroni, and F. Vinai
Reversible and Irreversible Processes of Structural Relaxation and Dynamic Young Modulus Behaviour in the $Fe_{40}Ni_{38}Mo_4B_{18}$ Amorphous Alloy
(2; 12.1; 21.1; 21.1.1)
(a) Vol. **88**, No. 2, p. 521–527 (1985).

K. Csach, J. Miškuf, and V. Karel
Angular Distribution of Emitted X-Rays from an Ion-Implanted Layer
(11 and 20; 21.1.1)
(a) Vol. **89**, No. 2, p. K153—K155 (1985).

A. N. Didenko, O. Kh. Asainov, V. P. Krivobokov, E. I. Logachev, A. D. Pogrebnyak, L. N. Puchkareva, G. E. Remnev, and Sh. M. Ruzimov
Positron Annihilation at Defects in Surface Layers of Steel Irradiated by Supercurrent Ion Beams
(10.1 and 11; 1.2; 21.1.1)
(a) Vol. **89**, No. 2, p. K167—K172 (1985).

A. L. Erzinkyan, V. V. Muraveva, V. P. Parfenova, and V. V. Turovtsev
Magnetic Hyperfine Interaction of ^{92m}Nb in $ZrFe_2$
(19; 21.1.1)
(a) Vol. **89**, No. 2, p. K201—K205 (1985).

I. G. Ratishvili
Annihilation of Frenkel Pairs in B.C.C. Metals (α-Iron). III. The Case of a Hydrostatically Compressed Lattice
(10.1; 11; 21.1.1)
(b) Vol. **130**, No. 1, p. 169—174 (1985).

W. Herremans, G. van Tendeloo, M. David, and R. Gevers
TEM Images of a Translation Interface on Edge
(1.4 and 10.1; 21.1.1)
(a) Vol. **90**, No. 1, p. 91—105 (1985).

J. A. Cusidó and A. Isalgué
A Simple Generalized Model for the Kinetics of Crystallization in Metallic Glasses
(1.2 and 2; 5; 8; 21.1.1)
(a) Vol. **90**, No. 1, p. 127—133 (1985).

W. Püschl
Reactions between Glide Dislocations and Forest Dislocations in Anistropic B.C.C. Metals
(10.1; 21; 21.1; 21.1.1; 21.2)
(a) Vol. **90**, No. 1, p. 181—189 (1985).

B. V. Petukhov
Theory of Solid Solution Softening in Comparison with Experiments on α-Fe
(10.1; 21.1.1)
(a) Vol. **90**, No. 1, p. 225—229 (1985).

A. I. Melker, S. N. Romanov, and V. S. Tokmakova
Displacement Spike Annealing in α-Iron Irradiated by PKA in the Energy Range 20 to 100 keV
(11; 10.1; 21.1.1)
(b) Vol. **130**, No. 2, p. 531—537 (1985).

K. Schröder
The Effect of a Cu Overlayer on the Magnetization of an Fe Film
(18.2.1; 21.1; 21.1.1)
(b) Vol. **130**, No. 2, p. 581—585 (1985).

HUEI LI HUANG and MING YEN LIN
Remanent Magnetization in Randomly Distributed Cubic Particles
(18; 21.1; 21.1.1)
(a) Vol. **91**, No. 1, p. 123–127 (1985).

I. N. SHABANOVA
Chemical Bond Peculiarities of Elements in Amorphous Alloy Surface Layers on the Basis of Fe
(1.6 and 2; 17; 21.1; 21.1.1)
(a) Vol. **91**, No. 1, p. K13–K17 (1985).

F. LECCABUE, J. L. SÁNCHEZ, L. PARETI, F. BOLZONI, and R. PANIZZIERI
Curie Temperatures of Some Substituted Nd_2Fe_{14} B Tetragonal Alloys
(18; 21.1.1)
(a) Vol. **91**, No. 1, p. K63–K65 (1985).

P. MUTH and V. CHRISTOPH
The Spontaneous Resistivity Anisotropy of Iron-Based Alloys
(13.1 and 14.1; 21.1.1)
(b) Vol. **131**, No. 2, p. 769–773 (1985).

H. OESTERREICHER
On the Spin Reorientation in $R_2Fe_{14}B$ Compounds
(13 and 18.2; 21.1.1)
(b) Vol. **131**, No. 2, p. K123–K127 (1985).

L. M. SANDLER, V. N. PUSHKAR, I. V. FILONCHUK, and T. S. KORSUNSKAYA
Crystallization of Amorphous Fe–Si Films
(1.2 and 2; 1.5; 21.1.1)
(a) Vol. **91**, No. 2, p. 371–378 (1985).

D. COLAITIS, D. VAN DYCK, and S. AMELINCKX
Electron Microscopic Study of the Domains and Interfaces in MV_4S_8 (M = Fe, Co, Ni)
(1.3 and 1.4; 21.1; 21.1.1)
(a) Vol. **91**, No. 2, p. 385–406 (1985).

J. ČERMÁK, M. ČERŇANSKÝ, P. DUHAJ, and J. HORVÁTH
Crystallization of the Metallic Glass $Fe_{40}Ni_{40}B_{20}$ during Isochronal Annealing in Argon and Hydrogen
(2 and 3; 21.1; 21.1.1)
(a) Vol. **91**, No. 2, p. 407–410 (1985).

M. FÄHNLE and A. SEEGER
On the Magnetic Contribution to the Free Enthalpy of Vacancy Formation in Ferromagnetic Crystals
(10.1 and 18.2; 21.1.1)
(a) Vol. **91**, No. 2, p. 609–617 (1985).

L. URAY
Investigation of Dilute Tungsten Alloys by Electrical Transport Measurements
(14.1 and 15; 1.3; 21.1; 21.1.1)
(a) Vol. **91**, No. 2, p. 729–735 (1985).

21.2 Alkali Metals

Original Papers and Short Notes

D. Sen and S. K. Sen
Transport in Simple Metals in the Framework of a Unified Pseudopotential Study
(8 and 14.1; 15; 21; 21.2)
(b) Vol. **127**, No. 1, p. 377–383 (1985).

H.-J. Fitting, K. König, and D. Hecht
Exoelectron Emission Due to Alkali Ions on Insulating Layers
(17; 1.5; 21.2; 22.1.2; 22.6)
(a) Vol. **87**, No. 1, p. K77–K79 (1985).

T. Soma, H. Ohsugi, and H.-Matsuo Kagaya
Third Order Contributions to Local and Band Modes of Lattice Dynamics for the Alkali Alloy Systems
(6; 21.2)
(b) Vol. **128**, No. 1, p. K5–K8 (1985).

N. J. Castellani, D. B. Leroy, and W. Lambrecht
Core-Hole Screening in Small Metallic Particles as a Function of Core-Hole Position
(13; 21; 21.2)
(b) Vol. **129**, No. 1, p. K69–K73 (1985).

L. J. Gallego, J. A. Somoza, and J. A. Alonso
Liquidus Curves of Eutectic Na–K and Na–Cs Systems from Semiempirical Theories of Mixtures
(2; 8; 21.2)
(a) Vol. **89**, No. 1, p. 73–78 (1985).

A. P. Zhernov, Kh. M. Pashaev, and A. M. Aliev
The Non-Equilibrium Electron Distribution Function in the Electrical Resistance Problem for Potassium Metal. Influence of N- and U-Processes
(14.1; 21.2)
(b) Vol. **129**, No. 2, p. 823–833 (1985).

P. Singh, Ch. Pal, and K. N. Khanna
Flory's Formula for the Entropy of Mixing of LiNa Alloy
(1.3 and 8; 2; 21.2)
(b) Vol. **129**, No. 2, p. K111–K115 (1985).

V. A. Polukhin, A. M. Bratkovskii, and V. G. Vaks
Molecular Dynamics Study of Thermodynamics and Structure of Liquid Sodium under Pressure
(2; 8; 21.2)
(b) Vol. **130**, No. 1, p. 87–94 (1985).

T. Soma, H.-Matsuo Kagaya, and H. Ohsugi
Thermal Properties for the Alloy System Composed of Alkali Metal Elements
(8; 6; 21.2)
(b) Vol. **130**, No. 1, p. K15–K18 (1985).

G. OOMI and S. B. WOODS
Pressure-Induced Martensitic Transformation in Li at Low Temperatures
(1.2 and 14.1; 21.2)
(b) Vol. **130**, No. 1, p. K77–K80 (1985).

W. PÜSCHL
Reactions between Glide Dislocations and Forest Dislocations in Anisotropic B.C.C. Metals
(10.1; 21; 21.1; 21.1.1; 21.2)
(a) Vol. **90**, No. 1, p. 181–189 (1985).

V. I. BOGDANOV and D. L. FUKS
Many-Body Contribution to the Energy of Metals with B.C.C. Lattice
(4; 1.3; 21.1; 21.2)
(b) Vol. **131**, No. 2, p. 495–503 (1985).

F. BOLDUAN, W. HÖNLE, H. D. HOCHREIMER, and W. HENKEL
Temperature and High Pressure Raman Study of M_3P_7 (M = Li, Na, K, Rb, Cs) Compounds
(1.2; 20.1; 21.2)
(b) Vol. **132**, No. 1, p. 41–50 (1985).

T. SOMA, H. HIGUCHI, and H.-MATSUO KAGAYA
Effect of Lattice Anharmonicity on Specific Heat of Alkali Alloys
(6; 4; 8; 21.2)
(b) Vol. **132**, No. 2, p. 383–388 (1985).

21.3 Alkali-Earth Metals

Original Papers and Short Notes

G. Neubert and S. Rogaschewski
Measurements of the Backscattering of 20 to 60 keV Electrons from Double Layers at Various Angles of Incidence
(11; 21; 21.3; 21.6)
(a) Vol. **87**, No. 2, p. 451–457 (1985).

A. Nohara
Dislocation-Structure Changes during the Dip Tests and the Dynamical Internal Stress in Al–Mg Alloys at High Temperatures
(10.1; 12.1; 21; 21.3)
(a) Vol. **88**, No. 1, p. 213–222 (1985).

S. Chakraborty, A. Manna, and A. K. Ghosh
Atomic Scattering Factors of Solid Elements
(13; 21; 21.1; 21.1.1; 21.3; 22.1.1; 22.1.2)
(b) Vol. **129**, No. 1, p. 211–220 (1985).

D. J. González, C. M. Zapata, and J. A. Alonso
Density Functional Pseudopotential Calculation of the Cohesive Properties of Disordered Solid Alloys of Alkaline-Earth Metals. Influence of the Ionic Pseudopotential
(1.3 and 13; 4; 21.3)
(b) Vol. **129**, No. 2, p. 483–488 (1985).

B. Dutta, S. K. Sarkar, and D. Roy
Three-Body Interaction in H.C.P. Metals
(4 and 6; 12.1; 21.3)
(b) Vol. **132**, No. 2, p. 345–354 (1985).

21.4 Rare-Earth Metals

Original Papers and Short Notes

E. A. GOREMYCHKIN, E. MÜHLE, B. LIPPOLD, O. D. CHISTYAKOV, and E. M. SAVITSKII
Crystal Field Study of $HoNi_5$ by Neutron Scattering
(13.4; 8; 18; 21.4)
(b) Vol. **127**, No. 1, p. 371–375 (1985).
Erratum
(b) Vol. **128**, No. 2, p. 807 (1985).

H. KIDO, T. HOSHIKAWA, M. SHIMADA, and M. KOIZUMI
Preparation of Magnetic Properties of $YMn_2(Si_{1-x}Ge_x)_2$ Compounds
(18.1 and 18.2; 1.3; 21.1; 21.4)
(a) Vol. **87**, No. 1, p. K61–K64 (1985).

S. DE, I. BOSE, and S. CHATTERJEE
Electronic Energy Bands of Pr under Pressure
(13.1; 21.4)
(b) Vol. **127**, No. 2, p. 605–610 (1985).

K. M. MUKIMOV, SH. M. SHARIPOV, and L. A. ERNAZAROVA
The Magneto-Optical Spectra and Behaviour of the Joint Density of States of the Intermetallic Compounds RFe_2 (R = Gd, Tb, Dy, Ho, Er) and YFe_2
(13.1 and 20.1; 21.1.1; 21.4)
(b) Vol. **127**, No. 2, p. K129–K131 (1985).

A. APOSTOLOV, CH. CHRISTOV, M. MIKHOV, and V. SKUMRIEV
Atomic Arrangement in Amorphous Er_4Co_3
(2; 21.1; 21.4)
(a) Vol. **88**, No. 1, p. K9–K13 (1985).

H. KIDO, T. HOSHIKAWA, M. SHIMADA, and M. KOIZUMI
Preparation and Magnetic Properties of YMnSi
(18.2 and 18.4; 18.1; 21.1; 21.4)
(a) Vol. **88**, No. 1, p. K39–K43 (1985).

BACH THANH CONG
On the Landau Phase Transition Theory for the Magnetic Itinerant Electron System Weakly Interacting with Localized Spins
(13 and 18; 21.1; 21.4)
(b) Vol. **128**, No. 2, p. 503–511 (1985).

B. KOZARZEWSKI, J. KURZYK, and J. DENISZCZYK
Indirect Magnetoelastic Interaction and Magnetostriction of Dilute Rare-Earth Alloys
(18; 21.4; 21.6)
(b) Vol. **129**, No. 1, p. 143–152 (1985).

N. Gopi Krishna, D. B. Sirdeshmukh, B. Rama Rao, B. J. Beaudry, and K. A. Gschneidner, Jr.
X-Ray Debye Temperature of Ytterbium
(6 and 8; 21.4)
(a) Vol. **89**, No. 1, p. K37–K38 (1985).

P. K. Srivastava and O. P. Kulshrestha
Lattice Dynamics of Lanthanum by Using a Model Pseudopotential Approach
(6; 21.4)
(b) Vol. **130**, No. 1, p. K23–K25 (1985).

J. Olejniczak, A. Rojek, and C. Sułkowski
Superconductivity and Resistivity of $LaRh_{1.1}Sn_4$
(14.1 and 14.2; 21.4; 21.7)
(b) Vol. **130**, No. 1, p. K73–K76 (1985).

G. Turilli, H. Szymczak, and H. K. Lachowicz
Elastic Constants of the $(RE)_2Fe_{14}B$ Tetragonal Compounds (RE = Y, Nd, and Pr)
(12.1; 21.1.1; 21.4)
(a) Vol. **90**, No. 2, p. K143–K146 (1985).

K. Witański
On the Magnetic Properties of $Gd_{1-x}RE_xAl_2$ Alloys
(18.2; 21.4)
(b) Vol. **131**, No. 1, p. 129–136 (1985).

G. S. Ameta and U. N. Upadhyaya
Electronic Properties of Ternary Superconductors
(14.2; 18.2; 21.4)
(b) Vol. **131**, No. 1, p. 369–386 (1985).

N. N. Delyagin, V. I. Krylov, G. T. Mujiri, V. I. Nesterov, and S. I. Reiman
The Magnetic Hyperfine Field for ^{119}Sn Atoms in GdMnSi Intermetallic Compound. Zero Contribution from 3d Moments
(5 and 18.3; 21.1; 21.4)
(b) Vol. **131**, No. 2, p. 555–561 (1985).

E. Bauer and J. Kołaczkiewicz
Low Energy Electron Exchange Excitation of 4f Transitions and Location of 4f Levels in Rare Earths
(13.1 and 13.5; 21.4)
(b) Vol. **131**, No. 2, p. 699–706 (1985).

Y. Q. Sheng, P. Munz, and R. Schultheiss
The Determination of the Thickness of Ultrathin Metal and Semiconductor Films on Conventional Semiconductor Substrates
(1.5; 14.3.1; 21; 21.1; 21.4; 21.6; 22.1.1)
(a) Vol. **92**, No. 1, p. 121–128 (1985).

M. S. Islam
Chemical Shifts and EXAFS of the L_{III} Absorption Discontinuity of Dysprosium in Some of Its Compounds
(20; 21.4)
(b) Vol. **132**, No. 2, p. 471–476 (1985).

21.5 Actinides

Short Notes

T. PALEWSKI
The Magnetic Properties of $U_xY_{1-x}As$ Solid Solution
(18.2; 21.5)
(a) Vol. **88**, No. 2, p. K149--K153 (1985).

R. CHAKRAVARTHY, S. K. PARANJPE, M. R. L. N. MURTHY, L. MADHAV RAO, and N. S. SATYA MURTHY
Magnetic Structure of $UCu_{5-x}Ni_x$
(18.4; 21.1; 21.5)
(a) Vol. **88**, No. 2, p. K155—K158 (1985).

21.6 Noble Metals

Review Article

K. H. FISCHER
Spin Glasses (II)
(18.2 and 18.4; 2; 5; 7; 8; 13; 14; 21; 21.1; 21.6; 22.6; 22.8)
(b) Vol. **130**, No. 1, p. 13–71 (1985).

Original Papers and Short Notes

M. TAKEDA and H. HASHIMOTO
Atomic Structure of CuAuII Studied by High Resolution Electron Microscopy
(1.4; 21.1; 21.6)
(a) Vol. **87**, No. 1, p. 141–149 (1985).

N. AJIKA, H. HASHIMOTO, and Y. TAKAI
Electron Microscope Observation of the Structure and Behaviour of Stacking Fault Tetrahedra and Single Vacacies in Gold Crystals Irradiated by 2 MeV Electrons
(1.4 and 10.1; 11; 21.6)
(a) Vol. **87**, No. 1, p. 235–252 (1985).

G. SEIFERT and H. ESCHRIG
LCAO-Xα Calculations of Transition Metal Clusters
(13; 21.1; 21.6)
(b) Vol. **127**, No. 2, p. 573–585 (1985).

G. NEUBERT and S. ROGASCHEWSKI
Measurements of the Backscattering of 20 to 60 keV Electrons from Double Layers at Various Angles of Incidence
(11; 21; 21.3; 21.6)
(a) Vol. **87**, No. 2, p. 451–457 (1985).

K. MIURA
Dissociation of Superdislocations and Damping in the Ll_2 Ordered Structure
(10.1; 21.1; 21.6)
(a) Vol. **88**, No. 1, p. 207–212 (1985).

T. BHATTACHARYA and A. K. GHOSH
Pair-Correlation Functions for Liquid Noble Metals through Percus-Yevick Theory
(2; 4; 21.1; 21.6)
(b) Vol. **128**, No. 2, p. 419–423 (1985).

K. MIURA and A. MARUYAMA
Equilibrium Superdislocation Damping in Cu_3Au near the Transformation Temperature
(10.1; 21.1; 21.6)
(a) Vol. **88**, No. 2, p. 529–532 (1985).

B. KOZARZEWSKI, J. KURZYK, and J. DENISZCZYK
Indirect Magnetoelastic Interaction and Magnetostriction of Dilute Rare-Earth Alloys
(18; 21.4; 21.6)
(b) Vol. **129**, No. 1, p. 143–152 (1985).

F. H. SÁNCHEZ, R. C. MERCADER, A. F. PASQUEVICH, A. G. BIBILONI, and A. R. LÓPEZ-GARCÍA
The Influence of Particle Sizes on the Oxidation Kinetics of AgSn Alloys Studies by Mössbauer Spectroscopy
(1.3 and 5; 21.6)
(a) Vol. **89**, No. 1, p. K17–K21 (1985).

R. STEARNS, R. BRAUNSTEIN, and L. MULDAWER
Optical Properties of β'-AuZn by Wavelength-Modulated Derivative Spectroscopy
(13.1 and 20.1; 21.1; 21.6)
(b) Vol. **129**, No. 2, p. 735–743 (1985).

M. KOBIYAMA and S. TAKAMURA
Internal Friction and Electrical Resistivity Measurements in Cu and Ag Alloys after Neutron Irradiation at Low Temperature I. Experiments
(10.1 and 14.1; 11; 21.1; 21.6)
(a) Vol. **90**, No. 1, p. 253–267 (1985).

S. TAKAMURA and M. KOBIYAMA
Internal Friction and Electrical Resistivity Measurements in Cu and Ag Alloys after Neutron Irradiation at Low Temperature. II. Discussion
(10.1 and 14.1; 11; 21.1; 21.6)
(a) Vol. **90**, No. 1, p. 269–276 (1985).

S. A. NEPIJKO, H. PAGNIA, and V. I. STYOPKIN
Electron Induced Off-Switching in Au Island Film Diodes
(11 and 14.1; 21.6)
(a) Vol. **90**, No. 1, p. K113–K116 (1985).

K. RAI, R. S. BHATTACHARYA, and M. H. RASHID
Ion Beam Mixing of Alternate Au–Ge Layer with GaAs
(11 and 14.3.4; 1.6; 21.6; 22.1.1; 22.2.1)
(a) Vol. **91**, No. 1, p. 287–293 (1985).

R. FROMAGEAU, A. MAGNOUCHE, and N. GERARD
The Resistivity of Dissolved Hydrogen in Palladium and Amorphous $Pd_{80}Si_{20}$
(2 and 14.1; 21.6)
(a) Vol. **91**, No. 1, p. K75–K80 (1985).

R. STEARNS, R. BRAUNSTEIN, and L. MULDAWER
Wavelength Modulated Spectra of α-$Ag_{0.7}Zn_{0.3}$ near the Optical Absorption Edge
(13.1 and 20.1; 21.1; 21.6)
(b) Vol. **131**, No. 2, p. 659–665 (1985).

21.7 Semi-Metals (Semiconducting Alloys See 22.7)

Original Papers and Short Notes

S. M. J. AKHTAR and E. E. KHAWAJA
A Study of the Resistivity and the Thermoelectric Power of Thin Films of Sb and Bi
(14.1 and 15; 21.7)
(a) Vol. **87**, No. 1, p. 335–340 (1985).

J. G. GASSER, J. L. BRETONNET, and A. BRUSON
Temperature Dependence of the Liquid Tin Resistivity
(2 and 14.1; 21.7)
(b) Vol. **128**, No. 2, p. 789–796 (1985).

M. MONDAL and K. P. GHATAK
Effect of Size Quantization on the Effective Electron Mass in Ultrathin Bi Films
(13.1; 21.7)
(b) Vol. **128**, No. 2, p. K133–K137 (1985).

J. OLEJNICZAK, A. ROJEK, and C. SUŁKOWSKI
Superconductivity and Resistivity of $LaRh_{1.1}Sn_4$
(14.1 and 14.2; 21.4; 21.7)
(b) Vol. **130**, No. 1, p. K73–K76 (1985).

B. K. DAS, A. K. TEWARI, and D. L. BHATTACHARYA
Study of Pressure on Metallic Tin Foil by Mössbauer Spectroscopy
(5 and 12.1; 21.7)
(b) Vol. **132**, No. 1, p. K23–K25 (1985).

Z. A. MATYSINA, A. M. OVRUTSKII, and L. M. CHUPRINA
Influence of Correlation and Anisotropy on the Molecular Roughness of the Crystal Face
(1.2 and 1.6; 21.7)
(a) Vol. **92**, No. 2, p. K93–K97 (1985).

22. Semiconductors and Ionic Crystals

Review Articles

YU. A. KATULEVSKII
Injection of Degenerate Electron–Hole Plasma
(14.3.4; 13.1; 13.4; 14.3.3; 22)
(a) Vol. **88**, No. 1, p. 11–42 (1985).

S. SUGAI
Lattice Vibrations in the Charge-Density-Wave States of Layered Transition Metal Dichalcogenides
(6 and 13; 1.2; 20.1; 22)
(b) Vol. **129**, No. 1, p. 13–39 (1985).

J. S. KANG and D. K. SCHRODER
The Pulsed MIS Capacitor. A Critical Review
(14.3.3; 22)
(a) Vol. **89**, No. 1, p. 13–43 (1985).

V. CHRISTOPH and G. RÖPKE
Theory of Inverse Linear Response Coefficients
(13 and 14; 15; 21; 22)
(b) Vol. **131**, No. 1, p. 11–42 (1985).

T. S. MOSS
Relations between the Refractive Index and Energy Gap of Semiconductors
(13.1 and 20.1; 22; 22.2.1; 22.2.3)
(b) Vol. **131**, No. 2, p. 415–427 (1985).

Original Papers and Short Notes

E. HEINER
A Dynamical Theory for a Disordered Hubbard Chain far from Equilibrium. I. Extended Master Equations
(13 and 14.3.4; 22)
(b) Vol. **127**, No. 1, p. 253–265 (1985).

G. I. PILIPENKO, D. V. OPARIN, O. I. TYUTYUNNIK, and F. F. GAVRILOV
Photostimulated Transformations of Complex Colour Centres in Lithium Hydride Single Crystals
(6 and 10.2; 20.1; 20.3; 22)
(b) Vol. **127**, No. 1, p. 299–306 (1985).

A. I. AGAFONOV, A. F. PLOTNIKOV, and V. N. SELEZNEV
Conductivity of Amorphous A_xB_{1-x} Semiconductors with Potential Fluctuations of the Mobility Edges
(13.4 and 14.3.4; 22)
(b) Vol. **127**, No. 1, p. 395–401 (1985).

P. ENDERS
Test of Two Relations between Tight-Binding Parameters for Tetrahedral Solids
(13.1; 22)
(b) Vol. **127**, No. 1, p. K51–K53 (1985).

K. R. ALLAKHVERDIEV, T. G. MAMEDOV, V. E. SHTEINSHREIBER, and I. K. EFENDIEVA
Fundamental Absorption Edge of TlSe and $TlInTe_2$ Crystals
(13.1 and 20.1; 13.5.1; 22; 22.8)
(b) Vol. **127**, No. 1, p. K55–K60 (1985).

P. R. VAYA, J. MAJHI, B. S. V. GOPALAM, and C. DATTATREYAN
Thickness Dependence of Hall Mobility of HWE Grown PbTe Films
(14.3.1; 22)
(a) Vol. **87**, No. 1, p. 341–350 (1985).

N. RYŠAVÁ, L. TICHÝ, Č. BARTA, A. TŘÍSKA, and H. TICHÁ
Kinetics of Recrystallization of Sb_2S_3 in the Glassy $(GeS_2)_{0.3}(Sb_2S_3)_{0.7}$
(2 and 3; 22)
(a) Vol. **87**, No. 1, p. K13–K17 (1985).

P. ENDERS
Structure and Bonding in Cubic IV–VI Crystals. IV. Evidence for Covalent Bonding in the IV–VI Family from the Extended Mulliken Scheme
(4 and 13; 1.2; 22)
(b) Vol. **127**, No. 2, p. 449–457 (1985).

E. GERLACH
Overshoot in Low-Conduction Semiconductors
(14.3; 22)
(b) Vol. **127**, No. 2, p. 727–733 (1985).

S. GUHA and C. BASU
Parametric Excitation of Electron-Acoustic Waves by Helicon Waves in Magnetised Semiconductor-Plasma
(13 and 20; 22)
(b) Vol. **127**, No. 2, p. K137–K142 (1985).

J. B. RADUNOVIĆ
The Influence of Large-Scale Fluctuations on the Anomalous Conductivity in Piezoelectric Semiconductors
(14.3; 6; 22)
(b) Vol. **127**, No. 2, p. K181–K185 (1985).

A. STRIGAZZI
A New Method for Measuring the Mixed Splay–Bend Elasticity and the Anchoring Strength in a Nematic Film
(1.5 and 12.2; 22)
(a) Vol. **87**, No. 2, p. 537–541 (1985).

I. L. Baginskii and E. G. Kostsov
Theory of Transient Currents in Dielectrics at a Limited Level of Single Injection. I. The Case of Strong Injection
(14.3.4; 14.4; 22)
(a) Vol. **88**, No. 1, p. 331 – 341 (1985).

R. Assenov, V. A. Moshnikov, and D. A. Yaskov
On the Behaviour of Iodine in PbTe and SnTe
(10.2; 14.3; 22)
(a) Vol. **88**, No. 1, p. K27 – K30 (1985).

P. I. Stoianova and R. E. Andreichin
Coordinate Sensitive Photoelement on the Basis of the Photoelectret Sate in Vitreous Semiconductors
(14.3.3 and 16; 2; 22)
(a) Vol. **88**, No. 1, p. K91 – K93 (1985).

V. G. Baryshevskii and S. V. Cherepitsa
Neutron Spin Precession and Spin Dichroism of Nonmagnetic Unpolarized Single Crystals
(1.1 and 11; 22)
(b) Vol. **128**, No. 2, p. 379 – 387 (1985).

A. Birczyński and Z. T. Lalowicz
^{1}H-NMR Spectra of Molecular Groups in a Paramagnetic Matrix
(19; 18.1; 18.4; 22)
(b) Vol. **128**, No. 2, p. 545 – 552 (1985).

Yu. S. Grushko, Yu. B. Paderno, K. Ya. Mishin, L. I. Molkanov, G. A. Shadrina, E. S. Konovalova, and E. M. Dudnik
A Study of the Electronic Structure of Rare Earth Hexaborides
(13.1; 22)
(b) Vol. **128**, No. 2, p. 591 – 597 (1985).

P. Enders
Electron–Phonon Interaction as Effective Electron–Electron Interaction
(13; 22)
(b) Vol. **128**, No. 2, p. 611 – 618 (1985).

Kaiyi Zhang and K. A. Chao
Nonlinear Thomas-Fermi Dielectric Function in Semiconductors
(14.4; 22)
(b) Vol. **128**, No. 2, p. 723 – 729 (1985).

O Ziep and R. Keiper
On the Theory of the Electron–Phonon Scattering Rate Including the Intra-Collisional Field Effect
(13 and 14.3.4; 22)
(b) Vol. **128**, No. 2, p. 779 – 787 (1985).

D. E. Jesson and M. P. Shaw
Dynamical Effects in Foil Thickness Determination Using Convergent-Beam Electron Diffraction
(1.5; 22)
(a) Vol. **88**, No. 2, p. 469 – 474 (1985).

I. A. CHAIKOVSKII, G. M. SHMELEV, and A. I. GERMAN
AC Conductivity of Highly Inhomogeneous Semiconductors
(14.3.4; 22)
(b) Vol. **129**, No. 1, p. 393–398 (1985).

U. HENNIGER and R. KEIPER
On Theoretical Modelling of Continuously Inhomogeneous Semiconductor Structures
(14.3; 22)
(b) Vol. **129**, No. 1, p. 399–406 (1985).

A. I. KASIYAN and P. I. RUSSU
Charge Carrier Mobility Caused by Scattering from Hybrid Plasmon–Phonon Modes
(6 and 13.5.2; 14.3.4; 22; 22.2.1)
(b) Vol. **129**, No. 1, p. 439–447 (1985).

T. R. GAWRON and J. MYCIELSKI
Erratum to: Shallow Impurity States in Strong Magnetic Field
(13.4; 20.1; 22)
(b) Vol. **129**, No. 1, p. 448 (1985).

S. CHANDRA and S. N. SAHU
Electrodeposited Tungsten Selenide Films. II. Optical, Electrical, Electrochemical Characterization and Photoelectrochemical Solar Cell Studies
(14.3.2; 16; 20.1; 22)
(a) Vol. **89**, No. 1, p. 321–331 (1985).

J. KRIŠTOFIK, J. J. MAREŠ, and V. ŠMÍD
The Effect of Pressure on Conductivity and Permittivity of As_2Te_3-Based Glasses
(14.3; 2; 22)
(a) Vol. **89**, No. 1, p. 333–345 (1985).

L. BENEŠ, J. VOTINSKÝ, P. LOŠŤÁK, J. KALOUSOVÁ, and J. KLIKORKA
Cobaltocene Intercalate of the Layered $SnSe_2$
(1.4; 22)
(a) Vol. **89**, No. 1, p. K1–K4 (1985).

C. H. LING, C. Y. KWOK, and K. PRASAD
Relative Hydrogen Content in Plasma-Enhanced CVD Silicon Nitride Films: Substrate Temperature Dependence and Effect of Thermal Annealing
(1.5 and 10.2; 22)
(a) Vol. **89**, No. 1, p. K39–K43 (1985).

YU. G. FARIVER, A. I. BAIRAMOV, T. D. DZHAFAROV, and M. S. SADIGOV
The Effect of Substrate on the Electrical Properties of As_2S_3 Films
(14.3.1 and 14.3.3; 22)
(a) Vol. **89**, No. 1, p. K95–K98 (1985).

W. G. MCMULLAN and J. C. IRWIN
An Interpolytypical Transition in NbS_2
(1.2; 22)
(b) Vol. **129**, No. 2, p. 465–469 (1985).

A. M. Andriesh, V. V. Bivol, V. I. Verlan, B. V. Efremushkin, M. S. Iovu, and E. G. Khanchevskaya
Influence of Electrode Material on the Electrical and Photoelectrical Properties of p-InP/As_2Se_3 Heterostructures
(14.3.3 and 16; 2; 14.3.1; 22; 22.2.2)
(a) Vol. **89**, No. 2, p. K231 – K234 (1985).

V. M. Polyanovskii
Harmonics of Cyclotron and Cyclotron–Phonon Resonances in Semiconductors with Superlattice
(6 and 13; 19; 20.1; 22)
(b) Vol. **130**, No. 1, p. 95 – 101 (1985).

J. Röseler
Application of the Density-Functional Formalism to Bound Polarons
(13.5.3; 22)
(b) Vol. **130**, No. 1, p. 303 – 314 (1985).

S. Lay, P. Delavignette, and J. Vicens
Analysis of Subgrain Boundaries in Tungsten Carbide Deformed at High Temperature
(1.4; 10.2; 22)
(a) Vol. **90**, No. 1, p. 53 – 68 (1985).

V. V. Vasilev and I. P. Mikhailovskii
Effect of High-Temperature Annealing on PL Spectra of Silicon Nitride
(20.3; 22)
(a) Vol. **90**, No. 1, p. 355 – 358 (1985).

K. Yasuda, A. Yoshida, M. Takeda, H. Masuda, and I. Akasaki
Boron Nitride Films Prepared by RF Sputtering
(2 and 3; 6; 20.1; 22)
(a) Vol. **90**, No. 1, p. K7 – K9 (1985).

T. Sato, H. Takizawa, and M. Shimada
Preparation and Electrical Properties of $(M_{1-x}M'_x)\,Si_2$ (M, M′ = Cr, Mn, Fe) ($0 \leqq x \leqq 0.3$)
(14.1 and 14.3; 1.3; 15; 21.1; 22)
(a) Vol. **90**, No. 1, p. K81 – K86 (1985).

R. M. Candea, D. Dărdâlat, R. Turcu, and E. Indrea
Properties of PbSe Films Prepared by Chemical "Anorganic" Deposition
(1.5 and 14.3.1; 16; 20.1; 22)
(a) Vol. **90**, No. 1, p. K91 – K95 (1985).

M. K. Agarwal, P. D. Patel, L. T. Talele, and D. Laxminarayana
Optical Band Gaps of Molybdenum Sulphoselenide (MoS_xSe_{2-x}, $0 \leqq x \leqq 2$) Single Crystals from Spectral Response
(14.3.3 and 16; 13.1; 22)
(a) Vol. **90**, No. 1, p. K107 – K111 (1985).

P. Enders
Structure and Bonding in Cubic IV–VI Crystals. VII. On the Influence of Defects on the Stability of the Cubic Structure
(4 and 13; 1.2; 10.2; 22)
(b) Vol. **130**, No. 2, p. 511 – 516 (1985).

R. ZIMMERMANN and H. STOLZ
The Mass Action Law in Two-Component Fermi Systems Revisited. Excitons and Electron–Hole Pairs
(13.5.1; 22)
(b) Vol. **131**, No. 1, p. 151–164 (1985).

F. M. GASHIMZADE and D. G. GULIEV
Continuity Chords of Bands and Band Structure of Semiconducting Compounds with TlSe-Type Structure
(13.1; 22)
(b) Vol. **131**, No.1, p. 201–206 (1985).

E. HEINER
A Dynamical Theory for a Disordered Hubbard Chain far from Equilibrium. II. The Low-Field Case
(13 and 14.3.4; 22)
(b) Vol. **131**, No. 1, p. 213–224 (1985).

P. LOŠŤÁK, J. HORÁK, and R. NOVOTNÝ
Some Optical Properties of $Bi_2Se_{3-x}S_x$ Single Crystals
(20.1; 22)
(b) Vol. **131**, No. 1, p. K67–K71 (1985).

H. STRAMSKA
Magneto-Optical Effects on the Generation of the Third Harmonic in Semiconductors
(18 and 20; 13; 22)
(b) Vol. **131**, No. 1, p. K77–K79 (1985).

K. NAKANO, K. SUZUKI, and S. TAKEUCHI
Dislocations in SmS Single Crystals
(10.2; 22)
(a) Vol. **91**, No. 1, p. 73–77 (1985).

A. K. ABASS, H. A. JASSIM, K. J. MAJEID, and R. H. MISHO
Optical Parameters of Chemically Deposited Tin Disulphide
(13.1 and 20.1; 22)
(a) Vol. **91**, No. 1, p. 129–133 (1985).

A. I. LEBEDEV and KH. A. ABDULLIN
Impurity Photoconductivity and Electrical Properties of $Pb_{1-x-y}Ge_xSn_yTe$ Doped with Indium
(14.3 and 16; 13.4; 22)
(a) Vol. **91**, No. 1, p. 225–234 (1985).

M. KITAO, K. YOSHII, and S. YAMADA
Thermoelectric Power of Glassy $As_{40}Se_{60-x}Te_x$
(14.3 and 15; 2; 22)
(a) Vol. **91**, No. 1, p. 271–277 (1985).

K.-P. MÖLLMANN and M. HAPP
Impurity Photoconductivity of Indium Doped Lead Tin Telluride
(16; 22)
(a) Vol. **91**, No. 1, p. K71–K74 (1985).

A. STERN and K. HÜBNER
The One-Band Tight-Binding Hamiltonian: Fivefold Rings and Band Edges
(13.1; 22)
(b) Vol. **132**, No. 1, p. K51 – K56 (1985).

S. E. ABDULGAFAROV, V. L. AVERYANOV, B. T. KOLOMIETS, and V. M. LYUBIN
Photoconductivity of Vitreous As_2Se_3 Films Modified with Nickel
(2 and 16; 22)
(a) Vol. **92**, No. 1, p. 257 – 262 (1985).

M. MATOS and B. MAFFEO
Non-Equivalence Effects in Cluster Self-Consistent Calculations of Ionic Solids
(13.1 and 13.4; 22)
(b) Vol. **132**, No. 2, p. 541 – 549 (1985).

V. D. ISKRA
On the Theory of Scattering by Short-Range Potentials
(13.4; 14.3; 22)
(b) Vol. **132**, No. 2, p. 581 – 590 (1985).

N. LAWERENCE and T. M. HARIDASAN
Mean Square Amplitude of Transition Metal Dichalcogenide 1T-$TiSe_2$ in the Normal and Commensurate Charge Density Wave Phases
(6; 1.2; 22)
(b) Vol. **132**, No. 2, p. K77 – K79 (1985).

C. SUGIURA and S. MURAMATSU
K-Edge X-Ray Absorption Spectra from Potassium Compounds
(20; 22)
(b) Vol. **132**, No. 2, p. K111 – K115 (1985).

P. R. VAYA
Magnetoresistance of PbTe Films
(14.3.1; 13.1; 22)
(a) Vol. **92**, No. 2, p. 555 – 561 (1985).

22.1 Elements

Original Papers and Short Notes

22.1.1 Germanium

Original Papers and Short Notes

H.-Matsuo Kagaya and T. Soma
Specific Heat of Si and Ge
(6 and 8; 22.1.1; 22.1.2)
(b) Vol. **127**, No. 1, p. K5–K7 (1985).

V. K. Arora
Intracollisional Field Effect in Quasi-Hot Electron Transport
(14.3.4; 6; 22.1.1; 22.2.1)
(b) Vol. **127**, No. 1, p. K95–K97 (1985).

V. M. Asnin, V. I. Stepanov, R. Zimmermann, and M. Rösler
Screening of the Spin–Orbit Exciton by the Electron–Hole Liquid in Germanium
(13.5.1 and 20.1; 13.5.2; 22.1.1)
(b) Vol. **127**, No. 2, p. 651–655 (1985).

F. R. Kessler, W. Köppen, and E. Gaulke
Spatiäl Resolved Ettingshausen Effect in i-Germanium
(15; 14.3; 22.1.1)
(b) Vol. **127**, No. 2, p. K175–K179 (1985).

D. A. Aronov, R. Mamatkulov, and V. V. Rubinov
The Criteria of Diffusionless Solution Validity in the Theory of Exclusion in "Super-Pure" Semiconductors
(14.3.4; 22.1.1)
(a) Vol. **87**, No. 2, p. 667–672 (1985).

Yu. K. Pozhela, E. V. Starikov, and P. N. Shiktorov
Far Infrared Absorption by Hot Holes in p-Ge under $\boldsymbol{E} \perp \boldsymbol{B}$ Fields
(14.3.4 and 20.1; 22.1.1)
(b) Vol. **128**, No. 2, p. 653–661 (1985).

I. V. Gutsul, I. G. Kirnas, P. G. Litovchenko, V. I. Marusyak, V. M. Nitsovich, S. E. Ostapov, and E. E. Petrosyan
Study of the Hall Coefficient in p-Type Germanium Taking into Account Anisotropic Charge Carrier Scattering
(14.3; 22.1.1)
(b) Vol. **128**, No. 2, p. 753–759 (1985).

A. Bath, J. Hugel, and R. Kleim
Thermoelectric Power of Liquid Germanium and Tin
(2 and 15; 14.1; 22.1; 22.1.1)
(b) Vol. **128**, No. 2, p. 761–767 (1985).

S. Chakraborty, A. Manna, and A. K. Ghosh
Atomic Scattering Factors of Solid Elements
(13; 21; 21.1; 21.1.1; 21.3; 22.1.1; 22.1.2)
(b) Vol. **129**, No. 1, p. 211–220 (1985).

G. CONTRERAS, L. TAPFER, A. K. SOOD, and M. CARDONA
Physical Properties of Ion-Implanted Laser Annealed n-Type Germanium
(6 and 13.4; 14.3; 20.1; 22.1.1)
(b) Vol. **131**, No. 2, p. 475–487 (1985).

F. BECHSTEDT, D. REICHARDT, and R. ENDERLEIN
Self-Consistent Tight-Binding Method for Total Energy Calculations of Tetrahedral Semiconductors Including Surfaces and Defects
(13.1; 4; 22.1; 22.1.1; 22.1.2)
(b) Vol. **131**, No. 2, p. 643–657 (1985).

W. FRENTRUP, M. GRIEPENTROG, and U. MÜLLER-JAHREIS
Negative Secondary Ion Emission Influenced by Alkali Atoms
(17; 1.6; 22.1.1; 22.1.2; 22.2.1; 22.2.2)
(a) Vol. **91**, No. 2, p. 447–452 (1985).

S. A. SHEVCHENKO, I. I. KHODOS, and I. I. SNIGHIREVA
Dislocation Dissociation and Electrical Properties of Plastically Deformed Germanium Single Crystals
(10.2; 14.3; 22.1.1)
(a) Vol. **91**, No. 2, p. 523–531 (1985).

E. ERBARUT and M. TOMAK
A Localized Orbital Description of Ge–Si Alloy
(13.1; 22.1.1; 22.1.2)
(b) Vol. **132**, No. 1, p. 173–177 (1985).

I. V. GUTSUL, I. G. KIRNAS, P. G. LITOVCHENKO, V. I. MARUSYAK, V. M. NITSOVICH, S. E. OSTAPOV, and E. E. PETROSYAN
Hall Factor in p-Ge with Anisotropic Energy Spectrum
(14.3; 22.1.1)
(b) Vol. **132**, No. 1, p. 271–276 (1985).

Y. Q. SHENG, P. MUNZ, and R. SCHULTHEISS
The Determination of the Thickness of Ultrathin Metal and Semiconductor Films on Conventional Semiconductor Substrates
(1.5; 14.3.1; 21; 21.1; 21.4; 21.6; 22.1.1)
(a) Vol. **92**, No. 1, p. 121–128 (1985).

V. V. KVEDER, R. LABUSCH, and YU. A. OSSIPYAN
Frequency Dependence of the Dislocation Conduction in Ge and Si
(10.2 and 14.3; 22.1.1; 22.1.2)
(a) Vol. **92**, No. 1, p. 293–302 (1985).

V. A. BESPALOV, P. K. KASHKAROV, V. F. KISELEV, V. A. MATVEEV, and G. S. PLOTNIKOV
The Study of Slow Relaxation of Charged Centres in Insulator–Semiconductor Structures Using Fluorescent Molecule Probes
(14.3.2; 13.3; 20.3; 22.1.1; 22.6)
(a) Vol. **92**, No. 1, p. 315–326 (1985).

22.1.2 Silicon

Original Papers and Short Notes

B. I. Reser
Mixed Basis Method with Overlapping Atomic Orbitals
(13.1; 22.1.2)
(b) Vol. **127**, No. 1, p. 267–272 (1985).

A. A. Andreev, A. V. Zherzdev, A. I. Kosarev, K. V. Kougia, and I. S. Shlimak
Photoluminescence and Recombination of Excess Carriers in Amorphous Hydrogenated Silicon
(13.4 and 20.3; 16; 22.1.2)
(b) Vol. **127**, No. 1, p. 273–278 (1985).

V. D. Gorobchenko and F. J. Litterst
Mössbauer Relaxation Spectra Due to Electric Field Gradient Reorientation and Nuclear Hopping for an $I = 5/2$ to $7/2$ Nuclear Transition
(5; 22.1.2)
(b) Vol. **127**, No. 1, p. 351–358 (1985).

H.-Matsuo Kagaya and T. Soma
Specific Heat of Si and Ge
(6 and 8; 22.1.1; 22.1.2)
(b) Vol. **127**, No. 1, p. K5–K7 (1985).

A. Armigliato, S. Solmi, C. Donolato, P. Negrini, E. Gabilli, A. Garulli, and M. Kittler
Influence of Chlorine Implantation on Phosphorus Diffusivity and Oxidation-Induced Defects in Silicon
(10.2 and 11; 14.3.3; 22.1.2)
(a) Vol. **87**, No. 1, p. 207–216 (1985).

F. Cembali, M. Servidori, E. Gabilli, and R. Lotti
Effect of Diffuse Scattering in the Strain Profile Determination by Double Crystal X-Ray Diffraction
(1.4; 11; 22.1.2)
(a) Vol. **87**, No. 1, p. 225–233 (1985).
Erratum
(a) Vol. **92**, No. 1, p. 336 (1985).

H. Toutah and G. Pananakakis
Study of Electronic States Located at the Si–SiO_2 Interface and within the Oxide Volume of VLSI MOS Capacitors Using Capacitance and Conductance Measurements
(13.3 and 14.3.3; 14.3.2; 22.1.2)
(a) Vol. **87**, No. 1, p. 311–317 (1985).

A. Yu. Leiderman
On the Generation–Recombination Currents in p–n Junctions of Semiconductors with Continuous Gap-State Spectrum
(2 and 14.3.3; 14.3.4; 22.1.2)
(a) Vol. **87**, No. 1, p. 363–372 (1985).

V. V. Kveder, Yu. A. Osipyan, I. R. Sagdeev, A. I. Shalynin, and M. N. Zolotukhin
The Effect of Annealing and Hydrogenation on the Dislocation Conduction in Silicon
(10.2 and 14.3; 13.4; 22.1.2)
(a) Vol. **87**, No. 2, p. 657–665 (1985).

A. C. Fiore de Mattos and G. Sarrabayrouse
The MISS Device. Modelling and Influence of Critical Parameters
(14.3.3; 14.3.4; 22.1.2)
(a) Vol. **87**, No. 2, p. 699–707 (1985).

B. B. Rao and W. A. Anderson
Comparing Furnace to Laser Annealing in Ion Implanted Surface Passivated MINP Solar Cells
(14.3.3 and 16; 22.1.2)
(a) Vol. **87**, No. 2, p. 735–739 (1985).

D. Stock, H.-D. Geiler, and K. Hehl
Theoretical Evidence for Opposite Moving Phase Fronts during Ultrafast Solidification Processes
(1.5 and 3; 11; 22.1.2)
(a) Vol. **87**, No. 2, p. K115–K118 (1985).

D. M. Vanderwalker
The Effect of Doping on the Motion of Partial Dislocations in Silicon
(10.2; 22.1.2)
(a) Vol. **87**, No. 2, p. K143–K148 (1985).

K. D. Glinchuk, N. M. Litovchenko, Z. A. Salnik, S. I. Skryl, and A. L. Troshin
Effect of Carbon on the Minority Carrier Lifetime in Heat-Treated Oxygen-Containing Silicon
(10.2 and 13.4; 22.1.2)
(a) Vol. **87**, No. 2, p. K149–K152 (1985).

V. S. Lysenko, V. I. Zimenko, I. P. Tyagulskii, I. N. Osiyuk, O. V. Snitko, and T. N. Sytenko
Flash-Lamp Annealing of Si–SiO_2 Transition Layer Defects
(11 and 13.3; 14.3.2; 22.1.2)
(a) Vol. **87**, No. 2, p. K175–K180 (1985).

B. V. Koba, S. M. Komarov, M. A. Kotov, N. G. Kryuchkov, V. L. Litvinov, V. V. Ogurtsov, N. A. Ukhin, I. B. Fedotov, and A. A. Filippov
Change in the Hall Coeeficient with Magnetic Field in Silicon, Irradiated by Neutrons
(14.3; 11; 22.1.2)
(a) Vol. **87**, No. 2, p. K195–K198 (1985).

B. V. Koba, S. M. Komarov, M. A. Kotov, N. G. Kryuchkov, V. L. Litvinov, V. V. Ogurtsov, N. A. Ukhin, I. B. Fedotov, and A. A. Filippov
The Physical and Geometrical Magnetoresistance of n-Si Irradiated by Neutrons
(14.3; 11; 22.1.2)
(a) Vol. **87**, No. 2, p. K199–K202 (1985).

S. A. Azimov, D. Isamukhamedova, P. M. Karageorgy-Alkalaev, A. Yu. Leiderman, and Yu. M. Yabov
On the Properties of Amorphous Silicon Films and Amorphous Si–Crystalline Si Heterojunction
(14.3.1 and 14.3.3; 16; 22.1.2)
(a) Vol. **87**, No. 2, p. K203–K208 (1985).

H. BERGNER, V. BRÜCKNER, F. KERSTAN, and W. NOWICK
Picosecond Processes of Laser-Excited Carriers in Silicon on Sapphire
(2 and 14.3.1; 16; 20.1; 22.1.2; 22.6)
(b) Vol. **128**, No. 2, p. 769—777 (1985).

V. I. KIRILLOV, D. I. MATERIKIN, E. I. PRIBYLOVA, YU. A. KAPUSTIN, and S. I. REMBEZA
On the Nature of Impurity Optical Absorption Bands in Silicon Doped by Noble Metals
(13.4 and 20.1; 22.1.2)
(b) Vol. **128**, No. 2, p. K163—K167 (1985).

FAN VAN AN, N. A. BULENKOV, and A. V. ANDREEVA
Structure of a Second-Order Twin Boundary in Silicon and Its Interaction with Thermally Generated Lattice Dislocations
(1.4 and 10.2; 22.1.2)
(a) Vol. **88**, No. 2, p. 429—441 (1985).

A. L. GOLOVIN, R. M. IMAMOV, and E. A. KONDRASHKINA
Potentialities of New X-Ray Diffraction Methods in Structural Studies of Ion-Implanted Silicon Layers
(1.6; 11; 22.1.2)
(a) Vol. **88**, No. 2, p. 505—514 (1985).

E. F. KRIMMEL, R. LANGFELD, and A. G. K. LUTSCH
Modelling of Dopant Profiles Modified by Diffusing Defects in Polycrystalline Silicon
(9 and 14.3.1; 11; 22.1.2)
(a) Vol. **88**, No. 2, p. 533—538 (1985).

S. E. ZABOLOTSKII, V. P. KALINUSHKIN, T. M. MURINA, M. G. PLOPPA, and K. TEMPELHOFF
The Influence of Growth Rate on the Impurity Cloud Parameters in Pure Silicon
(3 and 10.2; 22.1.2)
(a) Vol. **88**, No. 2, p. 539—542 (1985).

K. YASUDA, T. WADA, H. MASUDA, M. TAKEDA, and S. IKUTA
Defect Distribution in Crystalline Silicon Irradiated with High-Energy Electrons
(11 and 13.4; 14.3; 22.1.2)
(a) Vol. **88**, No. 2, p. 543—551 (1985).

S. I. KIRILLOVA, V. E. PRIMACHENKO, and O. V. SNITKO
Nonequilibrium Depletion Relaxation in Strong Electric Fields under Various Conditions at the Silicon Surface
(14.3.2 and 14.3.4; 13.3; 22.1.2)
(a) Vol. **88**, No. 2, p. 647—654 (1985).

A. ROMANOWSKI
The Impact of the Heat Generation of the Primary Scanning Beam on the SEM-EBIC Characteristics
(14.3.3; 22.1.2; 22.2.1)
(a) Vol. **88**, No. 2, p. 663—672 (1985).

A. V. MUDRYI, A. L. PUSHKARCHUK, and A. G. ULYASHIN
Noble Gas Atoms in Si: Energy Spectrum Distortion of Vacancy Complexes
(10.2 and 13.4; 22.1.2)
(b) Vol. **129**, No. 1, p. K83–K87 (1985).

A. STERN and K. HÜBNER
Electronic Density of States of a Valence-Alternation-Pair Defect at the Si–SiO_2 Interface
(13.3; 22.1.2; 22.6)
(b) Vol. **129**, No. 1, p. K89–K93 (1985).

D. STOCK, H.-D. GEILER, and K. HEHL
A Model of Crystallization Processes Controlled by Temperature Pulses in Amorphous Semiconductors.
I. Step-Like Temperature Pulse
(3; 1.2; 2; 22.1.2)
(a) Vol. **89**, No. 1, p. 57–63 (1985).

V. L. LITVINOV and N. A. UKHIN
Fast-Neutron Radiation Damages in Heavily Doped p-Silicon
(11; 14.3; 22.1.2)
(a) Vol. **89**, No.1, p. 95–104 (1985).

F. H. M. SPIT, H. ALBERS, A. LUBBES, Q. J. A. RIJKE, L. J. v. RUIJVEN, J. P. A. WESTERVELD, H. BAKKER, and S. RADELAAR
Diffusion of Antimony (^{125}Sb) in Polycrystalline Silicon
(9; 10.2; 22.1.2)
(a) Vol. **89**, No. 1, p. 105–115 (1985).
Erratum
(a) Vol. **91**, No. 1, p. 302 (1985).

W. R. FAHRNER, J. R. LASCHINSKI, D. BRÄUNIG, M. KNOLL, and A. NEIDIG
Damage Profiles after 50 to 500 MeV Ion Implantation as Deduced from Thyristor Leakage Currents
(11 and 14.3.3; 13.4; 22.1.2)
(a) Vol. **89**, No. 1, p. 347–351 (1985).

K. SCHMALZ, F.-G. KIRSCHT, S. NIESE, I. BABANSKAJA, M. KITTLER, H. RICHTER, J. SCHÖNEICH, and W. SEIFERT
On the Intrinsic Gettering in Cu-Contaminated Cz-Si
(10.2 and 13.4; 13.3; 22.1.2)
(a) Vol. **89**, No. 1, p. 389–401 (1985).

K. K. SHARMA and G. P. SRIVASTAVA
Barrier Height and Its Instability in Al–Ultrathin SiO_2–n/p-Si Devices
(14.3.3; 22.1.2)
(a) Vol. **89**, No. 1, p. 403–413 (1985).
Erratum
(a) Vol. **91**, No. 1, p. 303 (1985).

J. KUBĚNA and J. HLÁVKA
Space Correlation of Microdefects with Recombination of Excess Carriers in CZ-Si
(1.4 and 16; 22.1.2)
(a) Vol. **89**, No. 1, p. K23–K25 (1985).

B. Olejníková
A Method of Determination of the Trap Depth of Deep Centres by Capacitance Measurements
(13.4 and 14.3.4; 22.1.2)
(a) Vol. **89**, No. 2, p. K197–K200 (1985).

J. Koshy
Measurements of the Rectifying Barrier Height of Sputter Deposited Bi_2Te_3 Contacts on p-Silicon
(14.3.4; 22.1.2)
(a) Vol. **89**, No. 2, p. K219–K222 (1985).

F. P. Korshunov, I. G. Marchenko, and V. T. Troshchinskii
Variation of Switching Time of Silicon p–n–p–n Structures Irradiated by Fast Electrons
(11 and 14.3.3; 22.1.2)
(a) Vol. **89**, No. 2, p. K227–K230 (1985).

P. Baeri, M. A. Harith, G. Russo, E. Rimini, A. Giulietti, and M. Vaselli
Free Carrier Dynamics and Energy Transfer to the Si Lattice during Pico and Nanosecond Nd Laser Pulse Irradiation
(13 and 20.1; 22.1.2)
(b) Vol. **130**, No. 1, p. 225–233 (1985).

P. T. Landsberg, A. Neugroschel, F. A. Lindholm, and C. T. Sah
A Model Band-Gap Shrinkage in Semiconductors with Application to Silicon
(13.1 and 13.4; 22.1.2)
(b) Vol. **130**, No. 1, p. 255–266 (1985).

Xie Ying, L. Snyder, S. N. Sahu, and J. W. Corbett
MNDO Calculation on Molecular Clusters Chosen to Model the Bonding of Oxygen in Silicon
(4 and 13; 22.1.2)
(b) Vol. **130**, No. 1, p. 333–338 (1985).

A. E. Belyaev, O. P. Gorodnichii, V. S. Vikhnin, and R. M. Vinetskii
The Nonlinear Effect of Uniaxial Stress for Shallow Donors in Silicon
(13.4; 22.1.2)
(b) Vol. **130**, No. 1, p. K47–K50 (1985).

Yu. V. Gorelkinskii, N. N. Nevinnyi, and A. A. Kim
New EPR Spectrum in Helium-Implanted Silicon
(19; 11; 22.1.2)
(b) Vol. **130**, No. 1, p. K51–K53 (1985).

V. V. Aristov, W. Graeff, V. I. Polovinkina, and A. A. Snigirev
Dynamical Diffraction of X-Ray Spherical Waves in Perfect Crystals Using a Narrow Slit as Radiation Source
(1.1; 10.2; 22.1.2)
(a) Vol. **90**, No. 1, p. 69–75 (1985).

A. Armigliato, A. Parisini, R. Hillebrand, and P. Werner
Computer Simulation of High Resolution Electron Microscopy Images of Small Precipitates in P-Supersaturates Silicon
(1.4; 22.1.2)
(a) Vol. **90**, No. 1, p. 115–126 (1985).

T. Takai, T. Halicioğlu, and W. A. Tiller
Calculation of Potential Energy Parameters for the Silicon–Carbon System
(1 and 4; 22.1; 22.1.2; 22.3)
(b) Vol. **130**, No. 2, p. 475–481 (1985).

F. Thuselt and M. Rösler
Gap Shift in Doped Semiconductors at Finite Temperatures
(13.5; 22.1.2; 22.2.1)
(b) Vol. **130**, No. 2, p. 661–673 (1985).

G. V. Dedkov and E. Kh. Shidov
Planar Channeling Radiation of Positrons and Electrons at Moderate Energies in Simple Crystals
(11; 13; 22.1; 22.1.1; 22.1.2)
(b) Vol. **130**, No. 2, p. 719–726 (1985).

H. K. Sy, D. K. Desai, and C. K. Ong
Electron Screening and Mobility in Heavily Doped Silicon
(14.3; 22.1.2)
(b) Vol. **130**, No. 2, p. 787–792 (1985).

H.-Matsuo Kagaya, Y. Kitani, and T. Soma
Calculation of Specific Heat for the Si–Ge System
(8; 6; 22.1.1; 22.1.2)
(b) Vol. **130**, No. 2, p. K95–K96 (1985).

F. Thuselt and M. Rösler
Universal Approximation Formulas for the Gap Shift in Doped Semiconductors
(13.1; 22.1.2; 22.2.1)
(b) Vol. **130**, No. 2, p. K139–K144 (1985).

A. M. Afanasev, P. A. Aleksandrov, R. M. Imamov, E. M. Pashaev, and V. I. Polovinkina
X-Ray Topography of Thin Subsurface Layers
(1.6; 22.1.2)
(a) Vol. **90**, No. 2, p. 419–423 (1985).

A. V. Maslov, E. Kh. Mukhamedzhanov, R. M. Imamov, L. I. Man, and Le Cong Qui
Application of X-Ray Photoelectron Energy Analysis to Depth-Selective Studies of the Subsurface-Layer Structure
(1.6 and 17; 22.1.2)
(a) Vol. **90**, No. 2, p. 439–444 (1985).

O. Bostanjoglo, E. Endruschat, and W. Tornow
Kinetics of Laser-Induced Liquid Metal Etching of a-Si Films
(1.2; 1.4; 22.1.2)
(a) Vol. **90**, No. 2, p. 457–462 (1985).

V. V. Emtsev, T. V. Mashovets, and N. A. Vitovskii
The Spatial Distribution of Frenkel Pair Components Created in Germanium and Silicon under Irradiation
(10.2 and 11; 22.1.1; 22.1.2)
(a) Vol. **90**, No. 2, p. 523–530 (1985).

H. Flietner
U-Shaped Distributions at Semiconductor–Insulator Interfaces and the Nature of the Related Defect Centres
(13.3; 4; 10.2; 22.1.2)
(a) Vol. **91**, No. 1, p. 153–164 (1985).

N. T. Bagraev, L. S. Vlasenko, and A. A. Lebedev
Gold Solid Solution Decay in Silicon. II. Optical Nuclear Polarization Data
(10.2 and 13.4; 20.1; 22.1.2)
(a) Vol. **91**, No. 1, p. 165–172 (1985).

N. A. Kornyushkin, N. I. Bryzgalova, and A. A. Frantsuzov
Relaxation of Tunnel Injection Limited Current in Case of Low Capture Probability
(13.4 and 14.3; 22.1.2)
(a) Vol. **91**, No. 1, p. 211–218 (1985).

P. I. Baranskii, V. V. Savyak, Yu. V. Simonenko, and L. A. Shcherbina
Effect of High-Temperature Annealing on Phonon–Phonon Relaxation in Transmutationally-Doped Silicon Crystals
(6; 15; 22.1.2)
(a) Vol. **91**, No. 1, p. K25–K29 (1985).

P. Zaumseil
On the Increased Sensitivity of X-Ray Rocking Curve Measurements by Triple-Crystal Diffractometry
(1.5 and 10.2; 22.1.2)
(a) Vol. **91**, No. 1, p. K31–K33 (1985).

A. Goltzene, B. Meyer, C. Schwab, X. H. Liu, K. Bohringer, and S. Kalbitzer
EPR Monitoring of Annealing Effects in F-Implanted Amorphous Silicon
(10.2 and 19; 22.1.2)
(a) Vol. **91**, No. 1, p. K45–K48 (1985).

R. Pflueger, J. C. Corelli, and J. W. Corbett
Radiation-Enhanced Oxygen-Related Thermal Donor Formation in Neutron-Transmutation-Doped Floating-Zone Silicon
(10.2 and 11; 22.1.2)
(a) Vol. **91**, No. 1, p. K49–K54 (1985).

T. S. Shi, L. M. Xie, G. R. Bai, and M. W. Qi
The Nature of Two Intense Si–H IR Stretching Bands in FZ-Si:H
(6; 10.2; 20.1; 22.1.2)
(b) Vol. **131**, No. 2, p. 511–517 (1985).

P. Gehrmann, K. Lenkeit, and R. Stolle
Measurements of Proton Channeling Energy Losses in Silicon in the Intermediate Energy Region
(11; 22.1.2)
(b) Vol. **131**, No. 2, p. 519–526 (1985).

V. S. LYSENKO, A. N. NAZAROV, and T. E. RUDENKO
Simple Method for the Determination of the Doping Profile in MIS Structures with Implanted Shallow p–n Junction
(11 and 14.3.3; 22.1.2)
(a) Vol. **91**, No. 2, p. K173–K178 (1985).

M. BALKANSKI, M. JOUANNE, and M. KANEHISA
Bound Phonons
(6 and 13.4; 22.1.2; 22.4.3)
(b) Vol. **132**, No. 1, p. 57–67 (1985).

E. DE CASTRO and P. OLIVO
Quantum Effects in Accumulation Layers of Si–SiO_2 Interfaces in the WKB Effective Mass Approximation
(13.4 and 14.3.2; 13.1; 22.1.2; 22.6)
(b) Vol. **132**, No. 1, p. 153–163 (1985).

E. ERBARUT and M. TOMAK
A Localized Orbital Description of Ge–Si Alloy
(13.1; 22.1.1; 22.1.2)
(b) Vol. **132**, No. 1, p. 173–177 (1985).

T. KATTERLE and W. LORENZ
Quantumchemical Study of Imperfect Semiconductor Surfaces Including Step and Kink Sites. Silicon (111) and Chemisorption of H, OH, and H_2O
(13.3; 13.1; 22.1.2)
(b) Vol. **132**, No. 1, p. 225–238 (1985).

J. KULDA and P. MIKULA
The Three-Beam Case of Neutron Diffraction in a Distorted Crystal
(1.1; 22.1.2)
(a) Vol. **92**, No. 1, p. 95–100 (1985).

L. N. ALEKSANDROV, A. N. KOGAN, and N. P. TIKHONOVA
Study of the Peculiarities of Film Growth from Molecular Beams by the Method of Simulation
(1.5 and 3; 22.1.2)
(a) Vol. **92**, No. 1, p. 109–113 (1985).

P. GAWORZEWSKI and E. HILD
The Effect of Pre-Heat-Treatments on the Formation Kinetics of Thermal Donors in Silicon
(10.2; 13.4; 22.1.2)
(a) Vol. **92**, No. 1, p. 129–135 (1985).

P. NEGRINI, L. PASSARI, A. POGGI, M. SERVIDORI, and E. SUSI
Lattice Disorder and Recombination Centres in Heat-Treated FZ Silicon
(10.2; 14.3; 22.1.2)
(a) Vol. **92**, No. 1, p. 177–185 (1985).

D. TSOUKALAS and P CHENEVIER
Boron Diffusion in Silicon by a Vacancy Mechanism
(9; 10.2; 22.1.2)
(a) Vol. **92**, No. 2, p. 495 – 501 (1985).

C TEMPLIER, B. BOUBEKER, H. GAREM, E. L. MATHÉ, and J. C. DESOYER
Precipitation of Implanted Xenon in Silicon
(1.4 and 10.2; 11; 22.1.2)
(a) Vol. **92**, No. 2, p. 511 – 516 (1985).

S. ALEXANDROVA and A. SZEKERES
On the Correlation of the Charged Centres in Al-Thermal SiO_2–Si Structures
(13.3 and 14.3.2; 22.1.2)
(a) Vol. **92**, No. 2, p. 589 – 594 (1985).

C. A. LONDOS
Room-Temperature Irradiation of p-Type Silicon
(13.4; 10.2; 14.3.3; 22.1.2)
(a) Vol. **92**, No. 2, p. 609 – 614 (1985).

I. P. MIKHAILOVSKII and A. E. EPOV
A Model of Field (Tunnel) Generation of Positive Charge in Thermal Silicon Dioxide of Si–SiO_2 Structures
(14.3.4; 22.1.2)
(a) Vol. **92**, No. 2, p. 615 – 622 (1985).

P. DANESH
Annealing Behaviour of a-Si:H:Cl Films
(14.3.1; 22.1.2)
(a) Vol. **92**, No. 2, p. 631 – 635 (1985).

R. BURKOVA, S. TAKOVA, and G. DIMITROV
S-Pits on (111) Silicon Surfaces
(1.6 and 10.2; 22.1.2)
(a) Vol. **92**, No. 2, p. K107 – K108 (1985).

H LEMKE
Impurity Reaction in Au-Doped Silicon Single Crystals
(13.4; 22.1.2)
(a) Vol. **92**, No. 2, p. K139 – K143 (1985).

JUH TZENG LUE
Temperature and Field Dependence on the Photoconductive Decay Measurements of RF Sputtered and CVD Grown a-Si
(16, 2; 13.4; 22.1.2)
(a) Vol. **92**, No. 2, p. K161 – K165 (1985).

W A. BADAWAY and M S. EL-DESSOUKI
A Study of the Stability of n-Si/SnO_2 Solar Cells under Normal Working Conditions
(14.3.3 and 16, 22.1.2)
(a) Vol. **92**, No. 2, p. K167 – K170 (1985).

22.1.3 Group VI Elements

Original Papers and Short Notes

22.2 III–V Compounds

Review Article

M. Godlewski
On the Application of the Photo-EPR Technique to the Studies of Photoionization, DAP Recombination, and Non-Radiative Recombination Processes
(13.4 and 19; 16; 22.2; 22.4.1; 22.5)
(a) Vol. **90**, No. 1, p. 11–52 (1985).

Original Papers and Short Notes

H.-Matsuo Kagaya and T. Soma
Calculation of Phonon Dispersion Curves for AlP, AlAs, and AlSb
(6; 12.2; 22.2)
(b) Vol. **127**, No. 1, p. 89–93 (1985).

V. D. Tkachev, V. B. Shepilo, and A. M. Zaitsev
Cathodoluminescence of Cubic Boron Nitride – Mössbauer-Type Spectra
(20.3; 22.2)
(b) Vol. **127**, No. 1, p. K65–K67 (1985).

U. Pietsch
Static and Dynamical Valence Charge and Bond Charge Properties of Zincblende Structure Compounds
(4; 13; 22.2)
(b) Vol. **128**, No. 2, p. 439–451 (1985).

E. Anastassakis and M. Cardona
Effects of Strains on the Dynamic Effective Charge of III–V Semiconductors
(6 and 12.2; 4; 22.2)
(b) Vol. **129**, No. 1, p. 101–116 (1985).

Anil Kumar and D. Pal
Lattice Thermal Conductivity of Boron Nitride Crystals at Temperatures 1.5 to 300 K
(6 and 8; 22.2)
(b) Vol. **129**, No. 1, p. K9–K12 (1985).

U. Rensch and G. Eichhorn
Investigation of the State of Strain and of the Electrical Properties of AlN Layers on Monocrystalline Si
(1.5 and 14.3.1; 22.1.2; 22.2)
(a) Vol. **90**, No. 1, p. 135–141 (1985).

R. Strehlow, M. Hanke, and W. Kühn
Charge Self-Consistent Tight-Binding Parameters. Application to III–V Compound Semiconductors
(13.1; 22.2)
(b) Vol. **131**, No. 2, p. 631–642 (1985).

22.2.1 Arsenides

Review Article

T. S. Moss
Relations between the Refractive Index and Energy Gap of Semiconductors
(13.1 and 20.1; 22; 22.2.1; 22.2.3)
(b) Vol. **131**, No. 2, p. 415–427 (1985).

Original Papers and Short Notes

H. Köster, Jr., O. V. Kurnosova, and I. N. Yassievich
Tunneling from Deep Levels of 1–c Type in Electric Fields
(13.4 and 14.3.4; 22.2.1)
(b) Vol. **127**, No. 1, p. 339–350 (1985).

A. G. Mironov, V. A. Morozova, and V. S. Vavilov
Some Peculiarities of the Electrooptical Spectra of n-GaAs(Cr) at Liquid Helium Temperatures (Wannier Levels?)
(13.4 and 20.1; 14.3.4; 22.2.1)
(b) Vol. **127**, No. 1, p. 359–363 (1985).

A. Bhattacharya, A. Ghosal, and D. Chattopadhyay
Small-Signal AC Response of Two-Dimensional Hot Electrons in Quantum-Well Structures of Polar Semiconductors
(13 and 14.3.4; 22.2.1)
(b) Vol. **127**, No. 2, p. 427–432 (1985).

V. K. Arora
Intracollisional Field Effect in Quasi-Hot Electron Transport
(14.3.4; 6; 22.1.1; 22.2.1)
(b) Vol. **127**, No. 1, p. K95–K97 (1985).

H.-G. Brühl and H. Rhan
On the Extension of the Theory of Umweganregung with Respect to the Use of Divergent White X-Ray Radiation
(1.1; 22.2.1)
(a) Vol. **87**, No. 1, p. 121–126 (1985).

U. Pietsch
X-Ray Structure Amplitudes for GaAs and InP
(1.1; 22.2.1; 22.2.2)
(a) Vol. **87**, No. 1, p. 151–156 (1985).

Ľ. Hrivnák
Band-Gap Narrowing in Semi-Insulating GaAs
(13.1 and 14.3; 13.4; 22.2.1)
(b) Vol. **127**, No. 2, p. K165–K170 (1985).

O. Z. Alekperov, V. G. Golubev, V. I. Ivanov-Omskii, and A. Sh. Mekhtiev
The Effect of Electron–Electron Interaction on the Cyclotron Resonance Linewidth in n-GaAs
(14.3.4 and 19; 22.2.1)
(b) Vol. **127**, No. 2, p. K171 – K174 (1985).

J. Jimenez, J. Bonnafe, P. Hernandez, and J. A. de Saja
Optically Induced Photomemory by 1 to 1.35 eV Photons in the Near-Intrinsic Spectral Region on Semiinsulating Bulk GaAs
(16; 13.4; 22.2.1)
(a) Vol. **87**, No. 2, p. 623 – 628 (1985).

W. Siegel, G. Kühnel, and H. A. Schneider
On the Determination of the Carrier Concentration in Large-Grain Polycrystalline InP, GaAs, and GaP by Hall Effect Measurements
(14.3; 22.2.1; 22.2.2)
(a) Vol. **87**, No. 2, p. 673 – 681 (1985).

J. Betko and K. Měřínský
Analysis of Galvanomagnetic Parameters in Semi-Insulating GaAs with Respect to the Three-Band Model
(14.3; 22.2.1)
(a) Vol. **87**, No. 2, p. 683 – 691 (1985).

I. Mojzes, Yu. Vaulin, N. Gerasimenko, V. Obodnikov, B. Kovácz, and R. Veresegyházy
Spatial Distribution of the Components in Au/GaAs System after Scanning Laser Annealing
(1.6 and 11; 22.2.1)
(a) Vol. **87**, No. 2, p. K127 – K130 (1985).

A. I. Kasiyan and P. I. Russu
The Effect of Collective Electron–Hole Plasma Modes on Charge Carrier Mobility in Polar Semiconductors
(6 and 13.5.2; 14.3.4; 22; 22.2.1)
(b) Vol. **128**, No. 1, p. 243 – 249 (1985).

O. K. Gorodnichenko and N. D. Vasilenko
The Strength and Plasticity of GaAs in Saturated Melts of Solvent Metals at the Temperature of Epitaxy
(10.2; 22.2.1)
(a) Vol. **88**, No. 1, p. 223 – 230 (1985).

K. Yasuda, M. Takeda, H. Masuda, and A. Yoshida
Radiation Effects in a Photo-Coupler with High-Energy Electrons and γ-Rays and Recovery by Current Annealing
(11 and 14.3.3; 22.1.2; 22.2.1)
(a) Vol. **88**, No. 1, p. 389 – 396 (1985).

Yu. B. Bolkhovityanov and R. I. Bolkhovityanova
Stress-Induced Isothermal Formation of Films during LPE Growth of III–V Compounds with $\Delta a/a \neq 0$
(3; 1.5; 22.2.1; 22.2.2; 22.2.4)
(a) Vol. **88**, No. 1, p. K5 – K8 (1985).

V. A. KULAKHOV and S. I. CHIKICHEV
The Influence of Isoelectronic Impurities on Intrinsic Deep Levels in Liquid Phase Epitaxial Gallium Arsenide
(13.4; 10.2; 22.2.1)
(a) Vol. **88**, No. 1, p. K59 – K61 (1985).
Erratum
(a) Vol. **89**, No. 1, p. K115 (1985).

T. LALINSKÝ, P. KORDOS, B. KOVÁCS, I. MOJZES, B. SZENTPÁLI, and A. TICHY-RÁCZ
Optimization of the Heat Treatment for Forming AuGe Based Contacts to n-GaAs
(14.3.4; 22.2.1)
(a) Vol. **88**, No. 1, p. K87 – K90 (1985).

R. GULATI, S. B. KAUSHIK, P. L. TRIVEDI, H. S. SHARMA, I. CHANDRA, and B. L. SHARMA
Optical Effect in Normally-Off GaAs MESFET
(14.3.3; 16; 22.2.1)
(a) Vol. **88**, No. 1, p. K99 – K103 (1985).

L. WENDLER
Phonon-Polaritons in Bilayer Systems. s-Polarized Guided Wave Phonon-Polaritons
(6 and 20.1; 13.5; 22.2.1; 22.2.2)
(b) Vol. **128**, No. 2, p. 425 – 437 (1985).

I. URSU, V. CRĂCIUN, L. NANU, D. CRĂCIUN, I. N. MIHĂILESCU, AL. POPA, A. PANTELICĂ, and M. ŞĂLĂGEAN
On the Behaviour of GaAs Single Crystals Irradiated with a Free-Running Ruby Laser
(1.6; 11; 22.2.1)
(a) Vol. **88**, No. 2, p. 415 – 419 (1985).

A. ROMANOWSKI
The Impact of the Heat Generation of the Primary Scanning Beam on the SEM-EBIC Characteristics
(14.3.3; 22.1.2; 22.2.1)
(a) Vol. **88**, No. 2, p. 663 – 672 (1985).

J. KRIŠTOFIK, J. J. MAREŠ, V. ŠMÍD, and J. ZEMAN
AC Conductivity in Semi-Insulating GaAs
(14.3; 22.2.1)
(a) Vol. **88**, No. 2, p. K187 – K190 (1985).

W. ULRICI and P. KLEINERT
Optical Absorption of Cr^{4+} in GaAs:Cr
(13.4 and 20.1; 6; 22.2.1)
(b) Vol. **129**, No. 1, p. 339 – 347 (1985).

A. I. KASIYAN and P. I. RUSSU
Charge Carrier Mobility Caused by Scattering from Hybrid Plasmon–Phonon Modes
(6 and 13.5.2; 14.3.4; 22; 22.2.1)
(b) Vol. **129**, No. 1, p. 439 – 447 (1985).

T. Soma and H.-Matsuo Kagaya
Mode Grüneisen Parameters for Long-Wave Phonons of III–V Covalent Compounds under Pressure
(6; 4; 12.2; 22.2.1; 22.2.3)
(b) Vol. **130**, No. 2, p. 497–501 (1985).

W. Wesch, E. Wendler, G. Götz, K. Unger, H. Röppischer, and Chr. Resagk
Laser Induced Defects in GaAs Layers
(10.2 and 11; 20.1; 22.2.1)
(b) Vol. **130**, No. 2, p. 539–546 (1985).

E. P. Pokatilov, S. I. Beril, V. M. Fomin, and G. A. Pogorilko
Wannier-Mott Exciton States in Two-Layer Periodic Structures
(13.5.1; 22.2.1; 22.2.4)
(b) Vol. **130**, No. 2, p. 619–628 (1985).

F. Thuselt and M. Rösler
Gasp Shift in Doped Semiconductors at Finite Temperatures
(13.5; 22.1.2; 22.2.1)
(b) Vol. **130**, No. 2, p. 661–673 (1985).

F. Thuselt and M. Rösler
Universal Approximation Formulas for the Gap Shift in Doped Semiconductors
(13.1; 22.1.2; 22.2.1)
(b) Vol. **130**, No. 2, p. K139–K144 (1985).

A. J. Willis and M. A. Marais
Deduction of Semiconductor Parameters from the Frequency Dependence of Free Carrier Reflectivity
(20.1; 22.2.1)
(a) Vol. **90**, No. 2, p. K193–K197 (1985).

U. Pietsch, J. Mahlberg, and K. Unger
Investigation of Dynamical Bond Charge Transfer in GaAs by Changing X-Ray Reflection Power under High Electric Field
(4; 13; 22.2.1)
(b) Vol. **131**, No. 1, p. 67–73 (1985).

N. B. Lukyanchikova and N. T. Dekhtyaruk
Noise Investigation of the Impact Ionization in GaAs
(14.3.3 and 14.3.4; 22.2.1)
(a) Vol. **91**, No. 1, p. 279–286 (1985).

A. K. Rai, R. S. Bhattacharya, and M. H. Rashid
Ion Beam Mixing of Alternate Au–Ge Layer with GaAs
(11 and 14.3.4; 1.6; 21.6; 22.1.1; 22.2.1)
(a) Vol. **91**, No. 1, p. 287–293 (1985).

G. Balchaitis, J. Kameneckas, G. Petrauskas, and A. Sakalas
Physical and Chemical Changes in the Surface Layer of GaAs Induced by Recoil Implantation of Al Atoms
(11 and 14.3.2; 22.2.1)
(a) Vol. **91**, No. 1, p. 295–300 (1985).

V. I. Vdovin, L. A. Matveeva, G. N. Semenova, M. Ya. Skorohod, Yu. A. Tkhorik, and L. S. Khazan
Mechanism of Misfit Dislocation Network Formation in Heteroepitaxial System Ge–GaAs (001)
(1.5 and 10.2; 22.1.1; 22.2.1)
(a) Vol. **92**, No. 2, p. 379–390 (1985).

P. Trung Dung and M. Láznička
Outdiffusion of Mn into GaAs Films Grown on Semi-Insulating GaAs Substrates by MBE
(9 and 20.3; 22.2.1)
(a) Vol. **92**, No. 2, p. K113–K116 (1985).

K. D. Glinchuk, V. I. Guroshev, A. V. Prokhorovich, and N. S. Zayats
Photoluminescence in Electron-Irradiated Unannealed and Annealed GaAs:Zn – Irradiation-Induced 1.26 and 1.39 eV Emission Bands
(11 and 20.3; 13.4; 22.2.1)
(a) Vol. **92**, No. 2, p. K145–K149 (1985).

22.2.2 Phosphides

Original Papers and Short Notes

U. Pietsch
X-Ray Structure Amplitudes for GaAs and InP
(1.1; 22.2.1; 22.2.2)
(a) Vol. **87**, No. 1, p. 151–156 (1985).

B. Rheinländer, G. Oelgart, H. Haefner, and R. Pickenhain
Degradation of Green Light Emitting Diodes LPE-GaP:N. II. Injection Luminescence Spectroscopy and Kinetics
(14.3.4 and 20.3; 22.2.2)
(a) Vol. **87**, No. 1, p. 373–381 (1985).

J. Fiddicke and G. Oelgart
The Importance of the Excitation Volume for the Determination of the Minority Carrier Diffusion Length
(14.3.3; 22.2.2)
(a) Vol. **87**, No. 1, p. 383–389 (1985).

R. Fremunt, Y. Hirayama, F. Arai, and T. Sugano
Activation of S^+-Implanted InP by Scanned CW Laser Irradiation
(11 and 14.3; 22.2.2)
(a) Vol. **87**, No. 1, p. K49–K52 (1985).

D. Stephan and V. Geist
Investigation of Lattice Strain in Proton-Irradiated GaP by a Modified Auleytner Technique
(1.1 and 11; 22.2.2)
(a) Vol. **87**, No. 2, p. 589–596 (1985).

W. Siegel, G. Kühnel, and H. A. Schneider
On the Determination of the Carrier Concentration in Large-Grain Polycrystalline InP, GaAs, and GaP by Hall Effect Measurements
(14.3; 22.2.1; 22.2.2)
(a) Vol. **87**, No. 2, p. 673–681 (1985).

N. B. Lukyanchikova, N. P. Garbar, and M. I. Lisyanskii
Study of Degradation of Diode Structures by the Use of Probes
(14.3.3 and 20.3; 22.2.2)
(a) Vol. **87**, No. 2, p. 693–698 (1985).

S. Fung and R. J. Nicholas
Double Source Photoconductivity of InP:Fe
(16; 22.2.2)
(a) Vol. **88**, No. 1, p. 347–353 (1985).

G. S. Korotchenkov, A. A. Gutkin, and A. M. Marinova
The Characteristics of Thin-Film Gallium Phosphide MOS Structures with a Tunnel Transparent Native Anodic Oxide
(14.3.3; 16; 22.2.2)
(a) Vol. **88**, No. 1, p. 355–363 (1985).

Yu. B. Bolkhovityanov and R. I. Bolkhovityanova
Stress-Induced Isothermal Formation of Films during LPE Growth of III–V Compounds with $\Delta a/a \neq 0$
(3; 1.5; 22.2.1; 22.2.2; 22.2.4)
(a) Vol. **88**, No. 1, p. K5–K8 (1985).

L. Wendler
Phonon-Polaritons in Bilayer Systems. s-Polarized Guided Wave Phonon-Polaritons
(6 and 20.1; 13.5; 22.2.1; 22.2.2)
(b) Vol. **128**, No. 2, p. 425–437 (1985).

R. Pickenhain, V. Geist, W. Seifert, and K. Jacobs
Green and Yellow Light Emitting Diodes Produced from Vapour Phase Epitaxial GaP:N. II. Defects Produced by 1 MeV Proton Irradiation
(14.3.3 and 20.3; 13.4; 22.2.2)
(a) Vol. **88**, No. 2, p. 721–731 (1985).

L. Wendler
Electron-Phonon Interaction in Dielectric Bilayer Systems. Effect of the Electronic Polarizability
(6 and 13; 13.5.3; 14.4; 22.2.1; 22.2.2)
(b) Vol. **129**, No. 2, p. 513–530 (1985).

C. Ascheron, C. Bauer, H. Sobotta, and V. Riede
Investigations of Hydrogen Implanted GaP Single Crystals by Means of Particle Induced γ-Spectroscopy, Infrared Spectroscopy, and Rutherford Backscattering Channelling Technique
(11; 20.1; 22.2.2)
(a) Vol. **89**, No. 2, p. 549–557 (1985).

V. Riede, H. Sobotta, and H. Neumann
Nitrogen Concentration in GaP:N Epitaxial Layers from Localized Mode Absorption Measurements
(10.2 and 20.1; 6; 22.2.2)
(a) Vol. **89**, No. 2, p. K147–K151 (1985).

A. M. Andriesh, V. V. Bivol, V. I. Verlan, B. V. Efremushkin, M. S. Iovu, and E. G. Khanchevskaya
Influence of Electrode Material on the Electrical and Photoelectrical Properties of p-InP/As_2Se_3 Heterostructures
(14.3.3 and 16; 2; 14.3.1; 22; 22.2.2)
(a) Vol. **89**, No. 2, p. K231–K234 (1985).

K. Löschke
Phase-Contrast Microscopy in Comparison with Other Light-Microscopic Methods for Direct Observation of Dislocation Lines in Single Crystals
(10.2; 22.2.2)
(a) Vol. **90**, No. 2, p. 573–579 (1985).

H. Bergner, V. Brückner, F. Kerstan, L. L. Kulyuk, S. I. Radautsan, and E. E. Strumban
Picosecond Spectroscopy and Photoconductivity of n-InP
(16 and 20; 22.2.2)
(a) Vol. **90**, No. 2, p. K205–K208 (1985).

S. Nojima, M. Oishi, H. Asahi, and H. Nagai
An Interfacial Property of the InGaAs/Semi-Insulating InP Structure Prepared by Metal Organic Vapor Phase Epitaxy
(14.3.2 and 14.3.4; 22.2.2; 22.2.4)
(a) Vol. **90**, No. 2, p. K215–K218 (1985).

W. Frentrup, M. Griepentrog, and U. Müller-Jahreis
Negative Secondary Ion Emission Influenced by Alkali Atoms
(17; 1.6; 22.1.1; 22.1.2; 22.2.1; 22.2.2)
(a) Vol. **91**, No. 2, p. 447–452 (1985).

V. A. Budyanu, S. N. Chechuy, I. A. Damaskin, S. A. Fedoseev, A. A. Nasakin, S. L. Pyshkin, M. I. Valkovskaya, and V. P. Zenchenko
Investigation of $A^{III}B^{V}$–Si Heterojunctions Grown by Laser Deposition
(14.3.1 and 14.3.3; 13.3; 22.1.2; 22.2.1; 22.2.2)
(a) Vol. **91**, No. 2, p. 737–744 (1985).

C. Ascheron, A. Schindler, and G. Otto
A Study of Proton Bombardment Induced Swelling of GaP Single Crystals
(11; 22.2.2)
(a) Vol. **92**, No. 1, p. 169–176 (1985).

22.2.3 Antimonides

Review Articles

S. Porowski and W. Trzeciakowski
The Effect of Pressure on Deep Impurity States with Large Lattice Relaxation
(13.4; 12.2; 14.3; 16; 22.2.3; 22.2.4; 22.4.3; 22.5; 22.5.2)
(b) Vol. **128**, No. 1, p. 11 – 22 (1985).

T. S. Moss
Relations between the Refractive Index and Energy Gap of Semiconductors
(13.1 and 20.1; 22; 22.2.1; 22.2.3)
(b) Vol. **131**, No. 2, p. 415 – 427 (1985).

Original Papers and Short Notes

S. Ghosh and S. Dixit
Effect of Relativistic Mass Variation of the Electron on the Parametric Instability of Acoustic Waves in Transversely Magnetised Piezoelectric Semiconducting Plasmas
(13.5.2; 7; 22.2.3)
(b) Vol. **127**, No. 1, p. 245 – 252 (1985).

S. Ghosh and R. B. Saxena
Stimulated Brillouin Scattering of High-Power Helicon Waves in Longitudinally Magnetized Cubic Piezoelectric Semiconducting Plasmas
(13.5.2; 7; 22.2.3)
(b) Vol. **127**, No. 2, p. 557 – 566 (1985).

Th. Schurig and R. Herrmann
Some Properties of n-Inversion Layers at Grain Boundaries in InSb Bicrystals
(13.3 and 14.3.4; 22.2.3)
(b) Vol. **127**, No. 2, p. K159 – K163 (1985).

R. Herrmann, W. Kraak, and G. Nachtwei
Electrical Properties of Grain Boundaries in InSb Bicrystals
(14.3; 13.3; 22.2.3)
(b) Vol. **128**, No. 1, p. 337 – 344 (1985).

S. S. Kubakaddi and B. G. Mulimani
Contribution of Electron–Two-Phonon Scattering to Free Carrier Absorption in n-GaSb
(6 and 20.1; 22.2.3)
(b) Vol. **128**, No. 2, p. K117 – K122 (1985).

S. Ghosh and M. S. Pathak
Transverse Phonon–Helicon Interactions in Semiconducting Plasmas
(6 and 13.5.2; 7; 22.2.3)
(b) Vol. **129**, No. 1, p. 81 – 87 (1985).

R. Herrmann, W. Kraak, G. Nachtwei, and Th. Schurig
Two-Dimensional Electron Gas in the n-Inversion Layer of p-Type InSb Bicrystals
(13.1 and 14.3; 22.2.3)
(b) Vol. **129**, No. 1, p. 415–426 (1985).

A. L. Golovin, R. M. Imamov, and E. A. Kondrashkina
Absolute Measurements of Lattice Spacings in Surface Layers of Crystals
(1.6; 22.2.3)
(a) Vol. **89**, No. 1, p. K5–K7 (1985).

S. Ghosh and D. K. Sinha
Stimulated Brillouin Scattering and Brillouin Instability in Dissipative Piezosemiconducting Plasmas
(7; 22.2.3)
(b) Vol. **129**, No. 2, p. K117–K122 (1985).

G. Gobsch, G. Paasch, H. Übensee, R. Herrmann, and W. Kraak
Calculation of the Subband Structure of Inversion Layers in p-InSb Bicrystals
(13.1; 13.3; 14.3.2; 22.2.3)
(b) Vol. **129**, No. 2, p. K167–K171 (1985).

S. Ghosh and S. Dixit
Effect of Relativistic Mass Variation of the Electron on the Modulational Instability of Laser Beams in Transversely Magnetised Piezoelectric Semiconducting Plasmas
(7 and 13.5.2; 22.2.3)
(b) Vol. **130**, No. 1, p. 219–224 (1985).

Nguyen Nhu Dat and Nguyen Van Hieu
On the Microscopic Theory of the Resonant Nonlinear Refractive Index of InSb
(13.1 and 20.1; 22.2.3)
(b) Vol. **130**, No. 1, p. 235–241 (1985).

H. Übensee, G. Paasch, J.-P. Zöllner, Th. Fiedler, and G. Gobsch
Self-Consistent Theory of the Electronic Structure of Inversion Layers. I. A New Method Using the Modified Local Density Approximation
(13.3; 13.1; 14.3.2; 22.2.1; 22.2.3)
(b) Vol. **130**, No. 1, p. 387–402 (1985).

M. S. Kushwaha
Helicon Wave Propagation in Semiconductor Magnetoplasma
(13 and 20; 22.2.3)
(b) Vol. **130**, No. 1, p. K37–K41 (1985).

N. A. Avdeev, Yu. E. Gardin, V. A. Gurtov, and S. N. Kuznetsov
Study of Charge Accumulation Kinetics in MOS Structures with Intrinsic Anodic Oxide on InSb by Measuring Transient Photocurrents
(13.3 and 14.3.3; 16; 22.2.3)
(a) Vol. **90**, No. 1, p. 393–400 (1985).

T. SOMA and H.-MATSUO KAGAYA
Mode Grüneisen Parameters for Long-Wave Phonons of III–V Covalent Compounds under Pressure
(6; 4; 12.2; 22.2.1; 22.2.3)
(b) Vol. **130**, No. 2, p. 497–501 (1985).

W. KRAAK, R. HERRMANN, and G. NACHTWEI
Two-Dimensional Properties of n-Inversion Layers in InSb Grain Boundaries under High Hydrostatic Pressure
(13.1 and 14.3; 22.2.3)
(b) Vol. **130**, No. 2, p. 781–785 (1985).

S. GHOSH and S. DIXIT
Stimulatd Raman Scattering and Raman Instability of an Intense Helicon Wave in Longitudinally Magnetized n-Type Piezolectric Semiconducting Plasma
(13.5.2; 6; 22.2.3)
(b) Vol. **131**, No. 1, p. 255–265 (1985).

M. M. ALEKSANDROVA, V. D. BLANK, V. I. LARCHEV, S. V. POPOVA, and G. G. SKROTSKAYA
Transition Temperature and Heat of Crystallization of Amorphous Bulk Gallium Antimonide Obtained by Rapid Quenching from the Melt at High Pressure
(1.2 and 2; 22.2.3)
(a) Vol. **91**, No. 1, p. K5–K8 (1985).

B. MENGEL, H.-U. MÜLLER, F. LUDWIG, V. L. KONSTANTINOV, and R. HERRMANN
Magneto-Optical Investigations in the Submillimeter Wavelengh Range on p-InSb Bicrystals
(18 and 20.1; 13.4; 22.2.3)
(b) Vol. **131**, No. 2, p. K145–K147 (1985).

F. LUDWIG, H.-U. MÜLLER, B. MENGEL, and R. HERRMANN
Magneto-Optical Investigations on n-InSb in the Far Infrared
(18 and 20.1; 13.4; 22.2.3)
(b) Vol. **131**, No. 2, p. K149–K152 (1985).

W. BRAUNE, M. SCHNÜRER, N. KUBICKI, and R. HERRMANN
Carrier Mobility Determination in Inversion Layers of Anodized InSb MOS Structures
(14.3.2; 22.2.3)
(b) Vol. **131**, No. 2, p. K181–K183 (1985).

W. KNAP, R. STĘPNIEWSKI, and E. FANTNER
Optically Induced Nernst-Ettinghausen Effect in the Far Infrared and Strong Magnetic Fields in HgTe and InSb
(13.1; 16; 22.2.3; 22.4.3)
(b) Vol. **132**, No. 1, p. 133–140 (1985).

H.-MATSUO KAGAYA and T. SOMA
Specific Heat at Constant Volume of III–V Covalent Crystals
(8; 6; 22.2.1; 22.2.3)
(b) Vol. **132**, No. 1, p. K11–K14 (1985).

22.2.4 Mixed Crystals

Review Article

S. Porowski and W. Trzeciakowski
The Effect of Pressure on Deep Impurity States with Large Lattice Relaxation
(13.4; 12.2; 14.3; 16; 22.2.3; 22.2.4; 22.4.3; 22.5; 22.5.2)
(b) Vol. **128**, No. 1, p. 11–22 (1985).

Original Papers and Short Notes

P. Kleinert
Self-Consistent Field Approximation for the Vibrational Spectra of Quaternary Alloys
(6; 20.1; 22.2.4)
(b) Vol. **127**, No. 1, p. 109–119 (1985).

G. P. Peka, O. A. Tokalin, and N. K. Dryapiko
Cathodoluminescence of Graded Band-Gap Semiconductor Structures
(13.1 and 20.3; 13.3; 22.2.4)
(a) Vol. **87**, No. 1, p. 293–303 (1985).

D. Weaire and J. P. Kermode
Electron States in the GaAs–$Ga_xAl_{1-x}As$ Superlattice
(13.3; 22.2.4)
(b) Vol. **127**, No. 2, p. K143–K145 (1985).

R. Stegmann, G. Oelgart, and F. Thuselt
The Influence of Coulomb Screening and of NN_i-Pairs on the Recombination of Bound Excitons in $GaAs_{1-x}P_x$:N Mixed Crystals
(13.5.1 and 20.3; 22.2.4)
(a) Vol. **87**, No. 2, p. 641–650 (1985).

H. Oshima, A. Tanaka, and T. Sukegawa
Calculation of the Ga–Al–Sb Phase Diagram Using the Redlich-Kister Expression
(1.2; 3; 22.2.4)
(a) Vol. **87**, No. 2, p. K131–K134 (1985).
Erratum
(a) Vol. **92**, No. 1, p. K87 (1985).

S. K. Pal and A. Kumar
Resonant Scattering of Phonons in GaInSb
(6 and 8; 22.2.4)
(b) Vol. **128**, No. 1, p. K15–K19 (1985).

Yu. B. Bolkhovityanov and R. I. Bolkhovityanova
Stress-Induced Isothermal Formation of Films during LPE Growth of III–V Compounds with $\Delta a/a \neq 0$
(3; 1.5; 22.2.1; 22.2.2; 22.2.4)
(a) Vol. **88**, No. 1, p. K5–K8 (1985).

E. Bedel, R. Carles, A. Zwick, M. A. Renucci, and J. B. Renucci
Resonant Raman Scattering in $Ga_{1-x}In_xP$ Solid Solutions
(6 and 20.1; 22.2.4)
(b) Vol. **130**, No. 2, p. 467–474 (1985).

P. Kleinert
A Theoretical Study of the Vibrational Properties of $In_xGa_{1-x}As_{1-y}Sb_y$ Quaternary Alloys
(6; 20.1; 22.2.4)
(b) Vol. **130**, No. 2, p. 489–495 (1985).

E. P. Pokatilov, S. I. Beril, V. M. Fomin, and G. A. Pogorilko
Wannier-Mott Exciton States in Two-Layer Periodic Structures
(13.5.1; 22.2.1; 22.2.4)
(b) Vol. **130**, No. 2, p. 619–628 (1985).

J. A. Lahtinen and T. Tuomi
Energy Band Structure of $In_{1-x}Ga_xAs_yP_{1-y}$ Lattice Matched to InP by Means of the Line Profile Analysis of the Electroreflectance Spectra
(13.1; 20.1; 22.2.4)
(b) Vol. **130**, No. 2, p. 637–641 (1985).

J. J. Sinai and S. Y. Wu
Mixed Mode Behavior of $Ga_{1-x}In_xAs$
(6; 22.2.4)
(b) Vol. **130**, No. 2, p. K91–K93 (1985).

S. Nojima, M. Oishi, H. Asahi, and H. Nagai
An Interfacial Property of the InGaAs/Semi-Insulating InP Structure Prepared by Metal Organic Vapor Phase Epitaxy
(14.3.2 and 14.3.4; 22.2.2; 22.2.4)
(a) Vol. **90**, No. 2, p. K215–K218 (1985).

F. Bechstedt and R. Enderlein
Dielectric Screening, Polar Phonons, and Longitudinal Electronic Excitations of Quantum Well Double Heterostructures. Application to Light Scattering from Charge Density Fluctuations
(6 and 13.3; 14.4; 20; 22.2.4)
(b) Vol. **131**, No. 1, p. 53–66 (1985).

G. Beister, P. Krispin, J. Maege, H.-J. Bachert, and St. Stoeff
Emission from a Semi-Insulating Layer in AlGaAs–GaAs Double Heterostructures
(14.3.3 and 20.2; 13.4; 22.2.4)
(a) Vol. **91**, No. 1, p. K93–K96 (1985).

K. Jóźwikowski, Z. Orman, and A. Rogalski
On the Performance of Non-Cooled (In, As)Sb Photoelectromagnetic Detectors for 10.6 μm Radiation
(16; 22.2.4)
(a) Vol. **91**, No. 2, p. 745–751 (1985).

22.3 IV–IV, II–V, and III–VI Compounds

Review Article

B. L. Gelmont, B. T. Kolomiets, and K. D. Tsendin
Impurity Conduction of Chalcogenide Vitreous Semiconductors
(2 and 14.3; 13.1; 13.4; 22.3; 22.8)
(a) Vol. **91**, No. 2, p. 319—337 (1985).

Original Papers and Short Notes

K. Kambas, A. A. Anagnostopoulos, S. Ves, B. Ploss, and J. Spyridelis
Optical Absorption Edge Investigation of $CdIn_2S_4$ and β-In_2S_3 Compounds
(20.1; 22.3; 22.8)
(b) Vol. **127**, No. 1, p. 201—208 (1985).

A. G. Guseinov, A. Yu. Yangirov, K. K. Mamedov, S. M. Atakishiev, and K. M. Mamedova
Heat Capacity and Thermodynamic Properties of Tl_2S at Low Temperatures
(8; 22.3)
(a) Vol. **87**, No. 1, p. K29—K33 (1985).

S. K. J. An-Ani and C. A. Hogarth
Optical Absorption Edge in Amorphous $In_{30}Se_{70}$ Films
(2 and 20.1; 22.3)
(a) Vol. **87**, No. 1, p. K65—K67 (1985).

Ç. Tuncay and M. Tomak
Empirical Tight-Binding Band Structure of Wurtzite Semiconductors SiC, ZnSe, and ZnTe
(13.1; 22.3; 22.4.2; 22.4.3)
(b) Vol. **127**, No. 2, p. 543—547 (1985).

W. Coene, H. Bender, F. C. Lovey, D. van Dyck, and S. Amelinckx
On the Influence of Crystal Orientation on the High Resolution Image Contrast of Polytypes
(1.1; 21.1; 22.3; 22.4.1)
(a) Vol. **87**, No. 2, p. 483—497 (1985).

B. G. Tagiev, G. M. Niftiev, and F. Sh. Aidaev
Photoluminescence of GaSe:Tm Single Crystals
(20.3; 13.4; 22.3)
(b) Vol. **128**, No. 1, p. K65—K68 (1985).

S. Shigetomi, T. Ikari, Y. Koga, and S. Shigetomi
Drift Mobility of Electrons Parallel to the *c*-Axis in InSe
(14.3; 13.4; 16; 22.3)
(a) Vol. **88**, No. 1, p. K71—K73 (1985).

A. G. Abdullaev, K. R. Allakhverdiev, T. D. Ibragimov, and R. M. Sardarly
Two-Phonon Processes in TlSe and $TlInSe_2$
(6 and 20.1; 22.3; 22.3.1)
(b) Vol. **128**, No. 2, p. 401—409 (1985).

H. Matsuda, O. Uemura, and T. Satow
The X-Ray Diffraction Study of Molton Tl_2X and $AgTl_3X_2$ (X: Se and Te)
(2; 18; 22.3; 22.8)
(a) Vol. **90**, No. 1, p. 77–82 (1985).

C. de Blasi, D. Manno, and S. Mongelli
The Stacking of Polytypes in InSe Crystals
(1.4; 22.3)
(a) Vol. **90**, No. 1, p. K5–K6 (1985).

S. Shigetomi, T. Ikari, Y. Koga, and S. Shigetomi
Annealing Behavior of Photoacoustic Spectra of Undoped InSe
(7 and 13.4; 16; 22.3)
(a) Vol. **90**, No. 1, p. K61–K64 (1985).

T. Takai, T. Halicioğlu, and W. A. Tiller
Calculation of Potential Energy Parameters for the Silicon–Carbon System
(1 and 4; 22.1; 22.1.2; 22.3)
(b) Vol. **130**, No. 2, p. 475–481 (1985).

B. Mari, A. Šegura, and A. Chevy
Free Carrier Absorption in n-Type Indium Selenide
(6 and 20.1; 14.3; 22.3)
(b) Vol. **130**, No. 2, p. 793–799 (1985).

G. Dräger, W. Czolbe, E. Schulz, A. Simunek, J. Drahokoupil, Yu. N. Kucherenko, and V. V. Nemoshkalenko
Valence Band Investigations of GaS by Means of Polarized X-Ray Emission Spectra
(13.1 and 20; 22.3)
(b) Vol. **131**, No. 1, p. 193–200 (1985).

S. K. Lilov
Calculation of the Fermi Level, Fermi Distribution, and Activity Coefficient for p-SiC: Sc
(13.4; 22.3)
(b) Vol. **131**, No. 1, p. K81–K85 (1985).

I. B. Bakhtiyarov, S. M. Nakhmetov, P. G. Rustamov, I. M. Niftiev, and O. B. Tagiev
Physical Properties of $Ga_2S_3 \cdot Eu_2O_3$-Based Glasses
(2; 14.3; 20.3; 22.3; 22.6)
(a) Vol. **92**, No. 1, p. K31–K35 (1985).

A. Z. Iskenderzade, O. M. Sadykhov, and A. Sh. Abdinov
Effect of Gadolinium Impurities on Injection-Induced Impurity Photoconductivity in InSe Single Crystals
(16; 13.4; 22.3)
(a) Vol. **92**, No. 1, p. K77–K80 (1985).

A. U. Sheleg and V. V. Zaretskii
The Relationship of the "Invar Effect" with Polytypism and Modulation in α-ZnP_2
(1.2; 22.3)
(a) Vol. **92**, No. 2, p. K99–K104 (1985).

22.3.1 Mixed Crystals

Original Papers and Short Notes

K. R. ALLAKHVERDIEV, U. A. ALESHENKO, N. A. BAKHYSHOV, L. K. VODOPYANOV, F. M. GASHIMZADE, R. M. SARDARLY, and V. YA. SHTEINSHRAIBER
Vibrational Spectrum of TlS_xSe_{1-x} Solid Solutions
(6 and 20.1; 22.3.1)
(b) Vol. **127**, No. 2, p. 459–464 (1985).

E. M. SMOLYARENKO, V. M. TRUKHAN, and A. P. BOLOGA
Phase Diagram of the ZnP_2–$ZnAs_2$ System
(1.2; 22.3.1)
(a) Vol. **87**, No. 2, p. K123–K126 (1985).

A. G. ABDULLAEV, K. R. ALLAKHVERDIEV, T. D. IBRAGIMOV, and R. M. SARDARLY
Two-Phonon Processes in TlSe and $TlInSe_2$
(6 and 20.1; 22.3; 22.3.1)
(b) Vol. **128**, No. 2, p. 401–409 (1985).

E. K. ARUSHANOV, L. L. KULYUK, L. N. LUKYANOVA, A. N. NATEPROV, S. I. RADAUTSAN, and A. A. SHTANOV
Stimulated Emission Spectra and Valence Band Structure in the $Cd_3(As_xP_{1-x})_2$ System
(13.1 and 20.3; 22.3.1)
(b) Vol. **128**, No. 2, p. 583–589 (1985).

G. D. GUSEINOV, E. F. BAGIRZADE, S. S. ABDINBEKOV, and N. Z. GASANOV
Modulation Spectra Peculiarities in $A^{III}B^{III}X_2^{VI}$ Anisotropic Semiconductors
(13.5.1 and 20.1; 14.3.4; 22.3.1)
(b) Vol. **128**, No. 2, p. K129–K131 (1985).

E. M. SMOLYARENKO, V. M. TRUKHAN, and A. P. BOLOGA
Phase Diagram of the ZnP_2–CdP_2 System
(1.2; 22.3.1)
(a) Vol. **91**, No. 2, p. K101–K102 (1985).

22.4 II–VI Compounds

Original Papers and Short Notes

22.4.1 Sulphides

Review Article

M. GODLEWSKI
On the Application of the Photo-EPR Technique to the Studies of Photoionization, DAP Recombination, and Non-Radiative Recombination Processes
(13.4 and 19; 16; 22.2; 22.4.1; 22.5)
(a) Vol. **90**, No. 1, p. 11–52 (1985).

Original Papers and Short Notes

C. TRALLERO GINER and O. SOTOLONGO COSTA
One-Phonon Resonance Raman Scattering
(6 and 20.1; 13.5.1; 22.4.1)
(b) Vol. **127**, No. 1, p. 121–130 (1985).

W. COENE, H. BENDER, F. C. LOVEY, D. VAN DYCK, and S. AMELINCKX
On the Influence of Crystal Orientation on the High Resolution Image Contrast of Polytypes
(1.1; 21.1; 22.3; 22.4.1)
(a) Vol. **87**, No. 1, p. 483–497 (1985).

A. P. ZDEBSKII, YU. A. OLIKH, and A. U. SAVCHUK
Saturation and Oscillation of Current in Semiconductors Subjected to Uniaxial Deformation
(14.3.4; 12.2; 22.4.1)
(a) Vol. **87**, No. 2, p. K209–K212 (1985).

A. MEMON and D. B. TANNER
Far-Infrared Dielectric Function of Zincblende ZnS
(6 and 20.1; 22.4.1)
(b) Vol. **128**, No. 1, p. 49–52 (1985).

I. A. ILIEV, T. N. DIMOV, and H. LANGE
Stress Dependence of Polarization Properties of Hexagonal CdS Single Crystals around the Isotropic Point
(20.1; 22.4.1)
(b) Vol. **128**, No. 1, p. K45–K48 (1985).

K. STAVREV, G. ST. NIKOLOV, and K. KYNEV
Orbital and Term Energies for Manganese(II) in Trigonal Fields of C_{3v} Symmetry
(13.4; 20.1; 22.4.1)
(b) Vol. **128**, No. 2, p. K69–K73 (1985).

RAVIENDRA, D. and J. K. SHARMA
n-CdS/p-AgInSe$_2$ Solar Cells by Electrodeposition
(14.3.3 and 20.1; 1.5; 22.4.1; 22.8)
(a) Vol. **88**, No. 1, p. 365–368 (1985).

S. A. Mirsagatov and K. Sabirov
Investigation of Illuminance–Voltage Characteristics for Thin-Film p-CdTe–n-CdTe–n-CdS Structures
(14.3.3 and 16; 14.3.1; 22.4.1; 22.4.3)
(a) Vol. **90**, No. 2, p. K219 – K223 (1985).

H. Rossmann and F. Henneberger
Resonatorless Optical Bistability of CdS. Reconsideration of Transient Effects Due to Lattice Heating
(13.5.1 and 20; 22.4.1)
(b) Vol. **131**, No. 1, p. 185 – 192 (1985).

D. Backs, R. Stösser, and M. Lieberenz
Multiple-Quantum Transitions and Optical Excitation Processes in the EPR of Mn^{2+} in ZnS
(19; 22.4.1)
(b) Vol. **131**, No. 1, p. 291 – 297 (1985).

J. A. Jakimavičius and R. P. Purlys
Compton Profile and Valence Electron Momentum Distribution of CdS
(13; 22.4.1)
(b) Vol. **131**, No. 1, p. K63 – K66 (1985).

K. Syassen
Pressure-Induced Structural Transition in SrS
(1.2; 12.2; 22.4.1)
(a) Vol. **91**, No. 1, p. 11 – 15 (1985).

A. A. Borshch, M. S. Brodin, and V. N. Semioshko
The Anisotropy of Refraction Nonlinearity and Vector Self-Diffraction in the Wide-Gap Semiconductors of CdS Type
(20.1; 22.4.1; 22.4.4)
(a) Vol. **91**, No. 1, p. 135 – 141 (1985).

V. May, F. Boldt, and K. Henneberger
Many-Body Theory for the Dense Exciton Gas of Direct Semiconductors. III. Response of the Many-Exciton System on an Externally Driven Light Field
(13.5.1; 22.4.1)
(b) Vol. **131**, No. 2, p. 605 – 614 (1985).

V. May
Light-Transmission at the Is-Exciton Resonance of CdS in the Presence of a Density Profile of Excitons
(13.5.1 and 20.1; 22.4.1)
(b) Vol. **132**, No. 1, p. 145 – 151 (1985).

P. N. Ram, R. D. Bijalwan, P. D. Semalty, and M. Israel
Resonant Scattering of Phonons in ZnS:Fe^{2+} and CdTe:Fe^{2+}
(6 and 8; 22.4.1; 22.4.3)
(b) Vol. **132**, No. 2, p. 365 – 373 (1985).

S. W. Biernacki
Parametrization of Crystal Field Spectra of Mn^{2+} in ZnSe and ZnS
(13.4; 22.4.1; 22.4.2)
(b) Vol. **132**, No. 2, p. 557 – 564 (1985).

22.4.2 Selenides

Original Papers and Short Notes

G. Hein
Analysis of Shubnikov-De Haas Oscillations in HgSe
(13.1; 18; 22.4.2)
(b) Vol. **127**, No. 1, p. 227–236 (1985).

P. V. Vodolazskii, B. R. Kiyak, M. G. Matsko, V. B. Nosov, G. T. Petrovskii, and A. V. Shatilov
Reversible and Irreversible Changes in Exciton Spectra of ZnSe Luminescence
(13.5.1 and 20.3; 22.4.2)
(a) Vol. **87**, No. 1, p. K69–K71 (1985).

Ç. Tuncay and M. Tomak
Empirical Tight-Binding Band Structure of Wurtzite Semiconductors SiC, ZnSe, and ZnTe
(13.1; 22.3; 22.4.2; 22.4.3)
(b) Vol. **127**, No. 2, p. 543–547 (1985).

I. Rückmann, K. Jarašiūnas, and E. Gaubas
Transient Light-Induced Gratings Due to Excitons in CdSe
(13.5.1 and 20.1; 22.4.2)
(b) Vol. **128**, No. 2, p. 627–641 (1985).

N. Sielaff, H. Nelkowski, H. Pradella, J. Sahm, and K. Zink
EPR Measurements on Fe–Cu Complexes in ZnSe
(19; 13.4; 22.4.2)
(b) Vol. **129**, No. 2, p. K177–K179 (1985).

V. I. Sokolov, T. P. Surkova, M. P. Kulakov, and A. V. Fadeev
New Experimental Evidence Concerning the Nature of L, M, and N Lines in ZnSe:Co
(13.4 and 13.5.1; 14.3.4; 20.1; 22.4.2)
(b) Vol. **130**, No. 1, p. 267–272 (1985).

S. W. Biernacki
The Splitting of Impurity d-States in Zincblende Crystals
(13.4; 20.1; 22.4.2)
(b) Vol. **130**, No. 1, p. K59–K63 (1985).

G. I. Ibaev, S. I. Kulieva, and T. G. Magerramov
Erratum to: Frequency Dependences of the Dynamic Current–Voltage Characteristic of the Se–CdSe Heterostructure Based on Single-Crystal Selenium Film
(14.3.1 and 14.3.3; 22.1.3; 22.4.2)
(a) Vol. **90**, No. 1, p. K117 (1985).
Erratum
(a) Vol. **92**, No. 2, p. K175 (1985).

K. D. Tovstyuk, V. G. Deybuk, S. V. Melnichuk, and N. K. Tovstyuk
3d-Impurity Levels in CdTe and ZnSe
(13.4; 22.4.2; 22.4.3)
(b) Vol. **130**, No. 2, p. K153–K156 (1985).

22.4.3 Tellurides

Review Article

S. Porowski and W. Trzeciakowski
The Effect of Pressure on Deep Impurity States with Large Lattice Relaxation
(13.4; 12.2; 14.3; 16; 22.2.3; 22.2.4; 22.4.3; 22.5; 22.5.2)
(b) Vol. **128**, No. 1, p. 11–22 (1985).

Original Papers and Short Notes

Y. Madhavan, K. Ramachandran, and T. M. Haridasan
Resonant Modes Associated with a Vacancy in CdTe
(6 and 10.2; 22.4.3)
(b) Vol. **127**, No. 1, p. K15–K18 (1985).

Ç. Tuncay and M. Tomak
Empirical Tight-Binding Band Structure of Wurtzite Semiconductors SiC, ZnSe, and ZnTe
(13.1; 22.3; 22.4.2; 22.4.3)
(b) Vol. **127**, No. 2, p. 543–547 (1985).

Yu. P. Gnatenko, Yu. I. Zhirko, N. P. Gavaleshko, and P. I. Babii
On the Possibility of Electron–Hole Drop Formation on Neutral Acceptors in CdTe Crystal
(13.5 and 20.1; 13.4; 22.4.3)
(b) Vol. **127**, No. 2, p. K125–K128 (1985).

S. Kuźmiński and A. T. Szaynok
Shallow Energy States in CdTe
(13.3 and 14.3.2; 16; 22.4.3)
(a) Vol. **89**, No. 2, p. 623–628 (1985).

M. Grinberg, S. Łęgowski, and H. Męczyńska
The Energy of Shallow Donor States in CdTe, GaAs, AgCl, AgBr Semiconductors
(13.4; 6; 22.2.1; 22.4.3; 22.5.1)
(b) Vol. **130**, No. 1, p. 325–331 (1985).

V. N. Vivcharuk, Yu. V. Gudyma, V. N. Ermakov, and V. V. Nitsovich
Exciton Mechanism of Exchange Interaction in Semiconductors with Magnetic Impurities
(13.4 and 13.5.1; 18; 22.4.3)
(b) Vol. **130**, No. 2, p. K131–K134 (1985).

K. D. Tovstyuk, V. G. Deybuk, S. V. Melnichuk, and N. K. Tovstyuk
3d-Impurity Levels in CdTe and ZnSe
(13.4; 22.4.2; 22.4.3)
(b) Vol. **130**, No. 2, p. K153–K156 (1985).

D. Imhoff, F. Gelsdorf, B. Pellissier, and J. Castaing
Study of an Experimental Technique for High-Temperature Deformation of Compound Semiconductors (Liquid Encapsulation)
(10.2; 22.4.3)
(a) Vol. **90**, No. 2, p. 537–543 (1985).

22.4.4 Mixed Crystals

Original Papers and Short Notes

M. N. Aliev and L. R. Tagirov
Theory of Temperature Dependent EPR Linewidth in Semimagnetic Semiconductors
(19; 22.4.4)
(b) Vol. **127**, No. 1, p. K61 – K64 (1985).

G. D. Khattak, A. Twardowski, and R. R. Gałązka
Magnetization of $Cd_{1-x}Mn_xTe$
(18; 22.4.4)
(a) Vol. **87**, No. 1, p. K57 – K60 (1985).

M. Mondal and K. P. Ghatak
Theoretical Analysis of Oscillatory Heat Capacity in *n*-Channel Inversion Layers on Ternary Semiconductors
(8; 13.3; 22.4.4)
(b) Vol. **128**, No. 1, p. K21 – K26 (1985).

K. Sierański and J. Szatkowski
Free-Carrier Absorption in Semimagnetic Semiconductors
(18.1 and 20.1; 13.1; 22.4.4)
(b) Vol. **128**, No. 1, p. K33 – K35 (1985).

V. M. Burlakov, A. P. Litvinchuk, V. N. Pyrkov, G. G. Tarasov, and N. I. Vitrikhovskii
Optical Properties of the Quaternary II–VI Mixed Crystals in the Far Infrared Region
(6 and 20.1; 22.4.4)
(b) Vol. **128**, No. 2, p. 389 – 400 (1985).

P. I. Baranskii, A. E. Belyaev, and A. V. Zhidkov
Extreme Disordering of Electron Scattering in a Gapless Semiconductor
(14.3 and 18; 22.4.4)
(b) Vol. **128**, No. 2, p. K171 – K174 (1985).

G.-U. Reinsperger, F. Schwabe, and B. Selle
Nondestructive Analysis of Thin $Zn_xCd_{1-x}S$ Films by Rutherford Backscattering and Optical Measurements
(1.5 and 20.1; 22.4.4)
(a) Vol. **88**, No. 2, p. 745 – 751 (1985).

O. Žižić, M. Stojić, and B. Stošić
Hall Effect and Electrical Resistivity in the Paramagnetic Region of $Hg_{1-x}Mn_xSe$ Crystals
(14.3; 18.1; 22.4.4)
(a) Vol. **88**, No. 2, p. K191 – K196 (1985).

D. Shaw
The Chemical Diffusion of In in $Hg_{0.8}Cd_{0.2}Te$
(9; 22.4.4)
(a) Vol. **89**, No. 1, p. 173 – 183 (1985).

T. M. Razykov, B. Kh. Kadyrov, and M. A. Khodyaeva
Energy Band Models of n-$Zn_xCd_{1-x}S$–p-CdTe and n-$Zn_xCd_{1-x}S$–p-Si ($0 \leqq x \leqq 1$) Heterojunctions
(13.1 and 14.3.3; 22.4.4)
(a) Vol. **91**, No. 1, p. K87–K91 (1985).

Yu. G. Arapov, A. B. Davydov, and M. L. Zvereva
Tellurium Precipitate Influence on the Electron Hall Mobility in $Hg_{1-x}Cd_xTe$ ($x \approx 0.2$) Crystals
(14.3; 10.2; 22.4.4)
(a) Vol. **92**, No. 1, p. K81–K85 (1985).

W. Giriat
Optical Absorption in $Mn_xZn_{1-x}Se$
(20.1; 22.4.4)
(b) Vol. **132**, No. 2, p. K131–K132 (1985).

22.5 Halides

Review Articles

S. Porowski and W. Trzeciakowski
The Effect of Pressure on Deep Impurity States with Large Lattice Relaxation
(13.4; 12.2; 14.3; 16; 22.2.3; 22.2.4; 22.4.3; 22.5; 22.5.2)
(b) Vol. **128**, No. 1, p. 11–22 (1985).

M. Godlewski
On the Application of the Photo-EPR Technique to the Studies of Photoionization, DAP Recombination, and Non-Radiative Recombination Processes
(13.4 and 19; 16; 22.2; 22.4.1; 22.5)
(a) Vol. **90**, No. 1, p. 11–52 (1985).

Original Papers and Short Notes

M. Herget, A. Hofstaetter, and A. Scharmann
EPR of the F_2^--Centre in $LiYF_4$
(19; 10.2; 22.5)
(b) Vol. **127**, No. 1, p. K83–K87 (1985).

S. C. Agrawal
Analysis of the Equation of State of Ammonium Halides
(4; 22.5)
(b) Vol. **128**, No. 1, p. 69–73 (1985).

U. Braatz, F. Fink, and P. Klinger
Optoelectronic Switches on the Basis of Mercuric Iodide (Red Phase, α-HgI_2)
(14.3.4; 16; 20.2; 22.5)
(a) Vol. **88**, No. 1, p. K95–K97 (1985).

J. Przedmojski, S. Gierlotka, and B. Pałosz
X-Ray Thermal Investigations of Cadmium Iodide Single Crystals
(1.1 and 8; 22.5)
(a) Vol. **89**, No. 1, p. K9–K11 (1985).

M. Piotrowski and A. Murasik
Exchange and Dipolar Interactions in TbF_3
(18; 1.2; 22.5)
(a) Vol. **89**, No. 2, p. 571–580 (1985).

I. O. Bashkin, Yu. M. Dergachev, and E. G. Ponyatovskii
Pressure-Induced Phase Transition in RbH
(1.2; 22.5)
(b) Vol. **130**, No. 1, p. 73–77 (1985).

E. Anastassakis, J. S. Raptis, and W. Richter
Raman and Infrared Spectra of Tellurium Sub-Bromide (Te_2Br)
(6 and 20.1; 22.5)
(b) Vol. **130**, No. 1, p. 161–168 (1985).

G. I. Pilipenko, V. G. Plekhanov, V. I. Tyutyunnik, O. I. Tyutyunnik, and F. F. Gavrilov
Direct Observation of Bound Excitons in the Spectra of Resonance Secondary Emission of LiH:Mg Crystals
(13.5.1 and 20.1; 20.3; 22.5)
(b) Vol. **130**, No. 1, p. 273–280 (1985).

V. A. Bibik, I. V. Blonskii, M. S. Brodin, and N. A. Davydova
Structural Phase Transition in the Layer Semiconductor PbI_2 Induced by Laser Irradiation
(1.2 and 13.5.1; 20.1; 22.5)
(a) Vol. **90**, No. 1, p. K11–K14 (1985).

R. Bacewicz, B. Pałosz, and S. Gierlotka
Optical Band Gap of CdI_2 Polytypes
(13.1 and 20.1; 22.5)
(b) Vol. **130**, No. 2, p. K135–K137 (1985).

E. G. Tsitsishvili
Piezo-Optical Activity of Cubic Crystals
(12.2 and 20.1; 13.1; 13.5.1; 22.5)
(b) Vol. **131**, No. 1, p. 279–290 (1985).

D. Neogy and T. Purohit
The Behavior of Active Centers in a Laser Host. A Crystal Field Investigation on Nd^{3+} and Er^{3+} in LaF_3 Single Crystals
(13.4; 8; 18; 19; 22.5)
(b) Vol. **131**, No. 1, p. 329–338 (1985).

A. S. Barrière, H. Beji, G. Couturier, and G. Gevers
Hopping Conduction in NiF_2 Thin Films as a Function of Their Crystallization Degree and Composition
(14.3.1 and 14.3.3; 22.5)
(a) Vol. **91**, No. 1, p. 257–270 (1985).

T. Yamamoto, S. Kikkawa, and M. Koizumi
The Effect of Dispersed PLZT Particles on the Electrical Conductivity of Lithium Iodide
(14.4; 22.5)
(a) Vol. **91**, No. 1, p. K67–K70 (1985).

M. S. Brodin, I. V. Blonskii, A. O. Gushcha, and V. V. Tishenko
Secondary Emission Spectra and Energy Relaxation of Polaritons in Layer Polar Semiconductors
(6 and 13.5.1; 20.1; 20.3; 22.5)
(b) Vol. **131**, No. 2, p. 615–620 (1985).

F. V. Kusmartsev
The Narrow Bands of Exciton Absorption in Polar Solids and Alloys. Urbach's Rule
(13.5.1 and 20.1; 22.5; 22.5.4)
(b) Vol. **131**, No. 2, p. 621–629 (1985).

22.5.1 Silver Halides

Original Papers and Short Notes

A. K. M. A. Islam and E. Huq
On Cohesive Energy and Grüneisen Parameter of Silver Halide Crystals
(4 and 8; 6; 22.5.1)
(b) Vol. **127**, No. 2, p. K109–K111 (1985).

Th. Kleinefeld and W. von der Osten
Localized and Free Biexcitons in Silver Bromide
(13.5.1 and 20.3; 22.5.1)
(b) Vol. **128**, No. 2, p. 808 (1985).

M. Grinberg, S. Łęgowski, and H. Męczyńska
The Energy of Shallow Donor States in CdTe, GaAs, AgCl, AgBr Semiconductors
(13.4; 6; 22.2.1; 22.4.3; 22.5.1)
(b) Vol. **130**, No. 1, p. 325–331 (1985).

I. J. Booth and C. F. Schwerdtfeger
Optically Detected Cyclotron Resonance in AgBr
(19 and 20.3; 13.4; 22.5.1)
(b) Vol. **130**, No. 2, p. 749–756 (1985).

L. van Robbroeck, E. Goovaerts, and D. Schoemaker
Electron-Spin-Resonance Study of Co^{2+} and Ni^{+} Centers in AgCl (Cu, Co, Ni)
(13.4 and 19; 22.5.1)
(b) Vol. **132**, No. 1, p. 179–187 (1985).

L. van Robbroeck, A. Bouwen, and D. Schoemaker
Electron Spin Resonance Study of Ni^{2+} Centers in AgCl (Cu, Ni)
(13.4 and 19; 22.5.1)
(b) Vol. **132**, No. 2, p. 565–572 (1985).

22.5.2 Alkali Halides

Review Article

S. Porowski and W. Trzeciakowski
The Effect of Pressure on Deep Impurity States with Large Lattice Relaxation
(13.4; 12.2; 14.3; 16; 22.2.3; 22.2.4; 22.4.3; 22.5; 22.5.2)
(b) Vol. **128**, No. 1, p. 11–22 (1985).

Original Papers and Short Notes

W. N. Bhende, P. S. Bakhshi, and J. Shanker
Thermal Energy and the Equation of State for Alkali Halide Crystals
(4; 22.5.2)
(b) Vol. **127**, No. 1, p. 73–77 (1985).

N. V. K. Prabhakar and R. K. Singh
Anharmonic Properties of Sodium Halides
(12.2; 22.5.2)
(b) Vol. **127**, No. 1, p. 95–101 (1985).

N. Dass and M. Kumari
Derivation of Some Equations of State for Solids. A New Approach
(8; 12.2; 22.5.2)
(b) Vol. **127**, No. 1, p. 103–108 (1985).

B. G. Dick
Electric Field-Induced First Order Raman Scattering in Alkali Halides
(6 and 20.1; 22.5.2)
(b) Vol. **127**, No. 1, p. 131–138 (1985).

Yu. A. Ekmanis and I. S. Radchenko
Light Extinction by Colloidal Thallium in Ionic Crystals
(20.1; 22.5.2)
(b) Vol. **127**, No. 1, p. 287–298 (1985).

L. E. Nagli
F-Centres in KCl–Tl Crystals
(10.2 and 20.3; 22.5.2)
(b) Vol. **127**, No. 1, p. 319–326 (1985).

Y. Nakaoka
Crystal Structure Dependence of Hole Stability in Ionic Crystals
(13.5.3; 6; 13.4; 22.5.2; 23)
(b) Vol. **127**, No. 1, p. 327–338 (1985).

A. B. Gerasimov, Z. G. Gogua, and A. A. Tsertsvadze
On the Space Dependence of the Permittivity in the F-Centre Theory
(13.4; 22.5.2)
(b) Vol. **127**, No. 1, p. K73–K76 (1985).

YU. S. BOYARSKAYA, D. Z. GRABKO, and D. S. PISHKOVA
Correlation between Yield Stress and Microhardness in Crystals Such as NaCl
(10.2 and 12.2; 22.5.2)
(a) Vol. **87**, No. 1, p. 175–179 (1985).

G. A. ANDREEV and M. HARTMANOVÁ
Effect of Impurity State on Deformation-Induced Density Change in NaCl:Ca Crystals
(10.2; 22.5.2)
(a) Vol. **87**, No. 1, p. 217–223 (1985).

R. K. GUPTA, A. SHIROMANY, P. S. BAKHSHI, and J. SHANKER
Crystalline Properties of Alkali Halides Using the Generalized Huggins-Mayer Potentials
(4 and 8; 12.2; 22.5.2)
(b) Vol. **127**, No. 2, p. 473–479 (1985).

A. AKILBEKOV, A. DAULETBEKOVA, and A. ELANGO
Photo- and Thermochemical Reactions with Participation of Br_3^- Centres in X-Rayed KBr
(10.2; 11; 20.1; 22.5.2)
(b) Vol. **127**, No. 2, p. 493–501 (1985).

S. V. NISTOR, D. SCHOEMAKER, I. URSU, and B. R. YANG
The Study of Tl_2^+ ⟨110⟩ Centres in NaCl and RbCl
(10.2 and 19; 11; 22.5.2)
(b) Vol. **127**, No. 2, p. 503–508 (1985).

B. R. YANG, A. BOUWEN, and D. SCHOEMAKER
Electron Spin Resonance Study of Fe^+ Centres in NaCl:Fe^{++} and Motion of the Associated Vacancies
(13.4 and 19; 10.2; 22.5.2)
(b) Vol. **127**, No. 2, p. 657–662 (1985).

A. MADUEMEZIA and F. J. VEGA-CATALAN
A Refinement of the Mollwo-Ivey Rule for F-Centres in Alkali Halides
(13.4 and 20.1; 22.5.2)
(b) Vol. **127**, No. 2, p. 689–694 (1985).

T. KURIHARA, I. KANAZAWA, H. KOIZUMI, H. MURAKAMI, and T. SUZUKI
Mobility and Trapping States of Positrons in NaCl Single Crystal Containing Ca^{2+} Ions
(10.2; 22.5.2)
(b) Vol. **127**, No. 2, p. K113–K116 (1985).

W. N. BHENDE, P. S. BAKHSHI, and J. SHANKER
Thermoelastic Behaviour and Higher-Order Elastic Constants of Ionic Crystals
(4 and 12.2; 6; 8; 22.5.2; 22.6)
(b) Vol. **128**, No. 1, p. 105–112 (1985).

K. IWAHANA
Theory for the Magnetic Circular Polarization of the F-Center Luminescence
(13.4 and 20.3; 18.1; 22.5.2)
(b) Vol. **128**, No. 1, p. 297–311 (1985).

R. C. Bhuniya, P. Sridaran, and V. V. Ratnam
Emission Processes in Luminescence and Thermally Stimulated Luminescence of Ag- and Ca-Doped KCl
(13.4; 20.1; 20.3; 22.5.2)
(b) Vol. **129**, No. 1, p. 331–338 (1985).

S. Bandyopadhyay and S. K. Deb
Activation Energy of Anion Defects in Sodium Halides
(10.2; 22.5.2)
(b) Vol. **129**, No. 1, p. K27–K31 (1985).

G. A. Andreev and B. I. Smirnov
Effect of Plastic Deformation on the Density of Mg-Doped LiF Crystals
(10.2; 22.5.2)
(a) Vol. **89**, No. 1, p. 185–189 (1985).

R. Cywiński, E. Mugeński, W. Nowy-Wiechula, and J. Wiechulá
Aggregation of Mn^{2+}-Cation Vacancy Dipoles in KCl Matrix
(10.2 and 20.3; 22.5.2)
(b) Vol. **129**, No. 2, p. 601–605 (1985).

A. A. O'Connel-Bronin
Resonance LO Phonon-Assisted Luminescence and Light and Heavy Excitons in NaI Crystals
(6 and 20.3; 13.5.1; 22.5.2)
(b) Vol. **129**, No. 2, p. 675–681 (1985).

G. Jacobs
Transient I_e Centres in Alkali Halides
(6 and 20.1; 13.4; 22.5.2)
(b) Vol. **129**, No. 2, p. 755–762 (1985).

R. K. Chaudhary, A. V. Krishna Rao, and K. V. Rao
Optical Anisotropy and Dichroism of A-Centres in NaCl and KCl Single Crystals Irradiated with X-Rays under DC Fields
(10.2 and 20.1; 22.5.2)
(b) Vol. **129**, No. 2, p. 785–788 (1985).

R. Capelletti, M. Manfredi, and M. Solzi
Energy Transfer Mechanisms in the NaCl:Pb^{++}, Mn^{++} System along Aggregation Processes
(13.4 and 20.1; 10.2; 22.5.2)
(b) Vol. **129**, No. 2, p. 789–798 (1985).

M. Kumari and N. Dass
On a Simple Relation for the Adiabatic Anderson-Grüneisen Parameter in Solids
(8; 6; 22.5.2)
(b) Vol. **129**, No. 2, p. K105–K109 (1985).

B. T. Deshmukh, B. H. Ghiya, and R. B. Pode
On the Dose Dependence of the Formation of Z_1 Centres in KCl:Ca
(10.2 and 20.3; 11; 20.1; 22.5.2)
(a) Vol. **89**, No. 2, p. 517–520 (1985).

A. V. EGRANOV and A. I. NEPOMNYACHIKH
Emission Spectra of NaCl: Mg Crystals
(13.4 and 20.3; 10.2; 22.5.2)
(a) Vol. **90**, No. 1, p. K65–K69 (1985).

M. BARTH, W. LORKE, and E. MOHLER
Uniaxial Stress Dependence of the Static Dielectric Constant of Some Alkali Halides
(14.4; 22.5.2)
(b) Vol. **130**, No. 2, p. 457–466 (1985).

C. VAROTSOS, M. LAZARIDOU, K. ALEXOPOULOS, and P. VAROTSOS
Point Defect Entropies and Enthalpies in KCl
(10.2; 8; 22.5.2)
(b) Vol. **130**, No. 2, p. K105–K107 (1985).

N. TAKEUCHI, K. OHTAKI, and S. NAKAMURA
Kinetics of Thermoluminescence in X-Irradiated KCl: Ag Crystals
(20.3; 11; 22.5.2)
(b) Vol. **130**, No. 2, p. K145–K148 (1985).

A. N. PAVLOV and I. S. SMIRNOV
Thermally Activated Change of the Point Defect State in KBr Crystals
(13.4 and 20.1; 10.2; 22.5.2)
(b) Vol. **130**, No. 2, p. K163–K167 (1985).

S. M. BABULANAM
AC and DC Superimposed AC Conduction in Sodium Fluoride Thin Films
(14.4; 22.5.2)
(a) Vol. **90**, No. 2, p. 755–762 (1985).

C. N. AFONSO, C. ORTIZ, and G. J. CLARK
Defect Formation in LiF by Low Energy Ion Implantation
(10.2 and 11; 20.1; 22.5.2)
(b) Vol. **131**, No. 1, p. 87–96 (1985).

R. C. BHUNIYA and S. E. RAO
Colour Centres and Luminescence in KCl: Zn Crystals
(20.1 and 20.3; 10.2; 13.4; 22.5.2)
(b) Vol. **131**, No. 1, p. 339–348 (1985).

W. KUCHARCZYK
On the Nature of the Quadratic Electrooptic Effect in LiF
(4 and 20.1; 22.5.2)
(b) Vol. **131**, No. 2, p. K107–K109 (1985).

L. BOSI and M. NIMIS
An Analysis of the Classical Arguments Concerning Forecasts for the Z_1 Centre Formation in Alkali Halides
(10.2 and 13.4; 22.5.2)
(b) Vol. **131**, No. 2, p. K111–K116 (1985).

S. Bandyopadhyay and S. K. Deb
Solution Enthalpies of Aliovalent Defects in Alkali Halides
(10.2; 22.5.2)
(b) Vol. **132**, No. 2, p. 415–424 (1985).

L. A. Bugaev and R. V. Vedrinskii
Hartree-Fock Approximation in the Scattering Wave Method. K-Absorption of Ionic Crystals in the Single-Scattering Formalism (XANES and EXAFS)
(13.1; 20; 22.5.2)
(b) Vol. **132**, No. 2, p. 459–464 (1985).

H. A. Kassim
Energy Levels of Z_2^+-Centres in KCl:Ca^{2+}. A Model Potential Perturbation Calculation
(10.2 and 13.4; 22.5.2)
(b) Vol. **132**, No. 2, p. 531–539 (1985).

M. L. Dalal, S. Sivaraman, and V. Y. G. S. Murti
Optical Absorption Studies on Colored KCl:Na:Pb Crystals
(13.4 and 20.1; 22.5.2)
(b) Vol. **132**, No. 2, p. 591–600 (1985).

R. Cywiński, E. Mugeński, W. Nowy-Wiechuła, and J. Wichuła
Aggregation and Precipitation of Mn^{2+}–Cation Vacancy Dipoles in NaCl Matrix
(10.2 and 20.3; 22.5.2)
(b) Vol. **132**, No. 2, p. K91–K94 (1985).

T. Tsuboi
Formation of the $Tl^0(1)$ Center in KCl
(10.2 and 20.1; 22.5.2)
(b) Vol. **132**, No. 2, p. K95–K98 (1985).

A. I. Bobrysheva and Yu. M. Shvera
The Influence of the Excitonic Longitudinal–Transverse Splitting on the Biexciton Binding Energy
(13.5.1; 22.4.1; 22.5.2)
(b) Vol. **132**, No. 2, p. K107–K110 (1985).

A. M. Afanasev, R. M. Imamov, E. Kh. Mukhamedzhanov, and Le Cong Qui
Photoemission Curves Obtained in Conditions of X-Ray Grazing-Incidence Laue Diffraction from Crystals with a Disturbed Surface Layer
(1.6; 22.5.2)
(a) Vol. **92**, No. 2, p. 355–360 (1985).

M. Georgieva, P. Mishev, and G. Tumbev
Doppler Broadening Measurements in γ Irradiated LiF
(10.2; 22.5.2)
(a) Vol. **92**, No. 2, p. 503–506 (1985).

22.5.3 Alkali-Earth Halides

Original Papers and Short Notes

N. Dutt, O. P. Sharma, and J. Shanker
Analysis of Photoelastic Behaviour of Fluorite-Structure Crystals
(4 and 12.2; 14.4; 22.5.3)
(b) Vol. **127**, No. 1, p. 67–71 (1985).

B. Berulava and R. Mirianashvili
Radio-Frequency Discrete Saturation Measurement of the Signs of Ligand Hyperfine Interaction Tensor Components for Ce^{3+} in CaF_2
(19; 22.5.3)
(b) Vol. **127**, No. 1, p. K69–K72 (1985).

P. J. Alonso and R. Alcalá
EPR Study of Mn^+ Ions in SrF_2
(11 and 19; 22.5.3)
(b) Vol. **127**, No. 1, p. K77–K79 (1985).

P. Pipinys, A. Rimeika, and A. Kiveris
On a Trap Ionization Mechanism by Electric Field in CdF_2
(13.4 and 20.3; 14.3.4; 22.5.3)
(a) Vol. **87**, No. 1, p. K89–K91 (1985).

R. Mićić and B. Drašković
Some Photoluminescence Properties of Mn^{2+} in Magnesium Bromide
(10.2 and 20.3; 22.5.3)
(b) Vol. **128**, No. 2, p. 489–494 (1985).

P. J. Alonso and R. Alcala
EPR Study of a New Trigonal Mn^{2+} Defect in SrF_2
(19; 13.4; 22.5.3)
(b) Vol. **128**, No. 2, p. K153–K155 (1985).

P. Varotsos, K. Alexopoulos, and M. Lazaridou
Erratum to: On the Proportionality of the Migration Volume and Its Migration Enthalpy in Fluorides
(10.2; 9; 22.5.3)
(b) Vol. **128**, No. 2, p. K191 (1985).

P. Varotsos, K. Alexopoulos, C. Varotsos, and M. Lazaridou
Interconnection of Point Defect Parameters in BaF_2
(10.2; 8; 9; 22.5.3)
(a) Vol. **88**, No. 2, p. K137–K140 (1985).

N. Dutt, G. G. Agrawal, and J. Shanker
Investigation of the Equation of State and the Grüneisen-Anderson Parameters for Fluorite-Structure Crystals
(4; 6; 8; 12.2; 22.5.3; 22.8.1)
(b) Vol. **129**, No. 1, p. 73–79 (1985).

A. Kessler
Bulk Conductivity, Polarization, and Thermal Stimulated Currents of Insulated Polar Crystals
(14.3.3 and 14.3.4; 10.2; 22.5.3; 22.5.4)
(a) Vol. **90**, No. 2, p. 715–736 (1985).

N. Dutt, G. G. Agrawal, and J. Shanker
Volume Dependence of Dielectric Properties of Some Fluorite-Structure Crystals
(4 and 6; 14.4; 22.5.3; 22.8.1)
(b) Vol. **131**, No. 2, p. 505–510 (1985).

T. Sh. Abesadze and M. Punkkinen
Discrete Saturation of the Optical Absorption Line of Pr^{3+} in CaF_2 under the Condition of Pumping
(13.4 and 20.1; 19; 22.5.3)
(b) Vol. **131**, No. 2, p. K153–K155 (1985).

V. K. Jain and M. Shah Jahan
Changes in the Spectrum of Thermally Stimulated Emission from CaF_2:Mn at Low Temperatures
(20.3; 22.5.3)
(b) Vol. **131**, No. 2, p. K161–166 (1985).

V. K. Jain and M. S. Jahan
Low-Temperature Thermoluminescent Characteristics of CaF_2:Mn
(13.4 and 20.3; 22.5.3)
(a) Vol. **92**, No. 1, p. 237–248 (1985).

J. P. Jouart, C. Bissieux, G. Mary, and M. Egée
Study of the Sensitization Effect of Eu^{2+} Ions on the Trivalent Rare-Earth Fluorescence in CdF_2 Crystals
(13.4 and 20.3; 22.5.3)
(b) Vol. **132**, No. 2, p. 619–625 (1985).

A. P. Shkadarevich, M. V. Nikanovich, V. M. Reiterov, D. S. Umreiko, and A. R. Yarmolkevich
Luminescence and Stimulated Emission of M_A Colour Centers in MgF_2:Li Crystal
(20.2 and 20.3; 13.4; 22.5.3)
(a) Vol. **92**, No. 2, p. K135–K137 (1985).

22.5.4 Mixed Crystals

Original Papers and Short Notes

A. Kessler
Bulk Conductivity, Polarization, and Thermal Stimulated Currents of Insulated Polar Crystals
(14.3.3 and 14.3.4; 10.2; 22.5.3; 22.5.4)
(a) Vol. **90**, No. 2, p. 715–736 (1985).

S. Ihara and K. Suzuki
Molecular Dynamical Study of β-$AgI_{1-x}Cl_x$
(10.2 and 14.4; 1.2; 22.5.4)
(b) Vol. **131**, No. 1, p. 97–103 (1985).

F. V. Kusmartsev
The Narrow Bands of Exciton Absorption in Polar Solids and Alloys. Urbach's Rule
(13.5.1 and 20.1; 22.5; 22.5.4)
(b) Vol. **131**, No. 2, p. 621–629 (1985).

E. A. Popov, M. M. Kotlyarskii, and I. S. Edelman
Temperature and Field Dependence of Exciton-Magnon Absorption of $NaMnCl_3$
(13.5.1 and 13.5.4; 18.2; 20.1; 22.5.4)
(b) Vol. **132**, No. 2, p. 465–470 (1985).

S. N. Lukin and O. P. Teslya
The Structure of the Magnetic Centre in Cs_3CoCl_5
(13.4; 19; 22.5.4)
(b) Vol. **132**, No. 2, p. 601–606 (1985).

22.6 Simple Oxides (CdO, ZnO See 22.4)

Review Articles

K. H. Fischer
Spin Glasses (II)
(18.2 and 18.4; 2; 5; 7; 8; 13; 14; 21; 21.1; 21.6; 22.6; 22.8)
(b) Vol. **130**, No. 1, p. 13–71 (1985).

E. P. Wohlfarth
Magnetic Materials for Recording
(18; 18.3; 21.1; 21.1.1; 22.6; 22.8.2)
(a) Vol. **91**, No. 2, p. 339–350 (1985).

Original Papers and Short Notes

N. A. Mironova and V. N. Skvortsova
Fine Structure of the NiO:Co Infrared Absorption Spectrum
(13.5.3 and 20.1; 22.6)
(b) Vol. **127**, No. 1, p. K81–K82 (1985).

R. P. Goyal, Raviendra, D., and B. R. K. Gupta
Electroless Deposition of In_2O_3 and In_2O_3:Sn (ITO)
(1.5 and 14.3.1; 20.1; 22.6)
(a) Vol. **87**, No. 1, p. 79–83 (1985).

V. Kraševec, A. Prodan, M. Hudomalj, and S. Sulčič
A New Form of SnO_2 Found during Oxidation of α-SnO Thin Films
(1.5; 22.6)
(a) Vol. **87**, No. 1, p. 127–133 (1985).

S. Åsbrink, M. Wołcyrz, and S.-H. Hong
X-Ray Bond-Type Diffractometer Investigations on V_3O_5 in the Temperature Interval 298 to 480 K Including the Phase Transition Temperature T_t = 428 K
(1.2; 8; 22.6)
(a) Vol. **87**, No. 1, p. 135–140 (1985).
Erratum
(a) Vol. **89**, No. 1, p. 415 (1985).

H.-J. Fitting, K. König, and D. Hecht
Exoelectron Emission Due to Alkali Ions on Insulating Layers
(17; 1.5; 21.2; 22.1.2; 22.6)
(a) Vol. **87**, No. 1, p. K77–K79 (1985).

M. Averous, J. P. Desfours, and J. R. Vellas
Long- and Short-Range Magnetic Interactions in EuO
(14.3 and 18.2; 22.6)
(b) Vol. **127**, No. 2, p. 523–529 (1985).

V. G. TSIRELSON, M. YU. ANTIPIN, R. G. GERR, R. P. OZEROV, and YU. T. STRUCHKOV
Ruby Structure Peculiarities Derived from X-Ray Diffraction Data. Localization of Chromium Atoms and Electron Deformation Density
(1.3 and 10.2; 13.4; 22.6)
(a) Vol. **87**, No. 2, p. 425–433 (1985).

M. BEN SALEM, B. YANGUI, G. SCHIFFMACHER, and C. BOULESTEIX
Twinning of the Hexagonal (A) Structure of Rare Earth Sesquioxides
(1.4; 22.6)
(a) Vol. **87**, No. 2, p. 527–536 (1985).

K. IWAUCHI and N. KOIZUMI
Magnetic and Dielectric Relaxations of Fe_3O_4 and Some Ferrites
(14.4 and 18.3; 22.6; 22.8.2)
(a) Vol. **87**, No. 2, p. 617–621 (1985).

V. S. KORTOV, T. S. BESSONOVA, M. S. AKSELROD, and I. I. MILMAN
Hole-Induced Exoelectron Emission and Luminescence of Corundum Doped with Mg
(17 and 20.3; 10.2; 13.4; 22.6)
(a) Vol. **87**, No. 2, p. 629–639 (1985).

H. PAGNIA, J. SCHNELLBÄCHER, and N. SOTNIK
A Regeneration Model for Conducting Filaments in MIM Diodes
(13.3 and 14.3.3; 22.6)
(a) Vol. **87**, No. 2, p. 709–717 (1985).

D. LAPRAZ and P. IACCONI
Photo-Induced Thermoluminescence of X-Irradiated α-Al_2O_3
(11 and 20.3; 22.6)
(a) Vol. **87**, No. 2, p. K167–K169 (1985).

S. MOCHIZUKI
Radiative Optical Modes of Vibration in CeO_2 Thin Films
(6 and 20.1; 13.5; 22.6)
(b) Vol. **128**, No. 1, p. 23–30 (1985).

W. N. BHENDE, P. S. BAKHSHI, and J. SHANKER
Thermoelastic Behaviour and High-Order Elastic Constants of Ionic Crystals
(4 and 12.2; 6; 8; 22.5.2; 22.6)
(b) Vol. **128**, No. 1, p. 105–112 (1985).

G. MARTENS, P. RABE, and P. WENCK
Destructive Interference and Multiple Scattering Effects Observed in Ca K-Edge EXAFS Spectra
(1.4 and 20; 22.6; 22.8.1)
(a) Vol. **88**, No. 1, p. 103–111 (1985).

N. CH. MITRA and A. L. BHATTACHARYYA
Dielectric Relaxations in Dy_2O_3
(10.2; 14.4; 22.6)
(a) Vol. **88**, No. 1, p. 175–179 (1985).

B. Himmel, Th. Gerber, and H.-G. Neumann
X-Ray Diffraction Investigations of Differently Prepared Amorphous Silicas
(2; 22.6)
(a) Vol. **88**, No. 2, p. K127–K130 (1985).

R. R. Atabekyan, R. K. Ezoyan, V. A. Gevorkyan, and V. L. Vinetskii
Photostimulated Electron Redistribution between F and F^+ Centres in Corundum
(11 and 13.4; 20.1; 22.6)
(b) Vol. **129**, No. 1, p. 321–329 (1985).

F. Bechstedt and R. Enderlein
Electron–LO-Phonon Interaction near Interfaces. Application to Scattering of Inversion Layer Electrons on SiO_2–Si Interfaces
(6 and 13.3; 22.1.2; 22.6)
(b) Vol. **129**, No. 1, p. 349–362 (1985).

P. Triggs and F. Lévy
Optical and Electrical Properties of Ruthenium-Doped TiO_2
(13.4 and 20.1; 14.3; 22.6)
(b) Vol. **129**, No. 1, p. 363–374 (1985).

A. Stern and K. Hübner
Electronic Density of States of a Valence-Alternation-Pair Defect at the Si–SiO_2 Interface
(13.3; 22.1.2; 22.6)
(b) Vol. **129**, No. 1, p. K89–K93 (1985).

K. Atobe, N. Nishimoto, and M. Nakagawa
Irradiation-Induced Aggregate Centers in Single Crystal Al_2O_3
(10.2 and 20.1; 11; 22.6)
(a) Vol. **89**, No. 1, p. 155–162 (1985).

J. A. García, A. Remón, and J. Piqueras
Annealing-Induced Changes in the Photoluminescence of Deformed CaO Single Crystals
(20.3; 22.6)
(a) Vol. **89**, No. 1, p. 237–241 (1985).

R. Oberschmid
Absorption Centers of $Bi_{12}GeO_{20}$ and $Bi_{12}SiO_{20}$ Crystals
(13.4 and 20.1; 1.1; 16; 22.6)
(a) Vol. **89**, No. 1, p. 263–270 (1985).

H. Krause
Trap Induction and Breakdown Mechanism in SiO_2 Films
(14.3.4; 13.4; 22.6)
(a) Vol. **89**, No. 1, p. 353–362 (1985).

S. A. Y. Al-Ismail, K. Arshak, and C. A. Hogarth
Electron Spin Resonance Measurements and Electrical Characteristics, before and after Electroforming, of Thin Films of SiO/Nb_2O_5 and Nb_2O_5
(14.3.1 and 14.3.4; 2; 19; 22.6; 22.6.1)
(a) Vol. **89**, No. 1, p. 363–374 (1985).

A. J. MUHAMMED, D. V. MORGAN, and A. E. GUILE
Electronic Conduction in $Nb/Nb_2O_5/In$ Structures and the Effect of Space-Charge Overlap
(14.3.2 and 14.3.4; 22.6)
(a) Vol. **90**, No. 1, p. 371–381 (1985).

P. D. CIKMACH, J. J. KLEPERIS, A. R. LUSIS, and G. M. RAMĀNS
Influence of Ordering of Structure on the Optical Properties of WO_3 Thin Film
(1.5 and 20.1; 22.6)
(a) Vol. **90**, No. 1, p. K1–K4 (1985).

Y. TAKANO, Y. TANDOH, H. OZAKI, and N. MORI
XPS Study on a-GeO_x ($1.3 < x < 2$) Films Prepared from GeO_2
(2; 17; 22.6)
(b) Vol. **130**, No. 2, p. 431–436 (1985).

MBUNGU-TSUMBU, A. VAN DEN BOSCH, D. SEGERS, M. DORIKENS, and L. DORIKENS-VANPRAET
Positron Annihilation and Magnetic Susceptibility Measurements on Annealed Fast-Neutron Irradiated Crystalline Quartz
(11; 18.1; 22.6)
(a) Vol. **90**, No. 2, p. 595–606 (1985).

H. PAGNIA, N. SOTNIK, and H. STAUSS
Influence of Organic Molecules on the Current–Voltage Characteristic of Planar MIM Diodes
(14.3.3; 22.6)
(a) Vol. **90**, No. 2, p. 771–778 (1985).

N. A. RAZIK
Precise Lattice Constant Determination of Hexagonal, Rhombohedral, and Tetragonal Crystals from X-Ray Powder Diffractometric Data
(1.1; 22.6)
(a) Vol. **90**, No. 2, p. K125–K128 (1985).

A. L. TRONCONI and O. A. C. NUNES
Spin-Flip Intraband Absorption of Intense Laser Field in the Europium Compounds
(13.5.4 and 20.1; 22.6)
(b) Vol. **131**, No. 1, p. 137–140 (1985).

P. I. BARDETSKII, M. I. SHMIGLYUK, and SH. D. TIRON
Non-Linear Optical Nutation on Interexciton Transitions in Cu_2O
(13.5.1; 22.6)
(b) Vol. **131**, No. 1, p. 235–241 (1985).

S. N. VAIDYA
Theory of Melting. III. Nature of the Molten Glasses
(1.2 and 6; 2; 8; 22.6)
(a) Vol. **91**, No. 1, p. 37–43 (1985).

K. N. JOG, S. P. SANYAL, and R. K. SINGH
Third-Order Elastic Constants for Divalent Metal Oxides from Model Potentials
(12.2; 22.6)
(a) Vol. **91**, No. 1, p. 45–50 (1985).

P. Seidel, H.-J. Köhler, B. Mikolajczak, F. Müller, and H.-G. Meyer
Explanation of the AC Josephson Effect in Nb–NbO_x–Pb Tunnel Junctions with the Stewart-McCumber Model
(14.2 and 14.3.3; 21; 22.6)
(a) Vol. **91**, No. 2, p. K165 – K168 (1985).

E. De Castro and P. Olivo
Quantum Effects in Accumulation Layers of Si–SiO_2 Interfaces in the WKB Effective Mass Approximation
(13.4 and 14.3.2; 13.1; 22.1.2; 22.6)
(b) Vol. **132**, No. 1, p. 153 – 163 (1985).

F. Campabadal and X. Aymerich-Humet
On the Effects of Localized States and Tunnel Charge in Thin Insulating Films
(13.3 and 14.3.2; 14.3.3; 22.6)
(b) Vol. **132**, No. 1, p. 305 – 315 (1985).

M. J. Springis and J. A. Valbis
Red Luminescence of Color Centres in Sapphire
(13.4 and 20.3; 10.2; 22.6)
(b) Vol. **132**, No. 1, p. K61 – K65 (1985).

G. J. Muench, S. Arajs, and E. Matijević
The Morin Transition in Small α-Fe_2O_3 Particles
(18.2; 18.4; 22.6)
(a) Vol. **92**, No. 1, p. 187 – 192 (1985).

V. A. Bespalov, P. K. Kashkarov, V. F. Kiselev, V. A. M[illegible], and G. S. Plotnikov
The Study of Slow Relaxation of Charged Centres in Insulator–Semiconductor Structures Using Fluorescent Molecule Probes
(14.3.2; 13.3; 20.3; 22.1.1; 22.6)
(a) Vol. **92**, No. 1, p. 315 – 326 (1985).

I. B. Bakhtiyarov, S. M. Nakhmetov, P. G. Rustamov, I. M. Niftiev, and O. B. Tagiev
Physical Properties of $Ga_2S_3 \cdot Eu_2O_3$-Based Glasses
(2; 14.3; 20.3; 22.3; 22.6)
(a) Vol. **92**, No. 1, p. K31 – K35 (1985).

R. R. Atabekyan, V. L. Vinetskii, V. A. Gevorkyan, L. P. Godenko, and R. K. Ezoyan
Photoconductivity of Corundum Irradiated with High Energy Electrons at Excitation of F and F^+ Centres
(16; 13.4; 22.6)
(b) Vol. **132**, No. 2, p. 635 – 640 (1985).

22.6.1 Mixed Crystals

Original Papers and Short Notes

22.7 Semiconducting Intermetallic Compounds

Original Papers and Short Notes

J. Horák, P. Lošťák, and M. Matyáš
Point Defects in Sn-Doped Sb_2Te_3 Single Crystals
(13.4 and 14.3; 15; 20.1; 22.7)
(b) Vol. **129**, No. 1, p. 381 – 386 (1985).

L. G. Khvostantsev, A. I. Orlov, N. Kh. Abrikosov, and L. D. Ivanova
Kinetic Properties and Phase Transitions in Sb_2Te_3 under Hydrostatic Pressure up to 9 GPa
(1.2 and 15; 14.3; 22.7)
(a) Vol. **89**, No. 1, p. 301 – 309 (1985).

P. Lošťák, R. Novotný, J. Horák, and J. Klikorka
Properties of Sb_2Te_3 Single Crystals Doped with Tl Atoms
(10.2 and 20.1; 14.3; 22.7)
(a) Vol. **89**, No. 1, p. K55 – K59 (1985).

H. Neumann, W. Matz, W. Hoyer, and M. Wobst
Investigation of the Structure of Molten $Ge_{15}Te_{85}$ in Dependence on Temperature
(2; 1.3; 22.7)
(a) Vol. **90**, No. 2, p. 489 – 498 (1985).

M. Di Giulio, G. Micocci, P. Siciliano, and A. Tepore
On the Analysis of Complex Thermally Stimulated Capacitance Curves
(14.3.3; 13.4; 22.7)
(a) Vol. **91**, No. 1, p. 185 – 189 (1985).

M. Schenk, H. Berger, C. Kleint, N. S. Golovanova, O. I. Tananaeva, and V. P. Zlomanov
The Influence of Bi on Lattice Parameter and Carrier Concentration of PbTe
(10.2; 1.1; 14.3; 22.7)
(a) Vol. **91**, No. 1, p. K35 – K39 (1985).

22.7.1 Mixed Crystals

22.8 Tertiary and Higher Compounds

Review Articles

A. A. Kaminskii
Achievements in the Field of Physics and Spectroscopy of Activated Laser Crystals
(20.2; 20.1; 20.3; 22.8)
(a) Vol. **87**, No. 1, p. 11–57 (1985).

K. H. Fischer
Spin Glasses (II)
(18.2 and 18.4; 2; 5; 7; 8; 13; 14; 21; 21.1; 21.6; 22.6; 22.8)
(b) Vol. **130**, No. 1, p. 13–71 (1985).

B. L. Gelmont, B. T. Kolomiets, and K. D. Tsendin
Impurity Conduction of Chalcogenide Vitreous Semiconductors
(2 and 14.3; 13.1; 13.4; 22.3; 22.8)
(a) Vol. **91**, No. 2, p. 319–337 (1985).

Original Papers and Short Notes

H. Rauh and R. Geick
Crystal-Independent Interionic Potentials and the Lattice Dynamics of the Perovskite-Type Layer Material K_2ZnF_4
(6; 4; 22.8)
(b) Vol. **127**, No. 1, p. 55–65 (1985).

K. Kambas, A. Anagnostopoulos, S. Ves, B. Ploss, and J. Spyridelis
Optical Absorption Edge Investigation of $CdIn_2S_4$ and β-In_2S_3 Compounds
(20.1; 22.3; 22.8)
(b) Vol. **127**, No. 1, p. 201–208 (1985).

H. Neumann, W. Kissinger, Falah S. Hasoon, B. R. Pamplin, H. Sobotta, and V. Riede
Infrared Reflectivity Spectra of $CuSi_2P_3$
(6 and 20.1; 22.8)
(b) Vol. **127**, No. 1, p. K9–K14 (1985).

Z. Onyszkiewicz
Theory of Heat Capacity of the Mn_3GaC Metamagnet
(8 and 18; 22.8)
(b) Vol. **127**, No. 1, p. K37–K40 (1985).

K. M. Paraskevopoulos
Electron Effective Mass Dependence on Carrier Concentration in $TlBiTe_2$ Monocrystals
(13.1 and 20.1; 14.3; 22.8)
(b) Vol. **127**, No. 1, p. K45–K50 (1985).

G. Blasee and G. J. Dirksen
The Influence of the Phase Transformation on the Luminescence of $NbPO_5$
(1.2 and 20.3; 22.8)
(a) Vol. **87**, No. 2, p. K181 – K184 (1985).

V. Kumar, R. M. Singru, O. Parkash, and D. Bahadur
Mössbauer Studies of the System $La_{1-x}Sr_xCo_{1-y}Ti_yO_3$
(5; 22.8)
(b) Vol. **128**, No. 1, p. 223 – 230 (1985).

H. D. Hochheimer, E. Gmelin, Ch. Mensing, L. Viczian, K. D. Kreuer, and H. G. von Schnering
AC Conductivity, Calorimetric, X-Ray, and Raman Study of the Hexagonal Phase $TlInS_2$-III
(1.2; 6; 8; 14.3; 20.1; 22.8)
(a) Vol. **88**, No. 1, p. 43 – 48 (1985).

J. W. Tomm, K. Herrmann, A. E. Yunovich, and S. I. Solotov
Photoluminescence Spectra of $Pb_{1-x}Sn_xTe$ Epitaxial Layers and Heterostructures up to Room Temperature
(20.3; 14.3.1; 22.8)
(a) Vol. **88**, No. 1, p. 277 – 281 (1985).

Raviendra D. and J. K. Sharma
n-CdS/p-$AgInSe_2$ Solar Cells by Electrodeposition
(14.3.3 and 20.1; 1.5; 22.4.1; 22.8)
(a) Vol. **88**, No. 1, p. 365 – 368 (1985).

V. V. Sobolev, V. E. Grachev, and A. D. Smirnova
The Band Structure and Reflectance Spectra of $CdSnP_2$ Ternary Compound
(13.1 and 20.1; 22.8)
(b) Vol. **128**, No. 2, p. K143 – K147 (1985).

S. P. Chernov, L. I. Devyatkova, O. N. Ivanova, A. A. Kaminskii, V. V. Mikhailin, S. N. Rudnev, and T. V. Uvarova
$5d^14f^{N-1}$–$4f^N$ Absorption and Luminescence of Ce^{3+}, Pr^{3+}, and Nd^{3+} Ions in BaY_2F_8 Single Crystals
(20.2 and 20.3; 13.4; 22.8)
(a) Vol. **88**, No. 2, p. K169 – K173 (1985).

A. Edgar, J. M. Spaeth, and G. J. Troup
EPR and Optical Fluorescence Spectra of Cr^{3+} Ions in Alexandrite
(19 and 20.3; 22.8)
(a) Vol. **88**, No. 2, p. K175 – K178 (1985).

I. A. Smirnov, A. V. Golubkov, E. V. Goncharova, V. M. Sergeeva, V. V. Popov, M. V. Romanova, B. M. Buttaev, and V. S. Oskhotskii
Electrical and Galvanomagnetic Properties of $Tm_{1-x}La_xS$
(14.3; 15; 22; 22.8)
(a) Vol. **88**, No. 2, p. K201 – K203 (1985).

J. Horák, P. Lošťák, L. Jansa, and P. Matoušek
Departures from Stoichiometry of BiTeI Crystals Grown from the Vapour Phase
(10.2; 14.3; 22.8)
(a) Vol. **89**, No. 2, p. 493—498 (1985).

W. Bremser, K. H. Herrmann, and F. Rietenbach
Minority Carrier Lifetimes in $PbS_{0.1}Se_{0.9}$ Determined from Diffusion Length by a Scanning Laser Microprobe
(14.3 and 16; 22.8)
(a) Vol. **89**, No. 2, p. 653—656 (1985).

R. Oberschmid
Conductivity Instabilities and Polarization Effects of $Bi_{12}(Ge, Si)O_{20}$ Single-Crystal Samples
(14.3.4; 16; 22.8)
(a) Vol. **89**, No. 2, p. 657—671 (1985).

U. K. Reddy and P. J. Reddy
I–U Characteristics of Non-Crystalline As–Se–Cd Thin Films
(14.3.1; 2; 14.3.4; 22.8)
(a) Vol. **89**, No. 2, p. 679—682 (1985).

Th. Moritz and P. Schäfer
Preparation of Pb(Te, Se) Films by Simultaneous Evaporation of PbTe and Se
(1.5 and 3; 22.8)
(a) Vol. **89**, No. 2, p. K119—K121 (1985).

J. E. MacDonald, G. A. Saunders, M. S. Omar, and B. R. Pamplin
Comparison of the Elastic Behaviour of $CuGe_4P_3$ with that of $CuGe_2P_3$
(12.2; 22.8)
(a) Vol. **89**, No. 2, p. K137—K141 (1985).

E. E. Matyas and T. E. Borisenko
Optical Properties of Films of the PbTe-SnS System
(20.1; 22.8)
(a) Vol. **89**, No. 2, p. K177—K179 (1985).

K. Čermák and G. Juška
Drift Mobility of Electrons in $TlSbS_2$
(14.3; 22.8)
(a) Vol. **89**, No. 2, p. K207—K209 (1985).

W. Wojdowski
Vibrational Modes in $Bi_{12}GeO_{20}$ and $Bi_{12}SiO_{20}$ Crystals
(6 and 20.1; 4; 22.8)
(b) Vol. **130**, No. 1, p. 121—130 (1985).

H. Matsuda, O. Uemura, and T. Satow
The X-Ray Diffraction Study of Molton Tl_2X and $AgTl_3X_2$ (X: Se and Te)
(2; 18; 22.3; 22.8)
(a) Vol. **90**, No. 1, p. 77—82 (1985).

V. GHIORDĂNESCU and L. DOLLINGER
Some Observations on the Absorption of Copper in Photochromic Glass
(20.1; 13.4; 22.8)
(a) Vol. **90**, No. 2, p. K199–K204 (1985).

K. R. ALLAKHVERDIEV, T. G. MAMEDOV, V. V. PANFILOV, M. M. SHUKYUROV, and S. I. SUBBOTIN
Influence of Hydrostatic Pressure on the Fundamental Absorption Edge of $TlGaSe_2$, $TlGaS_2$, and $TlInS_2$ Crystals
(12.2 and 20.1; 1.2; 13.1; 22.8)
(b) Vol. **131**, No. 1, p. K23–K28 (1985).

G. D. GUSEINOV, V. M. BURLAKOV, T. M. TURSUNKULOV, and N. D. DZHURAEV
Special Features of Vibrational Spectra of Mixed $TlIn_{1-x}Fe_xS_2$ Crystals
(6 and 20.1; 1.2; 22.8)
(b) Vol. **131**, No. 1, p. K29–K33 (1985).

M. A. ALDZHANOV, K. K. MAMEDOV, and A. A. ABDURRAGIMOV
Heat Capacity and Moments of the Phonon Spectrum in $TlGaTe_2$
(6 and 8; 22.8)
(b) Vol. **131**, No. 1, p. K35–K37 (1985).

H. TAGUCHI and M. SHIMADA
X-Ray Photoelectron Spectroscopy Study of $CaMnO_{3-\delta}$
(4 and 13.4; 17; 22.8)
(b) Vol. **131**, No. 1, p. K59–K62 (1985).

K. ČERMÁK and G. JUŠKA
Drift Mobility of Holes in $TlSbS_2$
(14.3.4; 22.8)
(a) Vol. **91**, No. 1, p. 219–223 (1985).

R. K. VELIEV and M. A. ALDZHANOV
Low-Temperature Heat Capacity of $ZnGa_2S_4$, $ZnGa_2Se_4$, and $ZnGa_2Te_4$ Thiogallates
(8; 22.8)
(a) Vol. **91**, No. 1, p. K23–K24 (1985).

T. SATO, N. MORITA, and M. SHIMADA
Synthesis and Electrical Properties of $Li_xMg_{1-x}M_2O_4$ (M = Ti, V) ($0 \leqq x \leqq 1.0$)
(1.3 and 14.3; 22.8)
(a) Vol. **91**, No. 1, p. K81–K85 (1985).

J. L. DORMANN, A. SAIFI, V. CAGAN, and M. NOGUES
Frequency Dependence of the AC Susceptibility of Reentrant System $CdCr_{2x}In_{2-2x}S_4$ ($0.85 < x \leqq 1$)
(18.2; 22.8)
(b) Vol. **131**, No. 2, p. 573–583 (1985).

M. AVEROUS, B. A. LOMBOS, C. FAU, E. ILBNOUELGHAZI, J. C. TEDENAC, G. BRUN, and M. A. BARTKOWSKI
$Pb_{1-x}Gd_xTe$ a New Semimagnetic Semiconductor with Rare Earths
(14.3 and 18.4; 19; 22.8)
(b) Vol. **131**, No. 2, p. 759–768 (1985).

E. Grilli, M. Guzzi, E. Camerlenghi, and F. Pio
Photoluminescence of $Zn_xCd_{1-x}In_2S_4$ Layered Alloys ($0.7 < x < 1$)
(13.4 and 20.3; 22.8)
(a) Vol. **92**, No. 1, p. 213—220 (1985).

A. N. Anagnostopoulos, C. Manolikas, and A. Karòutis
Electrical Effects Associated with the Ordering Process in $CdInGaS_4$ Crystals. II. Electron Traps Determined with Conductivity Measurements
(13.4 and 14.3.4; 22.8)
(a) Vol. **92**, No. 1, p. 231—235 (1985).

S. K. Plachkova
Thermoelectric Power of GeTe-Rich $(GeTe)_{1-x}(AgBiTe_2)_x$ Solid Solutions
(14.3 and 15; 22.8)
(a) Vol. **92**, No. 1, p. 273—277 (1985).

J. P. Leyris and J. P. Aicardi
Photoconductivity Measurements on $AgGaS_2$ Crystals
(16; 20.3; 22.8)
(a) Vol. **92**, No. 1, p. 279—284 (1985).

V. P. Meiklyar, A. E. Usachev, Yu. V. Yablokov, and V. A. Shustov
EPR and ENDOR in $CsCaCl_3:Gd^{3+}$ Crystals
(1.2 and 19; 22.8)
(b) Vol. **132**, No. 2, p. K73—K76 (1985).

Yu. V. Sukhetskii, A. V. Soldatov, A. N. Gusatinskii, K. Žičkus, and A. Audzijonis
Energy Band Structure of $SbS_xSe_{1-x}I$
(13.1; 20; 22.8)
(b) Vol. **132**, No. 2, p. K103—K106 (1985).

V. P. Seth, V. K. Jain, A. Yadav, and R. S. Bansal
ESR Study of Vanadyl Ion in Potassium Succinate
(19; 13.4; 22.8)
(b) Vol. **132**, No. 2, p. K139—K142 (1985).

M. Quintero and J. C. Woolley
Crystallography and Optical Energy Gap Values for $Cd_{2x}(AgIn)_yMn_{2z}Te_2$ Alloys
(1.1; 13.1; 20.1; 22.8)
(a) Vol. **92**, No. 2, p. 449—456 (1985).

L. Tarricone, E. Grilli, M. Guzzi, and L. Zanotti
Optical Absorption and Recombination Processes in $CdGaInS_4$ Layered Crystals
(13.4 and 20.3; 16; 20.1; 22.8)
(a) Vol. **92**, No. 2, p. 595—602 (1985).

E. M. Kostova and M. K. Kostov
Experimental Study of Electrogyration in $Li_2Gd_4(MoO_4)_7$ Crystals
(12.2 and 14.4; 22.8)
(a) Vol. **92**, No. 2, p. K105—K106 (1985).

22.8.1 Salts of Oxy-Acids (Carbonates, Phospates, Silicates, Including Mica. etc.)

Review Article

Original Papers and Short Notes

A. REQUARDT and G. LEHMANN
Impurity-Related Electron Centres in α-Berlinite ($AlPO_4$)
(10.2 and 19; 11; 22.8.1)
(b) Vol. **127**, No. 2, p. K117–K120 (1985).

R. BLINC, P. CEVC, and N. I. SILKIN
EPR Study of the Pseudo-One-Dimensional Ferroelectric Phase Transition in Tl^{2+} Doped CsH_2PO_4
(14.4.1 and 19; 22.8.1)
(b) Vol. **127**, No. 2, p. K147–K149 (1985).

H. CERVA and W. GRAEFF
Contrast Investigations of Surface Acoustic Waves by Stroboscopic Topography. II. Wavefield Deviation Contrast
(7; 14.4.1; 22.8.1)
(a) Vol. **87**, No. 2, p. 507–516 (1985).

R. STYRKOWIEC
The Interaction between Moving Domain Walls in Rochelle Salt Crystals
(14.4.1; 22.8.1)
(a) Vol. **87**, No. 2, p. K135–K138 (1985).
Erratum
(a) Vol. **89**, No. 1, p. K117 (1985).

G. A. PETRAKOVSKII, G. S. PATRIN, and N. V. VOLKOV
Change of Magnetic Resonance Parameters in $FeBO_3$:NiO Induced by Radiation ($\lambda = 1.06$ μm)
(10.2 and 19; 22.8.1)
(a) Vol. **87**, No. 2, p. K153–K156 (1985).

B. OREL, M. KLANJŠEK, V. MOISEENKO, M. VOLNYANSKII, N. LE CALVÉ, B. PASQUIER, and A. NOVAK
Isotopic Impurity Modes and Polariton Dispersion in Li_2GeO_3 Crystals
(6 and 13.5; 20.1; 22.8.1)
(b) Vol. **128**, No. 1, p. 53–67 (1985).

F. BREHAT and B. WYNCKE
Soft Mode Spectroscopy in Potassium Dihydrogen Phosphate
(6 and 20.1; 14.4.1; 22.8.1)
(b) Vol. **128**, No. 1, p. 83–92 (1985).

G. ÜFFINGER and A. KASTEN
Specific Heat and Magnetocaloric Effects in $TbPO_4$. I. Experiments
(1.2 and 18.4; 8; 22.8.1)
(b) Vol. **128**, No. 1, p. 201–207 (1985).

A. A. ALYBAKOV, O. M. ARBOTOEV, V. A. GUBANOVA, and K. KUDABAEV
Cr^{5+} Paramagnetic Centers in X-Irradiated KH_2PO_4:Cr^{3+} Single Crystals
(13.4; 11; 19; 20.1; 22.8.1)
(b) Vol. **128**, No. 1, p. K93–K96 (1985).
Erratum
(b) Vol. **130**, No. 1, p. K89 (1985).

S. C. SABHARWAL and B. GHOSH
Electrical Conductivity of $NH_4H_2PO_4$ Single Crystal
(11 and 14.4; 22.8.1)
(a) Vol. **89**, No. 1, p. K83–K87 (1985).

B. WYNCKE, F. BREHAT, and G. V. KOZLOV
Low-Frequency Dielectric Response Function of Sodium Nitrate
(6 and 20.1; 14.4; 22.8.1)
(b) Vol. **129**, No. 2, p. 531–538 (1985).

R. KAWASHIMA, K. KATSUKI, and K. SUZUKI
Temperature Dependence of Optical Absorption near the Structural Phase Transition of Potassium Nitrate
(1.2 and 13.1; 14.4.1; 20.1; 22.8.1)
(b) Vol. **129**, No. 2, p. 697–705 (1985).

I. H. ISMAILZADE and O. A. SAMEDOV
The Nature of the Intermediate Phase of $PbZrO_3$
(1.2 and 14.4.1; 22.8.1)
(a) Vol. **89**, No. 2, p. K133–K136 (1985).

WANG HUAFU, SHI GUOTONG, and WU ZHONGKANG
Photovoltaic Effect in $LiNbO_3$:Mg
(16; 14.4.1; 22.8.1)
(a) Vol. **89**, No. 2, p. K211–K213 (1985).

F. FISCHER, J. PRAKASH, and W. SCHOBER
Thermally Stimulated Depolarization of $TlNO_2$ near the Glass Transformation
(14.4; 22.8.1)
(a) Vol. **89**, No. 2, p. K215–K218 (1985).

I. V. STASYUK and S. S. KOTSUR
On the Theory of Electrogyration in Dielectric Crystals. Sheelite-Type Crystals
(14.4 and 20; 22.8.1)
(b) Vol. **130**, No. 1, p. 103–113 (1985).

P. J. BECKER, H. G. KAHLE, and E. KELLER
The Phase Diagram of $TbPO_4$ Measured by Linear Optical Birefringence
(1.2 and 18.4; 20; 22.8.1)
(b) Vol. **130**, No. 1, p. 191–196 (1985).

J. BERDOWSKI and M. HAJDUK
The Influence of Isotope Effects on the Interaction of Light and Surface Acoustic Waves in DADA Crystals
(7 and 20.1; 14.4.1; 22.8.1)
(a) Vol. **90**, No. 1, p. 157–165 (1985).

R. POPRAWSKI, J. SHALDIN, and S. MATYJASIK
Pyroelectric and Optical Properties of α-$LiIO_3$ Crystals
(14.4 and 20; 22.8.1)
(a) Vol. **90**, No. 1, p. 167–171 (1985).

R. Blinc, P. Cevc, J. Seliger, and M. Koren
EPR of Tl^{2+} Doped Incommensurate K_2SeO_4
(13.5.4 and 19; 22.8.1)
(b) Vol. **131**, No. 2, p. K167–K172 (1985).
Erratum
(b) Vol. **132**, No. 2, p. K143 (1985).

V. V. Krasnogolovets and P. M. Tomchuk
The Anharmonicity Influence on the Proton Conductivity of an Hydrogen-Bonded Chain
(14.4; 6; 22.8.1)
(b) Vol. **131**, No. 2, p. K177–K179 (1985).

G. van Tendeloo, S. Amelinckx, C. Manolikas, and Wen Shulin
The Direct Observation of "Discommensurations" in Barium Sodium Niobate (BSN) and Its Homologues
(1.2 and 1.4; 22.8.1)
(a) Vol. **91**, No. 2, p. 483–501 (1985).

R. W. Chantrell and E. P. Wohlfarth
Rate Dependence of the Field-Cooled Magnetisation of a Fine Particle System
(18; 22.8.1)
(a) Vol. **91**, No. 2, p. 619–626 (1985).

Hj. Bernhardt
Investigations of the Orange Luminescence of $PbMoO_4$ Crystals
(20.1 and 20.3; 13.4; 22.8.1)
(a) Vol. **91**, No. 2, p. 643–648 (1985).

A. H. Piekara
A Theoretical Model of Ferroelectrics Exhibiting One, Two, or Three Curie Points
(14.4.1; 22.8.1)
(b) Vol. **132**, No. 1, p. 277–282 (1985).

Pan Xia-qing, Hu Mei-shen, Yao Ming-hui, and Feng Duan
A TEM Study of the Incommensurate Phase and Related Phase Transitions in Barium Sodium Niobate
(1.2 and 1.4; 22.8.1)
(a) Vol. **92**, No. 1, p. 57–68 (1985).

Minh Hoang, J.-C. Niepce, and J.-C. Mutin
Evidence of the Chemical Role Played by Niobium Oxide Nb_2O_5 during Sintering of Mixtures of $BaTiO_3$ and $CaZrO_3$ Perovskites
(1.3 and 14.4; 22.8.1)
(a) Vol. **92**, No. 1, p. 83–93 (1985).

Ye Ming, E. Krätzig, and R. Orlowski
Photorefractive Effects in $LiNbO_3$:Cr Induced by Two-Step Excitation
(20.1; 13.4; 16; 22.8.1)
(a) Vol. **92**, No. 1, p. 221–229 (1985).

22.8.2 Spinels, Garnets, Ferrites

Review Article

E. P. Wohlfarth
Magnetic Materials for Recording
(18; 18.3; 21.1; 21.1.1; 22.6; 22.8.2)
(a) Vol. **91**, No. 2, p. 339–350 (1985).

Original Papers and Short Notes

M. Ogata, H. Kozaka, and N. Uryû
Magnetic Interactions in Spinel Compound $CoGa_2O_4$
(18.4; 22.8.2)
(b) Vol. **127**, No. 1, p. K33–K36 (1985).

E. M. Smokotin, O. P. Kvashnina, and A. M. Kapitonov
Acoustic Properties of Terbium Ferrite Garnet in the Region of the Compensation Temperature Point
(7 and 18.3; 22.8.2)
(a) Vol. **87**, No. 1, p. K53–K56 (1985).

K. Iwauchi and N. Koizumi
Magnetic and Dielectric Relaxations of Fe_3O_4 and Some Ferrites
(14.4 and 18.3; 22.6; 22.8.2)
(a) Vol. **87**, No. 2, p. 617–621 (1985).

V. K. Vlasko-Vlasov and A. F. Khapikov
Reorientation of Domain Walls in a Garnet Film under the Action of an AC Magnetic Field
(18.3; 22.8.2)
(a) Vol. **87**, No. 2, p. K157–K162 (1985).

P. C. Riedi and L. Zammit-Mangion
Difference between Domain Wall Motion in Single Crystal and Polycrystalline YIG Observed by Pulsed NMR
(18.3 and 19; 22.8.2)
(a) Vol. **87**, No. 2, p. K163–K166 (1985).

D. Enakshi and K. V. Rao
Dielectric Properties of Tourmaline
(10.2 and 14.4; 14.3.4; 22.8.2)
(a) Vol. **87**, No. 2, p. K185–K189 (1985).

P. Colombet and W. S. Glaunsinger
Anomalous High-Temperature EPR Linewidths in the Spinel Solid Solution $Cu_{2x}Cr_{2x}Sn_{2-2x}S_4$
(19; 18.2; 18.4; 22.8.2)
(b) Vol. **128**, No. 1, p. 141–149 (1985).

V. Narasimhan, H. V. Keer, and D. K. Chakrabarty
Structural and Electrical Properties of the $LaCo_{1-x}Mn_xO_3$ and $LaCo_{1-x}Fe_xO_3$ Systems
(1.2 and 14.3; 15; 22.8.2)
(a) Vol. **89**, No. 1, p. 65–71 (1985).

A. A. Sidorov, V. P. Suetin, V. G. Pokazaniev, and B. V. Karpenko
Determination of Effective Indirect Exchange Interaction Integrals in Ferrimagnets
(18.3; 22.8.2)
(b) Vol. **129**, No. 2, p. 625–629 (1985).

C. Yu. Shunyaev, M. T. Varshavskii, and A. N. Men
Influence of Short-Range Order on the Magnetic Moment of Quasibinary Granat-Type Solid Solutions with Sublattice Substitution
(1.4 and 18; 22.8.2)
(b) Vol. **129**, No. 2, p. K131–K134 (1985).

S. García and E. Altshuler
Mössbauer Study of the Reaction Kinetics of Hexagonal M-Phase Ferrites
(1.2 and 5; 22.8.2)
(a) Vol. **89**, No. 2, p. 427–436 (1985).

K. Pátek and I. Tomáš
A Quantitative Characterization of Magnetic Defects by Domain Nucleation
(18.2; 22.8.2)
(a) Vol. **89**, No. 2, p. 595–599 (1985).

I. O. Troyanchuk, L. A. Bashkirov, L. V. Balyko, V. M. Pavlov, A. N. Pirogov, A. P. Vokhmyanin, and O. A. Zalukovskaya
Magnetic Ordering in the System $Ca(Cu_{1-x}Mn_x)_3Mn_4O_{12}$
(18.3; 22.8.2)
(a) Vol. **89**, No. 2, p. 601–608 (1985).

C. Michalk and W. Thiel
Garnet Formation of Al-Substituted Yttrium Iron Garnet Ceramics Studied by Mössbauer Effect and Magnetization Measurements
(1.2 and 5; 18.3; 22.8.2)
(a) Vol. **90**, No. 1, p. 325–330 (1985).

E. Z. Katsnelson and A. G. Karoza
Direct Observation of Ferrite Photomagnetization
(18.3; 22.8.2)
(a) Vol. **90**, No. 1, p. K49–K53 (1985).

S. R. Murthy and T. S. Rao
Magnetostriction of Ni–Zn and Co–Zn Ferrites
(18.3; 22.8.2)
(a) Vol. **90**, No. 2, p. 631–635 (1985).

M. R. CHAVES, J. L. RIBEIRO, A. SELMI, and P. GIBART
Electrical Resistivity and Seebeck Effect in $HgCr_2Se_4$
(14.3; 15; 22.8.2)
(a) Vol. **92**, No. 1, p. 263–271 (1985).

P. V. REDDY and T. S. RAO
Thermoelectric Power Studies of Li–Ni Mixed Ferrites
(15; 14.3; 22.8.2)
(a) Vol. **92**, No. 1, p. 303–307 (1985).

K. IWAUCHI
Dielectric Relaxation in Fe_3O_4
(14.4; 22.8.2)
(a) Vol. **92**, No. 1, p. 309–314 (1985).

R. Z. SADYKHOV, L. M. VALIEV, and A. I. AKHMEDOV
Magnetic Properties of the $Ni_{0.25}Cu_xFe_{0.75-x}Cr_2S_4$ ($0 \leqq x \leqq 0.25$) System
(18.2; 22.8.2)
(a) Vol. **92**, No. 1, p. K69–K72 (1985).

YU. G. CHUKALKIN, V. V. PETROV, V. R. SHTIRTS, and B. N. GOSHCHITSKII
Effects of Radiation Disorder in Chromium Spinels
(1 and 11; 22.8.2)
(a) Vol. **92**, No. 2, p. 347–354 (1985).

C. OTERO AREÁN and M. C. TROBAJO-FERNANDEZ
Cation Distribution in $Mg_xNi_{1-x}Ga_2O_4$ Oxide Spinels
(1.1; 1.3; 22.8.2)
(a) Vol. **92**, No. 2, p. 443–447 (1985).

V. F. KITAEVA, E. V. ZHARIKOV, and I. L. CHISTYI
The Properties of Crystals with Garnet Structure
(12.2; 7; 22.8.2)
(a) Vol. **92**, No. 2, p. 475–488 (1985).

Z. KLEM and J. KACZÉR
The FMR of a Close-Packed Bubble Lattice
(18.2; 19; 22.8.2)
(a) Vol. **92**, No. 2, p. 525–532 (1985).

S. A. MAZEN and A. H. WAFIK
Barkhausen Jumps of Single and Polycrystalline Ferrites
(18.3; 22.8.2)
(a) Vol. **92**, No. 2, p. 539–543 (1985).

K. IWAUCHI and Y. IKEDA
Magnetic and Dielectric Properties of $Ba_2Ti_4Fe_2O_{13}$–$K_2Ti_6O_{13}$
(14.4; 18.3; 22.8.2)
(a) Vol. **92**, No. 2, p. 623–629 (1985).

22.9 Organic Semiconductors

Review Article

H. W. STREITWOLF
Physical Properties of Polyacetylene
(13.5 and 13.5.3; 6; 18; 19; 20.1; 22.9)
(b) Vol. **127**, No. 1, p. 11–54 (1985).

Original Papers and Short Notes

P. REINEKER, J. KÖHLER, and U. SCHMID
Triplet Excitions in Non-Resonant Molecular Pairs. Saturation Effects in Electron Spin Resonance
(13.5.1; 19; 22.9)
(b) Vol. **127**, No. 1, p. 237–244 (1985).

PREM CHAND and M. UMAR
Electron Paramagnetic Resonance Study of Copper(II) Ions in Single Crystal of Strontium Formate Dihydrate
(13.4 and 19; 22.9)
(b) Vol. **127**, No. 1, p. 279–286 (1985).

M. M. SZOSTAK, A. V. BELUSHKIN, and I. NATKANIEC
Neutron Spectroscopy and Model Calculations of Phonon Dynamics in Crystalline p-Chloroaniline
(6; 1.2; 22.9)
(b) Vol. **127**, No. 1, p. K1–K4 (1985).

R. VERMA, K. N. LAKSHMINARAYAN, and K. K. SRIVASTAVA
Dielectric Parameters for Films and Sheets at Microwave Frequencies
(14.4; 22.9)
(a) Vol. **87**, No. 1, p. 323–333 (1985).

B. KOSTUREK
Depolarization Field Effect on Domain Wall Motion in TGS Crystals
(14.4.1; 22.9)
(a) Vol. **87**, No. 1, p. K93–K95 (1985).

B. J. SCHMID and P. REINEKER
Erratum to: The Mixing of Frenkel and Charge-Transfer Excited States. II. Numerical Treatment of Non-Symmetric Dimers
(13.5.1; 22.9)
(b) Vol. **127**, No. 2, p. 735 (1985).

W. SCHEIBNER and M. JÄCKEL
Thermal Conductivity and Specific Heat of an Epoxy Resin/Epoxy Resin Composite Material at Low Temperatures
(8; 2; 22.9)
(a) Vol. **87**, No. 2, p. 543–547 (1985).

K. IWAHANA, P. KNOLL, and H. KUZMANY
Effect of Irradiation with Intense Laser Light on Trans-Polyacetylene
(11; 20.3; 22.9)
(a) Vol. **128**, No. 2, p. 735–741 (1985).

G. S. MINGALEEV, A. P. TYUTNEV, A. V. VANNIKOV, V. I. ARKHIPOV, A. I. RUDENKO, V. M. LOGIN, and V. A. URSU
On the Influence of Neutral Contacts on the Steady-State Radiation-Induced Conductivity in Polymers
(11 and 16; 22.9)
(a) Vol. **88**, No. 2, p. 655–662 (1985).

A. P. TYUTNEV, V. N. ABRAMOV, V. S. SAENKO, P. I. DUBENSKOV, A. V. VANNIKOV, and E. D. POZHIDAEV
Radiation-Induced Conductivity in Foamed Dielectrics
(11 and 14.3; 22.6; 22.9)
(a) Vol. **88**, No. 2, p. 673–680 (1985).

A. K. ABASS, F. Y. M. AL-EITHAN, and R. H. MISHO
Indirect Electronic Transitions in Single Crystals of Triglycine Sulfate
(20.1; 6; 13.1; 22.9)
(a) Vol. **89**, No. 1, p. 225–229 (1985).

R. SIKDAR and A. K. PAL
EPR Studies Dimensionality in Copper Calcium Acetate Hexahydrate
(19; 18.1; 22.9)
(a) Vol. **89**, No. 1, p. 243–248 (1985).

V. I. ARKHIPOV, V. M. LOGIN, and S. YA. SHEVCHENKO
Effect of Multiple Trapping on Photoreceptor Discharge Characteristics under the Condition of Surface Carrier Generation
(14.3.2 and 16; 2; 22.9)
(a) Vol. **89**, No. 1, p. 293–299 (1985).

A. P. TYUTNEV, V. S. SAENKO, V. N. ABRAMOV, V. P. SICHKAR, A. I. KARPECHIN, E. D. POZHIDAEV, and A. V. VANNIKOV
Dose Effects in Transient Radiation-Induced Conductivity in Polymers
(11 and 14.3; 22.9)
(a) Vol. **89**, No. 1, p. 311–320 (1985).

A. V. IVANOV and A. I. MELKER
Interatomic Bond Rupture in the System of Two Coupled Anharmonic Chains with a Weak Interaction
(2 and 10.2; 4; 22.9)
(a) Vol. **89**, No. 1, p. K51–K54 (1985).

M. MÜLLER, TH. FRAUENHEIM, and C. HAMANN
Peierls Transition and Fluctuation Conductivity in Thin Lead-Phthalocyanine (PbPc) Films
(1.2 and 14.3.1; 22.9)
(a) Vol. **89**, No. 1, p. K89–K94 (1985).

A. Kuczkowski
Photoelectrochemical Solar Cells with Semiconducting Polymers Prepared by Modification of Polyvinylchloride PVC and Polytetrafluoroethylene PTFE
(14.3.3; 16; 22.9)
(a) Vol. **89**, No. 1, p. K109–K113 (1985).

M. Petrov, L. Komitov, P. Simova, and G. Hauck
Intermediate Smectic A-State at Nematic–Smectic C Phase Transition at Tilted Orientation
(a) Vol. **89**, No. 2, p. 451–456 (1985).

M. G. Kaplunov, E. B. Yagubskii, L. P. Rosenberg, and Yu. G. Borodko
Optical Properties of the Two Crystal Modifications of the Organic Conductor (BEDT-TTF)$_2$I$_3$
(20.1; 6; 13; 22.9)
(a) Vol. **89**, No. 2, p. 509–516 (1985).

H. Sodolski
Ionic Conduction in Rhodamine-Doped Polyester Polymer
(14.4; 22.9)
(b) Vol. **89**, No. 2, p. 647–652 (1985).

I. V. Brovchenko, N. I. Ostapenko, M. T. Shpak, L. L. Valkunas, and G. B. Vektaris
Resonance Interaction between Impurities and Crystal Molecules. Application to Benzselenophene in Naphthalene
(13.4; 20.1; 20.3; 22.9)
(b) Vol. **130**, No. 1, p. 281–291 (1985).

K. Ikezaki
Thermally Stimulated Current Transport Peak in High-Density Polyethylene
(14.4; 22.9)
(a) Vol. **90**, No. 1, p. 383–391 (1985).

M. K. Kerimov
Electron Paramagnetic Resonance of Stabilized Electrons in PTFE Coronoelectrets
(14.4.1 and 19; 22.9)
(a) Vol. **90**, No. 1, p. K71–K74 (1985).

H. von Berlepsch
Electric Field Dependent Trapping Time in HDPE
(14.3.1; 22.9)
(a) Vol. **90**, No. 1, p. K97–K100 (1985).

G. D. Khattak, M. S. Hussain, G. W. Hunter, and L. E. Wenger
Specific Heat and Magnetic Susceptibility Studies on Bromobis (1,4-Diazacycloheptane) Copper(II) Perchlorate, [Cu(dach)$_2$Br](ClO$_4$), at Low Temperatures
(8 and 18.2; 22.9)
(b) Vol. **130**, No. 2, p. 587–593 (1985).

R. Somashekar, Ch. Satyanarayana, and M. S. Madhava
Mesomorphic Behaviour of Mixtures of Cetyl Alcohol and Dimethyl Sulfoxide
(1 and 2; 22.9)
(a) Vol. **90**, No. 2, p. 449–456 (1985).

M. Krause, C. Michalk, M. Benedix, and J. Suwalski
Low-Temperature Mössbauer Measurement of a Polymeric Oxalate (2,2′-Bipyridine) Iron Compound
(5; 18; 22.9)
(a) Vol. **90**, No. 2, p. 657–660 (1985).

T. Słupkowski
Electrical Conductivity of Polyester Polymer Containing Carbon Black
(14.3; 22.9)
(a) Vol. **90**, No. 2, p. 737–741 (1985).

W. Scheibner, R. Häszler, M. Stephan, and M. Jäckel
On the Filler Induced Modification of the Glass Transition Temperature and Hardness of an Epoxy Resin Composite Material
(12.2; 2; 8; 22.9)
(a) Vol. **90**, No. 2, p. K139–K141 (1985).

S. L. Drechsler and M. Bobeth
Dielectric Properties of Trans-Polyacetylene. I. The Macroscopic Longitudinal Dielectric Function of a Peiers-Fröhlich Semiconductor of Commensurability 2 within RPA
(14.3; 13.5.2; 22.9)
(b) Vol. **131**, No. 1, p. 267–278 (1985).

Prem Chand, G. C. Upreti, M. Umar, and R. J. Singh
EPR of Cu^{2+} in Barium Formate Single Crystal
(13.4 and 19; 22.9)
(b) Vol. **131**, No. 1, p. 357–367 (1985).

B. Martin, J. M. Pastor, and J. A. de Saja
Ferroelectric Behaviour of TGS and TGSe from Microhardness Measurements
(12.2; 1.2; 22.9)
(b) Vol. **131**, No. 1, p. K5–K9 (1985).

J. M. Keller and S. C. Datt
Thermally Stimulated Depolarization Currents in Polypropylene Foils
(14.4; 22.9)
(a) Vol. **91**, No. 1, p. 205–210 (1985).

S. Schrader and M. Kryszewski
Multiple Trapping Approach to the TSC Drift Experiment
(13.4 and 14.3.4; 16; 22.9)
(a) Vol. **91**, No. 1, p. 243–255 (1985).

A. I. Melker and A. V. Ivanov
Fracture Model for a Strained Anharmonic Chain Having Atoms with Two Degrees of Freedom
(10.2; 22.9)
(a) Vol. **91**, No. 1, p. K41–K44 (1985).

H. von Berlepsch
Field-Dependent Injection in HDPE
(14.3.4; 13.3; 22.9)
(a) Vol. **92**, No. 1, p. 285–292 (1985).

J. Hönig
Thickness-Dependent Absorption in Lead Phthalocyanine Thin Films
(1.5 and 20.1; 22.9)
(a) Vol. **92**, No. 1, p. K7–K10 (1985).

A. K. George
Anomalous Ultrasonic Absorption in a Nematic and a Polymesomorphic Liquid Crystal
(2 and 7; 22.9)
(a) Vol. **92**, No. 1, p. K15–K17 (1985).

S. L. Chaplot
Phase Transition in Tetracyanoethylene at High Pressure by X-Ray Diffraction
(1.2; 12.2; 22.9)
(a) Vol. **92**, No. 1, p. K23–K27 (1985).

J. Ganster and D. Geiss
Polycrystalline Simple Average of Mechanical Properties in the General (Triclinic) Case
(1 and 12.2; 22.9)
(b) Vol. **132**, No. 2, p. 395–407 (1985).

A. B. Henriques and J. P. von der Weid
Decay Kinetics of Triplet Molecular Excitons in Alkali Cyanides
(13.5.1 and 20.3; 22.9)
(b) Vol. **132**, No. 2, p. 627–633 (1985).

S. Götz
Measurement of Anomalous Dispersion Effects around an Absorption Edge in the Soft X-Ray Region
(1.1 and 20; 22.9)
(a) Vol. **92**, No. 2, p. 407–412 (1985).

R. Mukhopadhyay, S. L. Chaplot, and K. R. Rao
Phase Transition in Tetracyanoethylene
(1.2; 22.9)
(a) Vol. **92**, No. 2, p. 467–474 (1985).

C. Dauwe and K. Motoko
Three-Quantum Annihilation in the Plastic Crystal Norbornane
(1.2 and 10.2; 22.9)
(a) Vol. **92**, No. 2, p. K117–K121 (1985).

M. Abd-Lefdil, M. Rolland, and M. Cadene
n^+–n Silicon–Trans$(CH)_x$ Schottky Diode
(14.3.1 and 14.3.3; 22.9)
(a) Vol. **92**, No. 2, p. K171–K174 (1985).

23. Solidified Gases

Original Papers and Short Notes

A. V. Eletskii and E. V. Stepanov
Mechanisms of Diffusion of a Small Impurity in Noble Gas Crystals
(9 and 10.2; 23)
(b) Vol. **130**, No. 2, p. 517–529 (1985).

M. Lazaridou and K. Eftaxias
On a Connection between Vacancy Formation Parameters and Melting Process in Rare Gas Solids
(8 and 10.2; 23)
(a) Vol. **90**, No. 2, p. K147–K149 (1985).

T. Schröder, O. Dössel, H. Nahme, and N. Schwentner
Transient Absorption of Xe and Kr Guest Atoms in Ar Crystals
(13.4 and 20.1; 23)
(b) Vol. **131**, No. 2, p. 741–746 (1985).

Author Index

International Classification System for Physics*)

(in use from January 1, 1986)

60. Condensed matter: structure, mechanical and thermal properties

61. Structure of liquids and solids; crystallography
    (see also 68.20. Solid surface structures, 71. Electron states)
    61.10. X-ray determination of structures
        (for specific determinations, see 61.55 to 61.80)
    61.12. Neutron determination of structures
        (for specific determinations, see 61.55 to 61.80)
    61.14. Electron determination of structures
        (for specific determinations, see 61.55 to 61.80)
    61.16. Other determination of structures
        (for specific determinations, see 61.55 to 61.80)
    61.20. Classical, semiclassical, and quantum theories of liquid structure
        (for electronic states, see 71; for liquid helium, see 67)
    61.25. Studies of specific liquid structures
    61.30. Liquid crystals
    61.40. Amorphous and polymeric materials
    61.50. Crystalline state (including molecular motions in solids)
        (for magnetic structure and spin systems, see 75.25)
    61.55. Specific structures of elements and alloys
    61.60. Specific structures of inorganic compounds
    61.65. Specific structures of organic compounds
    61.70. Defects in crystals
        (see also 61.80. Radiation damage, 62. Mechanical and acoustical properties,
        71.55. Impurities and defect levels, 76.30. EPR of color centers and other defects,
        78.50. Impurity and defect absorption in solids)
    61.80. Radiation damage and other irradiation effects
        (for techniques of structure determination, see 61.10 to 61.16; for electron and ion
        impact phenomena, see 79.20)
    61.90. Other topics in structure of liquids and solids

62. Mechanical and acoustical properties of condensed matter
    (see also 61.70. Defects in crystals, 68.30. Surfaces and interfaces)
    62.10. Mechanical properties of liquids
        (for viscosity of liquids, see also 66.20)
    62.20. Mechanical properties of solids (related to microscopic structure)
    62.30. Mechanical and elastic waves
    62.40. Anelasticity, internal friction, and mechanical resonances
        For thermomechanical effects, see 65.70
        For magnetomechanical effects, see 75.80
        For piezoelectric effects, see 77.60
        For elasto-optical effects, see 78.20
    62.50. High-pressure and shock-wave effects in solids
    62.60. Acoustical properties of liquids
        For lattice dynamics, phonons, see 63
        For second sound in quantum fluids, see 67.40
    62.65. Acoustical properties of solids
        For magnetoacoustic effects in metals, see 75.80
        For acoustoelectric effects, see 72.50
        For acousto-optical effects, see 78.20

---

*) Excerpt, reproduced with permission of International Council for Scientific and Technical Information (ICSTI).

62.80. Ultrasonic relaxation
(see also 74.30. Ultrasonic attenuation in superconductors)
62.90. Other topics in mechanical and acoustical properties of condensed matter

63. Lattice dynamics and crystal statistics
(see also 65. Thermal properties, 66.70. Thermal conduction, 68.30. Dynamics of surface and interface vibrations, 78.30. Infrared and Raman spectra)
63.10. General theory
63.20. Phonons and vibrations in crystal lattices
63.50. Vibrational states in disordered systems
63.70. Statistical mechanics of lattice vibrations
(see also 65. Thermal properties of condensed matter, and 66.70. Thermal conduction)
63.75. Statistical mechanics of displacive phase-transitions
For order–disorder and statistical mechanics of model systems, see 64.60
For crystallographic aspects of polymorphic and order–disorder transformations, see 61.50
63.90. Other topics in lattice dynamics and crystal statistics

64. Equations of state, phase equilibria, and phase transitions
64.10. General theory of equations of state and phase equilibria
64.30. Equations of state of specific substances
(see also 65.70. Thermal expansion)
64.60. General studies of phase transitions
(for magnetic, superconducting, and quantum fluid critical phenomena, see 75, 74, and 67, respectively)
64.70. Phase equilibria, phase transitions, and critical points of specific substances
64.75. Solubility, segregation, and mixing
64.80. Other phase properties of systems
64.90. Other topics in equations of state, phase equilibria, and phase transitions

65. Thermal properties of condensed matter
(see also 63. Lattice dynamics; for thermodynamic properties of quantum fluids, see 67.40; for thermal properties of solid helium, see 67.80)
65.20. Heat capacities of liquids
65.40. Heat capacities of solids
(for specific heat of superconductors, see 74.30; for specific heat of magnetic systems, see 75.40)
65.50. Thermodynamic properties and entropy
65.70. Thermal expansion and thermomechanical effects
(see also 64.30. Equations of state)
For thermal conduction in nonmetallic liquids, see 66.60; for nonmetallic solids, see 66.70
For electronic thermal conduction, see 72.10 and 72.20
For thermal conductivity of superconductors, see 74.30
For pyroelectric and electrocaloric effects, see 77.70
65.90. Other topics in thermal properties of condensed matter

66. Transport properties of condensed matter (nonelectronic)
66.10. Diffusion and ionic conduction in liquids
66.20. Diffusive momentum transport
(for viscosity of liquids, see also 62.10)
66.30. Diffusion in solids
66.60. Thermal conduction in nonmetallic liquids
(for thermal conduction in liquid metals, see 72.15)
66.70. Nonelectronic thermal conduction and heat-pulse propagation in nonmetallic solids
(for thermal conduction in solid metals, see 72.15; for statistical mechanics of lattice vibrations, see 63.70)
66.90. Other topics in nonelectronic transport properties

67. Quantum fluids and solids; liquid and solid helium
    67.20. Quantum effects on the structure and dynamics of nondegenerate fluids
    67.40. Boson degeneracy and superfluidity of helium-4
    67.50. Fermi fluids; liquid helium-3
    67.60. Mixed systems; liquid helium-3, -4 mixtures
    67.70. Films (including physical adsorption)
    67.80. Solid helium and related quantum crystals
    67.90. Other topics in quantum fluids
    (e.g., neutron-star matter)

68. Surfaces and interfaces; thin films and whiskers
    (for impact phenomena, see 79; for crystal growth, see 61.50)
    68.10. Fluid surfaces and fluid–fluid interfaces
    68.15. Liquid thin films
    68.20. Solid surface structures
    68.25. Mechanical and acoustical properties of solid surfaces and interfaces
    68.30. Dynamics of solid surfaces and interface vibrations
    68.40. Surface energy of solids; thermodynamic properties
    68.45. Solid–fluid interface processes
    68.48. Solid–solid interfaces (including bicrystals)
    (for grain boundaries, see 61.70)
    68.55. Thin film growth, structure, and epitaxy
    68.60. Physical properties of thin films: nonelectronic
    68.70. Whiskers and dendrites: growth, structure, and nonelectronic properties
    68.90. Other topics in the structure and nonelectronic properties of surfaces and thin films

70. Condensed matter: electronic structure; electrical, magnetic, and optical properties

71. Electronic states
    (see also 63. Lattice dynamics, 73. Electronic structure and electrical properties of surfaces, interfaces, and thin films)
    71.10. General theories and computational techniques
    71.20. Electronic density of states determinations
    (including energy states of liquid semiconductors)
    (see also 65.40. Electronic heat capacity)
    71.25. Nonlocalized single-particle electronic states
    71.30. Metal–insulator transitions
    71.35. Excitons and related phenomena (including electron–hole drops)
    71.36. Polaritons
    (including photon–phonon and photon–magnon interactions)
    71.38. Polarons and electron–phonon interactions
    (see also 63.20. phonon–electron interactions in lattices)
    71.45. Collective effects
    71.50. Localized single-particle electronic states (excluding impurities)
    71.55. Impurity and defect levels
    71.60. Positron states
    (see also 78.70. Positron annihilation)
    71.70. Level splitting and interactions
    (see also 75.10. — in magnetic phenomena, 75.30. Exchange and superexchange interactions, 73.20. Electronic surface states)
    71.90. Other topics in electron states

72. Electronic transport in condensed matter
    (for surfaces, interfaces, and thin films, see 73)
    72.10. Theory of electronic transport; scattering mechanism
    72.15. Electronic conduction in metals and alloys
    72.20. Conductivity phenomena in semiconductors and insulators
        (for nonelectronic conduction, see 66.70)
    72.30. High-frequency effects; plasma effects
    72.40. Photoconduction and photovoltaic effects; photodielectric effects
    72.50. Acoustoelectric effects
    72.55. Magnetoacoustic effects
    72.60. Mixed conductivity and conductivity transitions
    72.70. Noise processes and phenomena
    72.80. Conductivity of specific semiconductors and insulators
    72.90. Other topics in electronic transport in condensed matter

73. Electronic structure and electrical properties of surfaces, interfaces, and thin films
    73.20. Electronic surface states
        (for emission and impact phenomena, see 79)
    73.25. Surface conductivity
    73.30. Surface double layers, Schottky barriers, and work functions
    73.40. Interfaces
    73.60. Electronic properties of thin films
    73.90. Other topics in electrical properties of surfaces, interfaces, and thin films

74. Superconductivity
    74.10. Occurrence, critical temperature
    74.20. Theory
    74.30. General properties
    74.40. Fluctuations and critical effects
    74.50. Tunneling phenomena, Josephson effect, and proximity effects
    74.55. Type-I superconductivity
    74.60. Type-II superconductivity
    74.70. Superconducting materials
    74.90. Other topics in superconductivity

75. Magnetic properties and materials
    75.10. General theory and models of magnetic ordering
        (see also 71.25. Nonlocalized single-particle electronic states, 71.70. Level splitting and interactions)
    72.20. Diamagnetism and paramagnetism
    75.25. Spin arrangements in magnetically ordered materials
        (neutron studies, etc.)
    75.30. Magnetically ordered materials, other intrinsic properties
        (for critical point effects, see 75.40)
    75.40. Critical-point effects, specific heats, short-range order
        (see also 65.40. Heat capacities)
    75.50. Studies of specific magnetic materials
    75.60. Domain effects, magnetization curves, and hysteresis
    75.70. Magnetic films and plates
    75.80. Magnetomechanical and magnetoelectric effects, magnetostriction
      For galvanomagnetic effects, see 72.15 and 72.20
      For magneto-optical effects, see 78.20
    75.90. Other topics in magnetic properties and materials

76. Magnetic resonances and relaxation in condensed matter; Mössbauer effect
   76.20. General theory of resonances and relaxation
   76.30. Electron paramagnetic resonance and relaxation
   76.40. Diamagnetic and cyclotron resonances
   76.50. Ferromagnetic, antiferromagnetic, and ferrimagnetic resonances; spin-wave resonance (see also 75.30. Spin waves)
   76.60. Nuclear magnetic resonance and relaxation
   76.70. Magnetic double resonances and cross effects
   76.80. Mössbauer effect; other γ-ray spectroscopy
   76.90. Other topics in magnetic resonances and relaxation

77. Dielectric properties and materials
   (for conductivity phenomena, see 72.20 and 72.80)
   77.20. Permittivity
   77.30. Polarization and depolarization effects
   77.40. Dielectric loss and relaxation
   77.50. Dielectric breakdown and space-charge effects
   77.55. Dielectric thin films
   77.60. Piezoelectricity and electrostriction
   (for piezo-optical effects, see 78.20)
   77.70. Pyroelectric and electrocaloric effects
   77.80. Ferroelectricity and antiferroelectricity
   77.85. Electrical resonances
   77.90. Other topics in dielectric properties and materials

78. Optical properties and condensed-matter spectroscopy and other interactions of matter with particles and radiation
   (for phonon spectra, see 63)
   78.20. Optical properties and materials
   78.30. Infrared and Raman spectra and scattering
   78.35. Brillouin and Rayleigh scattering
   78.40. Visible and ultraviolet spectra
   78.45. Stimulated emission
   78.50. Impurity and defect absorption in solids
   78.55. Photoluminescence
   78.60. Other luminescence and radiative recombination
   For photoconduction and photovoltaic effects, see 72.40
   78.65. Optical properties of thin films
   78.70. Other interactions of matter with particles and radiation
   78.90. Other topics in optical properties of condensed matter and other interactions of matter with particles and radiation

79. Electron and ion emission by liquids and solids; impact phenomena
   79.20. Impact phenomena, including electron spectra and sputtering
   79.40. Thermionic emission
   79.60. Photoemission and photoelectron spectra
   79.70. Field emission and field ionization
   79.75. Exoelectron emission
   79.80. Resonance tunneling
   79.90. Other topics in emission and impact phenomena in condensed matter

Substance Classification (according to physica status solidi)

S1. Transition metals and their alloys and compounds
(for oxides, see S.10)
- S1.1. Fe
- S1.2. Other 3d metals
- S1.3. Noble metals
- S1.4. Lanthanides
- S1.5. Actinides
- S1.61. Carbides, silicides, germanides
- S1.62. Nitrides, phosphides, arsenides
- S1.63. Sulfides, selenides, tellurides

S2. Alkali metals and their alloys and compounds
(for alkali halides, see S9.11)

S3. Alkaline earth metals and their alloys and compounds
(for alkaline earth halides, see S9.12)

S4. Other metals and their alloys

S5. Nonmetallic elements
- S5.11. Silicon
- S5.12. Germanium
- S5.2. Solidified Gases

S6. Group IV compounds
(for transition metal compounds, see S1.61)

S7. Group V compounds
(for transition metal compounds, see S1.62)
- S7.1. III–V compounds
- S7.11. Phosphides
- S7.12. Arsenides
- S7.13. Antimonides
- S7.15. Mixed crystals
- S7.16. Higher compounds

S8. Chalcogenides
(for transition metal compounds, see S1.63)
- S8.1. II–VI compounds
- S8.11. Sulfides
- S8.12. Selenides
- S8.13. Tellurides
- S8.15. Mixed crystals
- S8.16. Higher compounds

S9. Halides
- S9.11. Alkali halides
- S9.12. Alkaline earth halides
- S9.15. Mixed crystals
- S9.16. Higher compounds

S10. Oxides
- S10.1. Simple oxides
- S10.15. Mixed crystals

S11. Other higher compounds
- S11.1. Salts of oxy-acids
- S11.2. Spinels, garnets, ferrites

S12. Organic compounds
- S12.1. Quasi-one-dimensional systems

19. Magnetic Resonance

20. Optical Properties
    20.1 Spectra. Optical Constants (X-Ray Spectra See 20)
    20.2 Lasers
    20.3 Luminescence (X-Ray Spectra See 20)

21. Metals. Alloys
    21.1 Transition Metals (Group Sc to Zn) and Their Alloys
        21.1.1 Fe and Fe Alloys
    21.2 Alkali Metals
    21.3 Alkali-Earth Metals
    21.4 Rare-Earth Metals
    21.5 Actinides
    21.6 Noble Metals
    21.7 Semi-Metals (Semiconducting Alloys See 22.7)

22. Semiconductors and Ionic Crystals
    22.1 Elements
        22.1.1 Germanium
        22.1.2 Silicon
        22.1.3 Group VI Elements
    22.2 III–V Compounds
        22.2.1 Arsenides
        22.2.2 Phosphides
        22.2.3 Antimonides
        22.2.4 Mixed Crystals
    22.3 IV–IV, II–V, and III–VI Compounds
        22.3.1 Mixed Crystals
    22.4 II–VI Compounds
        22.4.1 Sulphides
        22.4.2 Selenides
        22.4.3 Tellurides
        22.4.4 Mixed Crystals
    22.5 Halides
        22.5.1 Silver Halides
        22.5.2 Alkali Halides
        22.5.3 Alkali-Earth Halides
        22.5.4 Mixed Crystals
    22.6 Simple Oxides (CdO, ZnO See 22.4)
        22.6.1 Mixed Crystals
    22.7 Semiconducting Intermetallic Compounds
        22.7.1 Mixed Crystals
    22.8 Tertiary and Higher Compounds
        22.8.1 Salts of Oxy-Acids (Carbonates, Phosphates, Silicates Including Mica etc.)
        22.8.2 Spinels, Garnets, Ferrites
    22.9 Organic Semiconductors

23. Solidified Gases

For comments on the use of the classification scheme see p. 6.

32 913/32 914

Manuscripts and letters for physica status solidi (b) — basic research and physica status solidi (a) — applied research

should be sent to

DDR-1086 Berlin, Leipziger Str. 3—4, P. B. 1233

or

Kunst und Wissen, Erich Bieber OHG, Postfach 330 104, D-1000 Berlin 33.

The object of "physica status solidi (b)" and "physica status solidi (a)" is to publish rapidly important and topical results concerning fundamental research in the field of solid state physics and aspects of application and experimental techniques of fundamental solid state research, respectively. The issues of "physica status solidi (b)" and "physica status solidi (a)" will be published on the 1st and the 16th day of each month, respectively. Either series will contain the following categories:

1. Review articles written by arrangement with the editors. They will be remunerated.
2. Original papers selected from those which have not been submitted for publication elsewhere, after being reviewed. (The latest date for acceptance of contributions is the 8th and the 23rd day of each month for "physica status solidi (b)" and "physica status solidi (a)", respectively, 50 days before the delivery of the journal.)
3. Short notes published in the off-set part of the journal. They are, if necessary after being reviewed, selected by the editorial staff. (The latest date for acceptance of manuscripts for this part is the 12th and the 27th day of each month for "physica status solidi (b)" and "physica status solidi (a)", respectively, 16 days before the delivery of the journal.)
4. Pre-printed titles of papers to be published in the next issue of both series.

Editorial Information

Manuscripts should be submitted preferentially in English. Besides, papers in German, French, and Russian are admitted. The final decision on the incorporation of the papers in the series (b) or (a) is taken by the editorial staff after consultation of the referee. Papers on liquids are only published in physica status solidi if the problems treated are also relevant for the solid state. Authors are asked to observe the following rules:

1. Manuscripts have to be written as concisely as possible. As most, 20 and 4 pages (28 lines per page; abstracts, tables, references, figures, and captions of figures included, where 3 figures are counted as about one page) are admitted for original papers and short notes, respectively. All manuscripts should be type-written on one side only. Two copies are required. Badly arranged manuscripts are returned.
2. The author's name, initials, and institution should be given.
3. One abstract of about ten lines in English and one in German, French, or Russian are required for original papers.
4. Because of lack of space conclusions (if any) must not have the character of a summary (i. e. substantially a repetition of the contents of the abstract).
5. Formulae should not be written too small and not with pencil. Separate lines for formulae are desirable. Indistinct writing will lead to a return of the manuscript.
6. Drawings should be enclosed on separate sheets of paper, not exceeding the maximum and minimum sizes of 20 × 30 and 10 × 15 cm², respectively. Photographs should be submitted, on white paper, only for review articles and original papers. Duplicates are required.
7. Captions for the figures should be submitted on separate sheets of paper, and special instructions with regard to the size of illustrations should be given on these.
8. References should be numbered in order of appearance and should be given, on a separate sheet, in the form used in physica status solidi.
9. SI-units should be used.
10. Each paper must contain a proposal for the subject classification according to the classification scheme appearing in the January issues of each annual set.

The authors will be sent galley proofs of review articles and original papers. Proof reading of short notes and pre-printed titles will be done by the editorial staff.

For each original paper, and short note, a total of 75 reprints is available free of charge. The number of reprints free of charge per review article is 25. Requests for additional reprints (against payment) should be given together with the submitted manuscript. **The deadline for orders of additional reprints is the 15th and the 30th of each month** for "physica status solidi (b)" and "physica status solidi (a)", respectively, two weeks before the delivery of the journal.

Bei Fragen zur Produktsicherheit wenden Sie sich bitte an:
If you have any questions regarding product safety,
please contact:

Walter de Gruyter GmbH
Genthiner Straße 13
10785 Berlin
productsafety@degruyterbrill.com